CARDIAC CELL BIOLOGY

Cardiac Cell Biology

Edited by

ELISSAVET KARDAMI

Institute of Cardiovascular Sciences
St. Boniface General Hospital Research Centre
Faculty of Medicine
University of Manitoba
Winnipeg, Canada

LARRY HRYSHKO

Institute of Cardiovascular Sciences
St. Boniface General Hospital Research Centre
Faculty of Medicine
University of Manitoba
Winnipeg, Canada

NASRIN MESAELI

Institute of Cardiovascular Sciences and
Division of Stroke and Vascular Disease
St. Boniface General Hospital Research Centre
and Department of Physiology
University of Manitoba,
Winnipeg, Canada

Reprinted from *Molecular and Cellular Biochemistry*, Volume 242 (2003)

Kluwer Academic Publishers

Dordrecht / Boston / London

Distributors for North, Central and South America:
Kluwer Academic Publishers
101 Philip Drive
Assinippi Park
Norwell, Massachusetts 02061 USA
Telephone (781) 871-6600
Fax (781) 681-9045
E-mail <kluwer@wkap.com>

Distributors for all other countries:
Kluwer Academic Publishers Group
Distribution Centre
Post Office Box 322
3300 AH Dordrecht, The Netherlands
Telephone 31 78 6392 392
Fax 31 78 6546 474
E-mail <services@wkap.nl>

Electronic Services <http://www.wkap.nl>

Library of Congress Cataloging-in-Publication Data

A C.I.P. Catalogue record for this book is available from the Library of Congress

ISBN 1-4020-7296-1

Printed on acid-free paper

Printed in The Netherlands

Molecular and Cellular Biochemistry:

An International Journal for Chemical Biology in Health and Disease

CONTENTS VOLUME 242, Nos. 1 & 2, January (I) 2003

CARDIAC CELL BIOLOGY

Drs. Elissavet Kardami, Larry Hryshko and Nasrin Mesaeli

Molecular and Cellular Biochemistry **242:** 1, 2003.

Preface

Cardiac cell biology has come of age. This was evident at the 2001 ISHR World Congress where many studies addressed fundamental questions of clinical cardiology at the cellular level. Papers included in this focused issue of MCB are based on selected abstracts that have made use of cutting edge technological advances and a plethora of specialized reagents to allow the study and understanding of heart cells to an unprecedented level.

Recognition of activated or modified signaling molecules by specific antibodies, new selective inhibitors and fluorescent fusion tags are but a few of the tools used to dissect signaling pathways and cross-talk mechanisms that may eventually allow rational drug design. Thus Markou *et al.* delineated the relationship between purinergic stimulation of adult cardiomyocytes and components of MAPK pathways, as well as their link to cPLA2 activation. Ellis *et al.*, in a series of elegant experiments, provided evidence that CaM kinase II C phosphorylates (and thus 'inactivates') the chaperone protein 14-3-3, resulting in the return of histone deacetylase in the nucleus, formation of a complex with the transcription factor MEF-2, and thus silencing its transcription in smooth muscle cells. Bayer *et al.* reported distinct autophosphorylation patterns for different PKC isoenzymes supporting the notion that each isozyme may play different roles in the development of left ventricular hypertrophy and heart failure.

Understanding the regulation of cardiac hypertrophy in all its complexity remains a fundamental goal of cardiac research. Tsoporis *et al.* have studied the hypertrophic response in cultured cardiomyocytes and the role of the calcium binding proteins S100A1 and S100B. They propose an intriguing model where down regulation of S100A1 is required to allow early induction of hypertrophic genes subsequent to injury, including S100B, an intrinsic negative regulator of the hypertrophic response. Using a genetic rat model (SHHF-fa^{cp}) Radin *et al.* report that obesity, secondary to leptin resistance in homozygous (cp/cp) rats resulted in increased salt sensitivity mediated by endothelin. An elegant analysis of the metabolic profile of the hypertensive heart is presented by Vincent *et al.* who by combining *ex vivo* semi-recirculating working perfused heart with ^{13}C substrates and mass isotopomer analysis by gas chromatography-mass spectrometry, describe specific metabolic changes that resemble metabolic alterations reported in cardiac patients.

Since the advancement of adenovirally mediated gene transfer, transfection efficiency is no longer a limiting factor in the study of cardiomyocytes. Fan *et al.* were thus able to deliver a mini-gene expressing mutated L-type calcium channel subunits to modulate EC-coupling in adult cardiomyocytes, while Tian *et al.* used expression of dominant-negative Ras gene to establish the link between ouabain, activation of p42/44 MAPKs and increasing $[Ca^{2+}]$.

A limiting factor in considering cell transplantation as a strategy to repair the damaged heart is cell availability at the right time. Studies such as those by Yokomuro *et al.*, who report on optimal conditions and feasibility of cardiac myocyte as well as cardiac tissue cryopreservation provide the groundwork for making heart cell transplantation a practical approach.

Cardiac gap junctions, intercellular communication channels that allow electrical and metabolic coupling and play an important role in arrhythmogenesis are now understood to be exquisite sensors of cardiac change. Thanks to the benefits of confocal microscopy this is beautifully illustrated by Kostin *et al.* who report that focal disorganization of gap junction distribution and downregulation of connexin-43 are typical features of myocardial remodeling in patients with dilated, ischemic and inflammatory cardiomyopathies. Comparative studies of the distribution of the major connexin species by Coppen *et al.* point to significant similarities between the developing mouse and human hearts and strengthen support for the use of transgenic mouse models to study anatomical substrates of arrhythmias. Finally, a non-channel dependent, growth regulating role for the C-terminal fragment of connexin-43 is proposed by Dang *et al.*, introducing the possibility of direct nuclear signaling by connexins.

I would like to take the opportunity to thank all the authors for their contribution and hope that you, the reader, will share in our enthusiasm for all the papers presented in the present issue.

Elissavet Kardami
Institute of Cardiovascular Sciences
Winnipeg, Manitoba
Canada

Molecular and Cellular Biochemistry **242:** 3–10, 2003.

Decoy calcium channel beta subunits modulate contractile function in myocytes

Q. Ivy Fan, Kathleen M. Vanderpool, Jessica O'Connor and James D. Marsh

Program in Molecular and Cellular Cardiology, Department of Internal Medicine, Wayne State University School of Medicine, Detroit, MI, USA

Abstract

To test the hypothesis that mutated β2-subunits of the L-type calcium channel could serve as a decoy and interdict calcium channel trafficking and function, we engineered a β2 subunit that contained the β interaction domain for α1c subunit interaction, but lacked N- and C-terminal domains that might be essential for sarcolemmal localization. An adenoviral vector was constructed containing the gene for the β-interaction domain (BID) fused to green fluorescence protein (GFP), using a vector containing only GFP as control. Freshly plated, dissociated adult rat myocytes were infected and expression and function were assessed at 60 h. Fluorescence microscopy confirmed GFP expression; immunoblot analysis confirmed dose-dependent GFP-BID expression. Mechanical properties of adult rat ventricular myocytes were evaluated using a video edge-detection system. Contractility analysis (optical/video, field stimulation) demonstrated that contracting cells decreased from 60 to 2%. Contractile amplitude (percent shortening) decreases significantly from 5.6 vs. 2.4% with no change in time to peak twitch. Recombinant adenovirus overexpressing mutated β2 subunits in adult mammalian myocytes can markedly alter excitation-contraction coupling. This paradigm may offer new approaches to understanding and modulating EC coupling. (Mol Cell Biochem **242**: 3–10, 2003)

Key words: calcium channel, beta 2 subunit, cardiac myocytes, contractility, adenoviral vector, fusion protein

Introduction

The L-type voltage-dependent calcium channel, also known as the DHP (dihydropyridine) receptor, in the cardiovascular system is an important component of excitation-contraction coupling of contracting cells. Numerous drugs have been developed to modify function of this channel for the treatment and prevention of coronary artery disease, cardiac arrhythmias, hypertension and stroke. The channel is a heteromultimeric protein that includes the α1, β- and α2δ subunits. Although the α1 subunit is the primary protein forming the ion pore of the channel, the other subunits are necessary for fully physiologically functional calcium channels. It has been widely reported that co-expression of the auxiliary subunits, β- and α2δ, promote the membrane localization of the calcium channel, and influence the gating properties of the channel [1, 2].

Co-expression of the β subunit with the α1 subunit increases current density and affects the kinetics of the current; the β subunit increases the number of channels in the sarcolemma as well as having modulatory effects on gating of the channel [3–5]. It is likely that the β subunit might serve as a chaperone protein and is necessary for correct folding of the α1 subunit and insertion into the cell membrane. Recently, it has been reported that in neurons there is an endoplasmic reticulum retention signal within the I–II loop of the α1 subunit that severely restricts the plasma membrane incorporation of α1 subunit. The β subunit reverses the inhibition imposed by the retention signal of the α1 subunit [6]. Thus, we postulate that structural alterations

Address for offprints: J.D. Marsh, Program in Molecular and Cellular Cardiology, Wayne State University School of Medicine, 421 East Canfield Avenue, Detroit, MI 48201, USA (E-mail: jmarsh@intmed.wayne.edu)

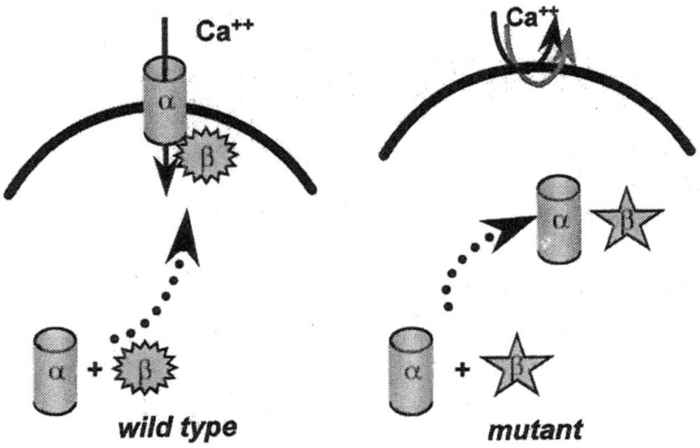

Fig. 1. Schematic diagram of the hypothesis that overexpression of the BID fragment will interfere with the endogenous β subunit interaction with the α1 subunit. This will reduce the number of functional calcium channels in the sarcolemmal membrane without interference with calcium channel kinetics.

of the β subunit that affect its chaperone function will subsequently affect the targeting of the calcium channel to the sarcolemma. However, to this point, the effect of structural alterations of β subunits on calcium channel function and excitation-contraction coupling has not been reported.

It has been thought that the predominant isoform of the β subunit in heart was β2. Colecraft *et al.* [7] have reported that in cardiac myocytes, β2 subunits localize to the plasma membrane. It has been reported that there is a protein domain within the β subunit, β-interacting domain (BID) [8], which physically interacts with α1 subunit. Other domains of the β subunit serve other functions [2, 9–12]. It appears that the amino acid sequences (Cys3Cys4 & SH3) of the β2A subunit are responsible for its membrane association characteristics. Mutation of these residues abolishes sarcolemmal membrane localization in tsA201 cell and COS 7 cell [13, 14]. Although β subunit function has been investigated in heterologous expression system, in model system and neurons, its function in cardiac myocytes is less well understood.

This study tests the hypothesis that overexpression of a mutated β2 subunit (BID fragment) will reduce the number of functional calcium channels localizing to the sarcolemmal membrane in adult cardiac myocytes (Fig. 1). We postulate that a decrease in the number of the calcium channels in the cell membrane will cause a decrease in the calcium influx that will subsequently reduce the cell's contraction in systole.

Materials and methods

Producing the adenoviral vector carrying the BID gene

Adenovirus preparation was based on the methods described by He *et al.* [15]. The cDNA for the β2A subunit was a generous gift of Dr. Perez-Reyes [18]. Briefly, cDNA encoding the BID fragment (bp 857–1177) was cloned into pEGFP-C1 vector (Clontech). The BID is in frame with green fluorescence protein (GFP), and in the fusion protein it is localized at the C-terminal of the GFP. A CMV promoter drives fusion protein expression (Fig. 2A). The CMV-GFP-BID DNA fragment was subcloned into the p-Shuttle vector. A CMV-GFP DNA fragment (lacking any β subunit construct) was also subcloned into a p-Shuttle vector to construct a control virus. After recombination of the shuttle vector and the pAdEasy viral backbone vector in BJ5183 cells, the viral DNA was transfected into HEK293 cells by using the FuGene 6 reagent (Roche). Subsequently, adenoviruses were amplified in HEK293 cells and purified by a CsCl2 gradient. The number of the viral particles was determined spectrophotometrically at 260 λ.

Isolation and culture of adult cardiac myocytes

Ventricular myocytes were isolated from adult male Sprague–Dawley (S/D) rats and plated on cover slips coated with laminin as described previously [16]. Briefly, young male S/D rats were anesthetized, the heart was dissected out and perfused with calcium-free solution and subsequently with trypsin and collagenase. The ventricle was then minced and further digested with collagenase (shaking at 37°C for 30 min). After passing the mixture through gauze mesh to eliminate any undigested tissue, the myocytes were gradually brought back to normal calcium concentration (10^{-3} M) and suspended in culture medium. The cells were plated on laminin-coated coverslips.

Immunoblot analysis

The immunoblots were carried out by conventional methods as we have previously described [16, 17]. The anti-GFP antibody was purchased from Invitrogen.

Myocyte contractility

Mechanical properties of cultured ventricular myocytes were assessed using a video based edge-detection system (Crescent Electronics, Sandy, UT, USA). In brief, coverslips with myocytes attached were placed in a chamber mounted on the stage of an inverted microscope and perfused with buffer (pH 7.4) containing 1 mM Ca^{2+}, other ions and glucose. The cells were field stimulated with a suprathreshold voltage at a frequency of 0.5 Hz for 3 msec. A video-based edge-detector was used to capture and convert changes in cell length during shortening and relengthening into an analogue voltage signal (IonOptix, Milton, MA, USA) which was subsequently

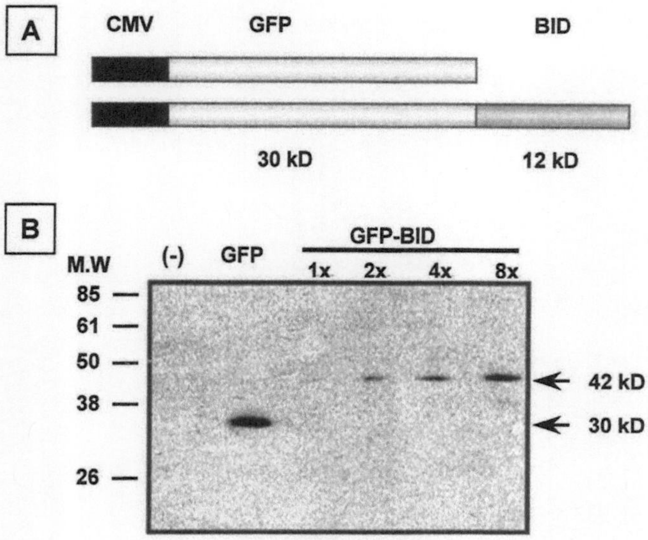

Fig. 2. Immunoblot analysis. (A) Schematic diagram of the constructs for GFP alone and the GFP-BID fusion protein. (B) Cardiac myocytes infected with adenovirus carrying GFP, or with increasing doses of virus carrying GFP-BID. One dose of GFP-BID equals 1×10^9 viral particles. Primary antibody is anti-GFP. The predicted molecular weight is 30 kD for GFP, and 42 kD for the fusion protein GFP-BID.

analyzed with pClamp software. Cell shortening and relengthening were assessed using the following indices: peak twitch amplitude (PTA), time-to-90% PTA (TPT90), time-to-90% relengthening (TR 90), maximal velocities of shortening (+dL/dt) and relengthening (–dL/dt).

Data analysis

Data are presented as mean ± S.E.M. Statistical significance for each variable was estimated by ANOVA followed by a Newman–Keuls test.

Materials

All chemicals and media used in this study were purchased from Sigma except as indicated. The restriction enzymes used for DNA cloning were purchased from Invitrogen and NEB.

Results

Production of adenoviruses that express a decoy β subunit

Ventricular myocytes from adult rat were used to express the control and decoy (experimental) genes (GFP or GFP-BID). Two h after isolation, the myocytes were infected with increasing amounts of an adenovirus carrying the gene for GFP or GFP-BID fusion protein. Adenovirus dosage was $1–8 \times 10^9$ particles for each well/6 well dish. Sixty h after infection, fluorescence microscopy and immunoblot analysis were used to confirm successful viral infection.

Immunoblot analysis was used to confirm expression of the specific protein; the anti-GFP antibody was used. The observed molecular weights of 30 and 42 kD are consistent with our predicted molecular weights for GFP and GFP-BID respectively. Additionally, there is a dose-dependent increase in protein expression with increased amount of the adenovirus used in the experiments (Fig. 2B). Using the anti-BID antibody, immunoblot analysis showed that the decoy β subunit was expressed at a much higher level compared to endogenous wild type β subunits in infected myocytes (data not shown). Thus, overexpression of a potentially dominant negative decoy protein was achieved.

To test our hypothesis, we need evidence that the truncated β subunit, BID, does not have the ability to localize to the cell membrane but rather remains in the cytoplasm. Immunocytochemisty with an anti-β antibody will show the location of the exogenous β mutant, but it will also recognize the endogenous wild type β subunit which might well be located in the cell membrane. To eliminate interference by the endogenous β subunit, we created a fusion protein by conjugating the EGFP cDNA with the cDNA fragment encoding BID. This allowed us to track the exogenous β subunit protein's intracellular location by looking for GFP-BID with fluorescence microscopy.

Fluorescence microscopy demonstrated that the fusion protein GFP-BID was restricted to the cytoplasmic compartment. This distribution pattern is similar to that of the control protein, GFP, that is known as a cytoplasmic protein [19]. Paralleling what we have observed from the immunoblot, there is a dose dependent increase in the number of the green cells as well as a dose dependent increase in the intensity of green color in individual cells (Fig. 3).

Effect of overexpression of GFP-BID on myocyte shortening

Myocytes were electrically stimulated to contract at 0.5 Hz under isotonic conditions in a physiological buffer. In the GFP control group, about 60% of the myocytes contract in response to the electrical stimulation. However, for myocytes infected with the vector expressing the GFP-BID construct there was a dose-dependent decrease in percent of the cells that could be stimulated to contract. At the highest dosage, there were only 2% cells still contracting regularly (Fig. 4).

The average resting cell length (CL) of ventricular myocytes was 105 ± 5 μm. In the myocytes infected with GFP virus, peak twitch amplitude (PTA) in response to electrical stimulation after normalization to CL (PTA/CL) was 5.6 ±

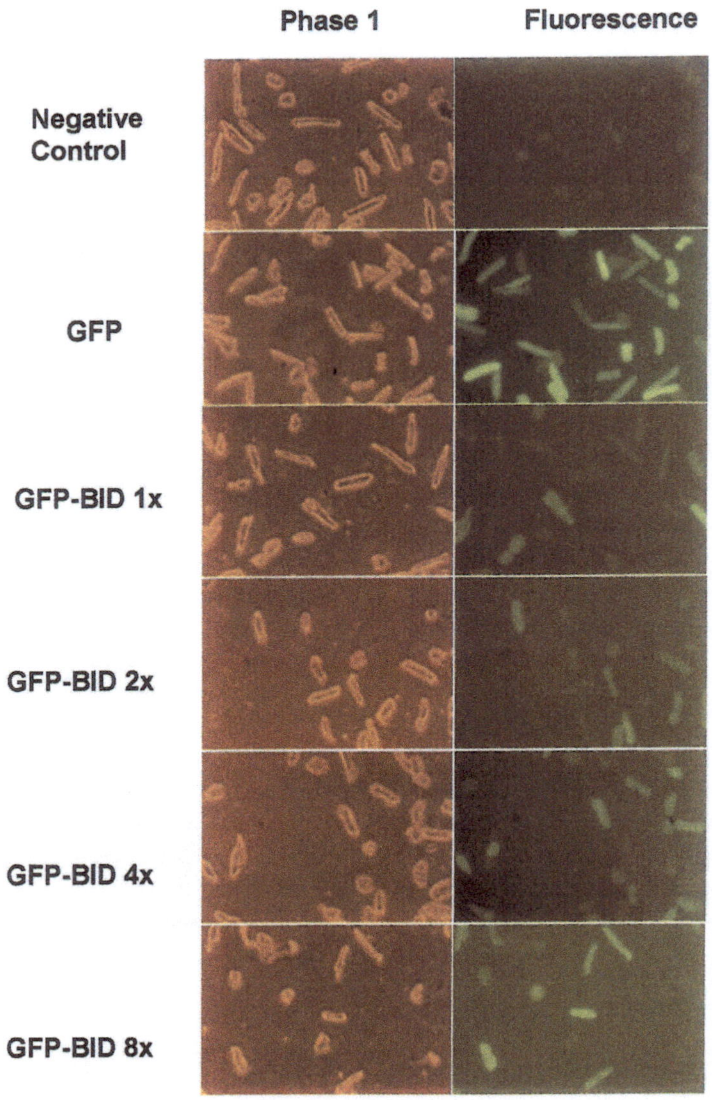

Fig. 3. Green fluorescence detection of gene expression in cultured ventricular myocytes. Photomicrographs were taken 60 h after infection with the adenovirus. Left panels show bright field view of cardiac myocytes and right panels show fluorescence image of the same cells (100×).

0.5%. Of note, PTA/CL is similar to that we have previously reported for noninfected cells, excluding a major contractility effect of GFP alone [20]. There is a dose-dependent decrease in PTA among the myocytes which have been infected with the GFP-BID virus. At the highest dose (8×10^9 viral particles), the PTA/CL was 1.9 ± 0.1%. Overall, there was a significant reduction of the shortening ability in myocytes infected with GFP-BID virus with 42 ± 3.8% ($p < 0.05$) in 1× dose virus group to 66 ± 5% ($p < 0.05$) in 8× dose group (Fig. 5A). Normal time-to-90% peak twitching (TPT90) (Fig. 5B) was observed. This suggests that although trigger calcium may be decreased, the kinetics of calcium release from the sarcoplasmic reticulum is not altered. With depressed PTA and normal TPT90, all groups showed a significant reduction in the maximum velocities of shortening except for the lowest dose group (Fig. 5C).

Relaxation kinetics was also altered by overexpression of the decoy β subunit. At the 4× and 8× dosage of virus infection, the time to 90% relaxation (TR90) was increased (Fig. 6A) and the maximal velocity of relaxation (–dL/dt) was decreased in a dose-dependent manner (Fig. 6B).

Discussion

There are several novel and important findings of the present study. First, we demonstrate that in cardiac myocytes it is possible to over express L-type calcium channel β subunits that

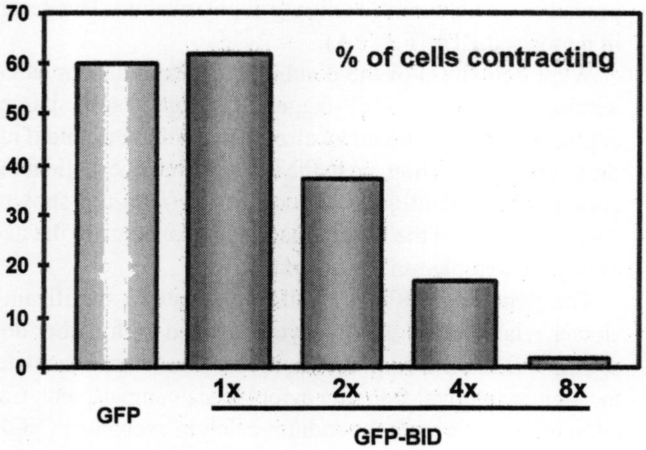

Fig. 4. Percentage of cells contracting in response to field stimulus after 60-h infection with adenoviral vector containing the gene for either GFP or GFP-BID fusion protein. The percentage of contracting cells expressing GFP-BID is significantly decreased in a dose-dependent manner when compared to cells expressing the control construct, GFP alone.

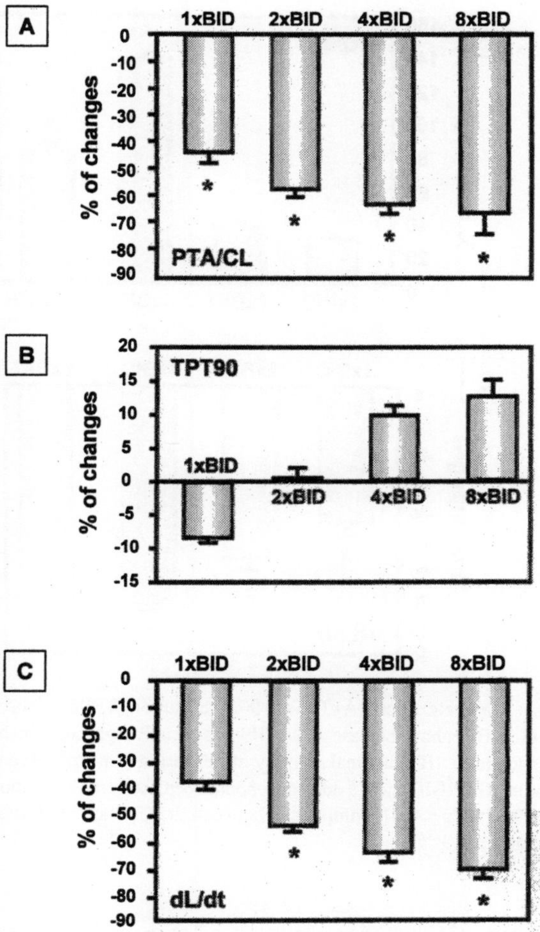

Fig. 5. Cell contractility. (A) Peak cell shortening after normalization by cell length (PTA/CL). Cells expressing GFP-BID show a significant dose-dependent decrease in contractility when compared to cells expressing only GFP. (B) Time to 90% peak twitch (TPT90). Cells expressing increasing levels of GFP-BID show no significant change in the time they take to reach 90% peak twitch when compared to cells expressing only GFP. (C) Maximal velocity of contraction (+dL/dt). Cells expressing GFP-BID show a significant, dose-dependent reduction in the velocity of contraction when compared to cells expressing GFP alone. $^*p < 0.05$; comparison between the GFP and individual GFP-BID group; n = 25–33 for each group.

are mutated and do not localize to the sarcolemmal membrane. The adenovirally-mediated delivery of subunits is efficient, with dose-dependent subunit and GFP expression in essentially 100% of adult myocytes. The second novel finding is that when the mutated subunit is overexpressed, there are important functional consequences. Normal excitation-contraction coupling is affected in a gene dose-dependent manner. There is a decline in cell contractility that is specific for the process dependent on calcium entry through the L-type channel; there is no disruption in kinetics of contraction or relaxation, which are largely dependent on function of proteins of the sarcoplasmic reticulum: the ryanodine receptor, sarco(endo)plasmic reticulum calcium ATPase and phospholamban. Thus the BID minigene is a very sharp probe of E-C coupling.

β Subunits are a heterogenous group of proteins; at least four genes have been identified encoding β subunits: β1 [21–23], β2 [16], β3 [24] and β4 [25]. All the β subunits share some degree of homology in their central core sequence, especially at the BID which interacts with the α1 subunit [8]. The unique N-terminal and C-terminal regions of each isoform make them functionally different from each other. Researchers in the Dolphin laboratory studying dorsal root ganglia have described the negative consequences of depletion of β subunits in neurons [26]. When a 26 bp antisense DNA for β subunits was microinjected into individual cells, there was a reduction in maximum calcium current amplitude of 47% and shift in the activation curves [26]. Furthermore, alteration of the expression of β subunits has been reported in human diseases. A recent study by Hullin's group showed that in cardiac allografts demonstrating diastolic failure related to alteration of the calcium handling, down regulated expression of the β subunit might have contributed [27]. Conversely, in hypertrophic obstructive cardiomyopathy where systolic function is supranormal, an increase in the expression of the β subunit has been reported [28].

Among all the β subunits, β2 subtype is the most abundant form in the heart. It is also the isoform that physically interacts with the α1c subunit to form the L-type calcium channel [29]. Alternative splicing of the β2 subunits has also been reported. At least two splicing forms of the β2 subunit have been cloned from heart tissue, β2A and β2B, and a third splic-

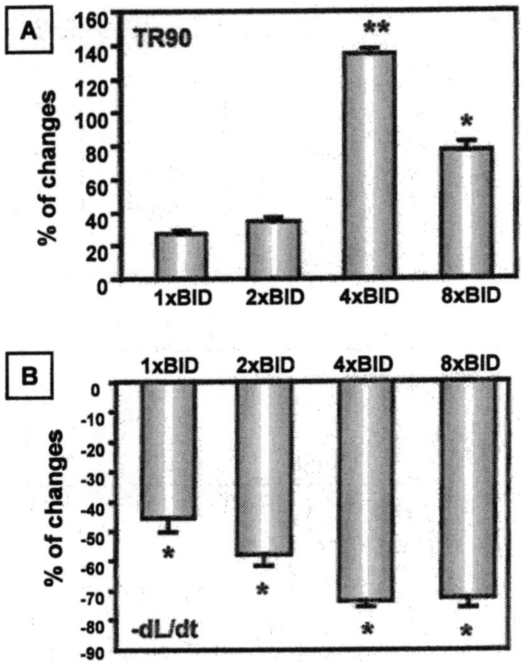

Fig. 6. Cell relaxation. (A) Time to 90% relaxation (TR90). Cells expressing GFP-BID show increase in the TR90, especially significant in the higher dosage group. (B) Maximal velocity of relaxation (–dL/dt). Myocytes expressing GFP-BID show a dose-dependent decrease in the relaxation speed. $*p < 0.05$; $**p < 0.01$, comparison between the GFP and individual GFP-BID group.

ing form β2C has been observed in brain and heart tissue. The two cardiac β2 isoforms are different in that β2B has an additional 28 amino acids at the N-terminal of the protein [29, 30]. Although structurally β2A subunit is predicted to be a hydrophilic protein, in tsA201 cells expression of β2A alone showed a membrane distribution [31]. A recent study showed that the predominant isoform of β subunit in rat heart is β2C, which differed from the β2A subunit by substitution of the first 16 amino acids with a different 67 amino acid sequence [32]. The alteration is due to alternative slicing of the gene. To date, membrane localization of the expressed β2 subunit has not been extensively studied in adult cardiac myocytes. As predicted from physiological studies we also observed similar phenomenon when we expression of GFP-β2A (wild type) protein in cardiac myocytes. The fusion protein locates to the membrane rather than remaining in the cytoplasmic compartment as the GFP-BID (data not shown).

The current study shows there is a dose-dependent increase in the GFP-BID fusion protein expression (Figs 2B and 3) in cardiac myocytes. The mutated β subunit competes with endogenous wild type β2 subunits and binds with the α1C subunit. Therefore, with more GFP-BID being expressed within the cells, less α1C subunit will be correctly folded and inserting into cell membrane (Fig. 1). Accordingly, we would predict less calcium influx during depolarization. This results in depressed PTA (Fig. 5A).

What domain(s) of the β subunit are essential for sarcolemmal localization? GFP-tagged full length β subunit overexpressed in myocytes do localize the sarcolemma, and if just the C-terminal portion, up to the BID is deleted, excellent sarcolemmal localization is also observed (data not shown). Thus, presence of the N-terminal domain appears to the necessary for sarcolemmal localization.

The changes in TR90 and –dL/dt indicated a significantly slower relaxation of the myocytes infected with higher dosage GFP-BID virus (Fig. 6). These two parameters are related to calcium removal from the cytoplasmic compartment. Calcium efflux is through the sodium-calcium exchanger (NCX) while calcium sequestration into the sarcoplasmic reticulum is carried out by sarcoplasmic reticulum Ca-ATPase (SERCA) [33–35]. The diminished relaxation velocity would be expected for myocytes with decreased cytosolic calcium concentration, as we and others have observed in the presence of blocked or down-regulated calcium channels [36, 37]. With diminished $[Ca^{2+}]_i$ there are altered gradients for both SERCA and for NCX. Altered expression of genes encoding these calcium regulatory proteins, which is an part calcium dependent, cannot be excluded.

Clinically, calcium overload has been described in several disease conditions such as in diastolic heart failure and myocardial ischemia, and organic calcium channel blockers have been widely used for the treatment of hypertension, cardiac arrhythmia, and myocardial ischemia. The molecular approach to modulation of calcium channel expression and function may offer a number of substantial advantages over pharmacological blockade of calcium entry. For instance, strategies can be developed to modify calcium channel expression and function to the immediate environment of specific myocytes in the myocardium or atrio-ventricular node. As an example, a hypoxia response element can be engineered into the promoter of a therapeutic gene so that gene expression could respond to conditions of hypoxia or ischemia, two important arrhythmogenic triggers.

In summary, we now show for the first time in an adult mammalian cardiac myocytes that an adenovirus can deliver a mini-gene encoding a β subunit that can function as a decoy and alter excitation-contraction coupling. The mutated β subunit localizes to the cytosol and its expression can be titrated by altering viral dose to produce a graded physiological effect.

References

1. Welling A, Bosse E, Cavalie A, Bottlender R, Ludwig A, Nastainczyk W, Flockerzi V, Hofmann F: Stable co-expression of calcium channel α1, β and α2/δ subunits in a somatic cell line. J Physiol 471: 749–765, 1993

2. Massa E, Kelly KM, Yule DI, MacDonald RL, Uhler MD: Comparison of fura-2 imaging and electrophysiological analysis of murine calcium channel alpha 1 subunits coexpressed with novel beta 2 subunit isoforms. Mol Pharmocol 47: 707–716, 1995
3. Perez-Garcia MT, Kamp TJ, Marban E: Functional properties of cardiac L-type calcium channels transiently expressed in HEK293 cells. Roles of $\alpha 1$ and β sububit. J Gen Physiol 105: 289–305, 1995
4. Josephson IR, Varadi G: The β subunit increases Ca currents and gating charge movements of human cardiac L-type Ca channel. Biophys J 70: 1285–1293, 1996
5. Cens T, Restituito S, Vallentin A, Charnet P: Promotion and inhibition of L-type calcium channel facilitation by distinct domains of the beta subunit. J Biol Chem 273: 18308–18315, 1997
6. Bichet D, Cornet V, Geib S, Carlier E, Volsen S, Hoshi T, Mori Y, DeWaard M: The I–II loop of the Ca^{2+} channel alpha1 subunit contains endoplasmic reticulum retention signal antagonized by the beta subunit. Neuron 25: 177–190, 2000
7. Colecraft HM, Wei SK, Johns DC, Yue DT: Overexpression of calcium channel beta3 subunits in adult myocytes reveals a primary role for beta subunits in PKA modulation of L-type calcium channels. Circulation 100: I-190, 1999
8. De Waard M, Scott VE, Pragnell M, Campbell KP: Identification of critical amino acids involved in alpha1-beta interaction in voltage-dependent Ca^{2+} channels. FEBS Lett 380: 272–276, 1996
9. Cens T, Restituito S, Charnet P: Regulation of Ca-sensitive inactivation of a L-type Ca^{2+} channel by specific domains of beta subunits. FEBS Lett 450: 17–22, 1999
10. Chien AJ, Zhao X, Shirokov RE, Puri TS, Chang CF, Sun D, Rios E, Hosey MM: Roles of a membrane-localized β subunit in the formation and targeting of functional L-type Ca^{2+} channels. J Biol Chem 270: 30036–30044, 1995
11. Qin N, Olcese R, Cabello OA, Birnbaumer L, Stefani E: Identification of a second region of the beta-subunit involved in regulation of calcium channel inactivation. Am J Physiol 271: c1539– c1545, 1996
12. Gerster U, Neuhuber B, Groschner K, Striessnig J, Flucher BE: Current modulation and membrane targeting of the calcium channel 1C subunit are independent functions of the beta subunit. J Physiol 517: 353–368, 1999
13. Chien AJ, Carr KM, Shirokov RE, Rios E, Hosey MM: Identification of palmitoylation sites within the L-type calcium channel beta2a subunit and effects on channel function. J Biol Chem 271: 26465–26468, 1996
14. Brice NL, Berrow NS, Campbell V, Page KM, Brickley K, Tedder I, Dolphin AC: Importance of the different beta subunits in the membrane expression of the alpha1A and alpha2 calcium channel subunits: studies using a depolarization-sensitive alpha1A antibody. Eur J Neurosci 9: 749–759, 1997
15. He TC, Zhou S, da Costa LT, Yu J, Kinzler KW, Vogelstein B: A simplified system for generating recombinant adenoviruses. Proc Natl Acad Sci USA 95: 2509–2514, 1998
16. Davidoff AJ, Maki TM, Ellingsen O, Marsh JD: Expression of calcium channels in adult cardiac myocytes is regulated by calcium. J Mol Cell Cardiol 298: 1791–1803, 1997
17. Towbin H, Taehelin T, Ordon J: Electrophoretic transfer of proteins from polyacrylamide gels to nitrocellulose sheets: procedure and some applications. Proc Natl Acad Sci USA 76: 4350– 4354, 1979
18. Perez-Reyes E, Castellano A, Kim HS, Bertrand P, Baggstrom E, Lacerda AE, Wei XY, Birnbaumer L: Cloning and expression of a cardiac/brain beta subunit of the L-type calcium channel. J Biol Chem 267: 1792–1797, 1992
19. Marshall J, Molloy R, Moss GWJ, Howe JR, Hughes TE: The jellyfish green fluorescent protein: A new tool for studying ion channel expression and function. Neuron 14: 211–215, 1995
20. Ellingsen O, Davidoff AJ, Prasad SK, Berger HJ, Springhorn JP, Marsh JD, Kelly RA, Smith TW: Adult rat ventricular myocytes cultured in defined medium: Phenotype and electromechanical function. Am Physiol 265: H747–54, 1993
21. Pragnell M, Sakamoto J, Jay SD, Campbell KP: Cloning and tissue-specific expression of the brain calcium channel beta-subunit. FASEB Lett 91: 253–258, 1991
22. Powers PA, Liu S, Hogan K, Gregg RG: Skeletal muscle and brain isoforms of a beta-subunit of human voltage-dependent calcium channels are encode by a single gene. J Biol Chem 267: 22967–22972, 1992
23. Gregg RG, Powers PA, Hogan K: Assignment of the human gene for the beta subunit of the voltage-dependent calcium channel to chromosome 17 using somatic cell hybrids and linkage mapping. Genomics 15: 185–187, 1993
24. Collin T, Lory P, Taviaux S, Courtieu C, Guilbault P, Berta P, Nargeot J: Cloning, chromosomal location and functional expression of the human voltage-dependent calcium-channel beta 3 subunit. Eur J Biochem 220: 257–262, 1994
25. Castellano A, Wei X, Birnbaumer L, Perez-Reyes E: Cloning and expression of a neuronal calcium channel beta subunit. J Biol Chem 268: 12359–12366, 1993
26. Berrow NS, Campbell V, Fitzgerald EM, Brixkley K, Dolphin AC: Antisense depletion of beta-subunit modulates the biophysical and pharmacological properties of neuronal calcium channels. J Physiol 482: 481–491, 1995
27. Hullin R, Asmus F, Ludwig A, Hersel J, Boekstegers P: Subunit expression of the cardiac L-type calcium channel is differentially regulated in diastolic heart failure of the cardiac allograft. Circulation 100: 155–163, 1999
28. Haase H, Kresse A, Hohaus A, Schulte H, Maier M, Osterziel K, Lange P, Morano I: Expression of calcium channel subunits in the normal and diseased human myocardium. J Mol Med 74: 99–104, 1996
29. Collin T, Wang JJ, Nargeot J, Schwartz A: Molecular cloning of three isoforms of the L-type voltage-dependent calcium channel β subunit from normal human heart. Circ Res 72: 1337–1344, 1993
30. Hullin R, Singer-Lahat D, Freichel M, Biel M, Dascal N, Hofmann F, Flockerzi V: Calcium channel beta subunit heterogeneity: Functional expression of cloned cDNA from heart, aorta and brain. EMBO J 11: 885–890, 1992
31. Chien AJ, Gao T, Perez-Reyes E, Hosey MM: Membrane targeting of L-type calcium channels. J Biol Chem 273: 23590–23597, 1998
32. Yamada Y, Nagashima M, Tsutsuura M, Kobayashi T, Seki S, Makita N, Horio Y, Tohse N: Cloning of a functional splice variant of l-type calcium channel beta 2 subunit from rat heart. J Biol Chem 276: 47163–47170, 2001
33. Lewartowski B, Wolska BM, Zdanowski K: The effects of blocking the Na-Ca exchange at intervals throughout the physiological contraction-relaxation cycle of single cardiac myocyte. J Mol Cell Cardiol 24: 967–976, 1992
34. Bers DM, Bassani JW, Bassani RA: Na-Ca exchange and Ca fluxes during contraction and relaxation in mammalian ventricular muscle. Ann NY Acad Sci *779*: 430–442, 1996
35. Gaughan JP, Furukawa S, Jeevanandam V, Hefner CA, Kubo H, Margulies KB, McGowan BS, Mattiello JA, Dipla K, Piacentino V III, Li S, Houser SR: Sodium/calcium exchange contributes to contraction and relaxation in failed human ventricular myocytes. Am J Physiol 77: H714–24, 1999

36. Weber CR, Ginsburg KS, Philipson KD, Shannon TR and Bers DM: Allosteric regulation of Na/Ca exchange current by cytosolic Ca in intact cardiac myocytes. J Gen Physiol 117: 119–131, 2001

37. Golden KL, Fan QI, Chen B, Ren J, O'Connor J, Marsh JD: Adrenergic stimulation regulates Na/Ca exchanger expression in rat cardiac myocytes. J Mol Cell Cardiol 32: 611–620, 2000

Molecular and Cellular Biochemistry **242:** 11–17, 2003.

Importance of Ca^{2+} influx by Na^+/Ca^{2+} exchange under normal and sodium-loaded conditions in mammalian ventricles

Hiroshi Satoh, Masaaki Mukai, Tsuyoshi Urushida, Hideki Katoh, Hajime Terada and Hideharu Hayashi
Division of Cardiology, Internal Medicine III, Hamamatsu University School of Medicine, Handayama, Hamamatsu, Japan

Abstract

Na^+/Ca^{2+} exchange (NCX) is a major Ca^{2+} extrusion system in cardiac myocytes, but can also mediate Ca^{2+} influx and trigger sarcoplasmic reticulum Ca^{2+} release. Under conditions such as digitalis toxicity or ischemia/reperfusion, increased $[Na^+]_i$ may lead to a rise in $[Ca^{2+}]_i$ through NCX, causing Ca^{2+} overload and triggered arrhythmias. Here we used an agent which selectively blocks Ca^{2+} influx by NCX, KB-R7943 (KBR), and assessed twitch contractions and Ca^{2+} transients in rat and guinea pig ventricular myocytes loaded with indo-1. KBR (5 μM) did not alter control steady-state twitch contractions or Ca^{2+} transients at 0.5 Hz in rat, but significantly decreased them in guinea pig myocytes. When cells were Na^+-loaded by perfusion of strophanthidin (50 μM), the addition of KBR reduced diastolic $[Ca^{2+}]_i$ and abolished spontaneous Ca^{2+} oscillations. In guinea pig papillary muscles exposed to substrate-free hypoxic medium for 60 min, KBR (10 μM applied 10 min before and during reoxygenation) reduced both the incidence and duration of reoxygenation-induced arrhythmias. KBR also enhanced the recovery of developed tension after reoxygenation. It is concluded that (1) the importance of Ca^{2+} influx via NCX for normal excitation-contraction coupling is species-dependent, and (2) Ca^{2+} influx via NCX may be critical in causing myocardial Ca^{2+} overload and triggered activities induced by cardiac glycoside or reoxygenation. (Mol Cell Biochem **242**: 11–17, 2003)

Key words: myocytes, Na^+/Ca^{2+} exchange, sarcoplasmic reticulum, arrhythmia

Introduction

Na^+/Ca^{2+} exchange (NCX) is the main mechanism of Ca^{2+} extrusion from cardiac myocytes [1]. Indeed, depending on the species, NCX can compete with the sarcoplasmic reticulum (SR) Ca^{2+}-ATPase to make a significant contribution to cardiac relaxation. For example, in rabbit ventricular myocytes Ca^{2+} extrusion by NCX is responsible for ~ 25% of twitch relaxation, but this number is only ~ 7–8% in rat ventricular myocytes. Compared to the sarcolemmal Ca^{2+}-ATPase, NCX is typically ~ 10 times more potent in extruding Ca^{2+} from the cell during both rest as well as during contraction and relaxation.

Less quantitative information is available concerning Ca^{2+} influx via NCX. Based on thermodynamic considerations and mathematical modeling, Ca^{2+} influx via NCX is most likely to occur during the very early phase of the action potential [1]. Under normal conditions, the total amount of Ca^{2+}, which enters NCX, is rather small. However, when intracellular Na^+ concentration ($[Na^+]_i$) increases during Na^+/K^+ ATPase inhibition or during ischemia/reperfusion, much greater amounts of Ca^{2+} can enter the cell via NCX [1, 2].

While the total amount of Ca^{2+} entry via NCX during the action potential may be small, there is evidence to suggest that this Ca^{2+} entry may be sufficient to trigger SR Ca^{2+} release [3–9]. This could be due to either the direct effect of depolarization on the driving force for Ca^{2+} entry via NCX [5–9] or secondary to Na^+ channel current (I_{Na}) and the local

Address for offprints: H. Satoh, Division of Cardiology, Internal Medicine III, Hamamatsu University School of Medicine, 1-20-1 Handayama, Hamamatsu, 431-3192, Japan (E-mail: satoh36@hama-med.ac.jp)

rise in subsarcolemmal $[Na^+]_i$ [3, 4]. However, the physiological relevance of this mode of SR Ca^{2+} release is controversial [10, 11].

KB-R7943 (KBR) is an agent which has been reported to preferentially block the Ca^{2+} influx mode of the cardiac NCX rather than the Ca^{2+} extrusion mode at the range of 1–10 μM [12–14]. We have clarified that KBR at the concentration of 5–10 μM can be used as a specific inhibitor of the Ca^{2+} influx mode of NCX in rat ventricular myocytes and in guinea pig papillary muscles [15, 16]. Although the many aspects of its inhibitory properties remain controversial, we examined the effects of KBR in mammalian ventricles to clarify the importance of Ca^{2+} influx via NCX in excitation-contraction (E-C) coupling under normal twitch contractions, and in Na^+-loaded condition induced by cardiac glycoside and hypoxia/reoxygenation.

Materials and methods

Preparation of ventricular myocytes

This investigation conforms with the *Guide for the Care and Use of Laboratory Animals* published by the US National Institutes of Health. Ventricular myocytes were isolated from male Sprague–Dawley rats (200 ~ 240 g) or female guinea pigs (300 ~ 400 g) and were loaded with indo-1 acetoxymethyl ester (indo-1 AM) at room temperature as described previously [2, 17]. A small aliquot of myocytes was placed in an experimental chamber mounted on the stage of an inverted microscope (TMD, Nikon, Tokyo, Japan) and perfused with the solution containing (in mM) NaCl 137, KCl 4, $MgSO_4$ 1.2, glucose 10, N-2-hydroxyethylpiperazine-N'-2-ethanesulfonic acid (HEPES) 10, $CaCl_2$ 1.5, with pH adjusted to 7.4 with NaOH. The myocytes received bipolar field stimulation through platinum electrodes at 0.5 Hz.

Apparatus

Intracellular $[Ca^{2+}]$ ($[Ca^{2+}]_i$) was measured from indo-1 fluorescence of myocytes excited at 340 nm and detected at 405 and 485 nm. The 405/480 nm fluorescence ratios after subtraction of corresponding backgrounds were used to indicate changes in $[Ca^{2+}]_i$. Cells were simultaneously trans-illuminated with red light (> 600 nm), and contractions were measured using a linear image sensor and the edge detection system (Hamamatsu Photonics, Hamamatsu, Japan).

Papillary muscle preparations

Papillary muscles were obtained from the right ventricles of guinea pig heart following the method previously reported [18]. Thin papillary muscles were dissected in oxygenated modified Krebs solution (pH 7.4) of the following composition (in mM): NaCl 113.1, KCl 4.6, $CaCl_2$ 2.45, $MgCl_2$ 1.2, $NaHCO_3$ 21.9, and glucose 10. The muscle was mounted in a Perspex bath perfused with oxygenated modified Krebs solution at 37 ± 0.2°C. The mural end of the muscles was clamped, and the tendinous end was tied by a short length of silk thread to a stainless steel rod to measure isometric tension. The length of the muscle was adjusted until the resting tension was 50–100 mg. Stimuli (1 Hz) were applied to the basal part of the preparation through a bipolar Ag-AgCl electrode. The tension was displayed on a storage oscilloscope (model 5113, Tektronix, Tokyo, Japan) and recorded on a pen recorder (model WS-641G, Nihon Kohden, Japan) and on a digital audiotape recorder (model RD-120T, TEAC, Japan).

Reagents and solutions

2-[2-[4-(4-nitrobenzyloxy)phenyl]ethyl] isothiourea methanesulfonate (KB-R7943) was generous gift from the New Drug Research Laboratories, Organon Ltd. (Osaka, Japan). Indo-1 AM was supplied by Molecular Probes. These reagents were used from stock solutions in ethanol or DMSO. All other chemicals were purchased from Sigma (St. Louis, MO). Final [DMSO] at 5 or 10 μM KBR was ≤ 0.1%.

Statistical analyses

Results were expressed as means ± S.E. for the indicated number (n) of isolated myocytes. Statistical analyses were performed using Student's *t*-test. The probability was considered significant at $p < 0.05$.

Results

Effects of KBR on steady-state (SS) twitches and Ca^{2+} transients in rat and guinea pig ventricular myocytes

Figure 1A shows simultaneous recordings of twitch cell shortening and Ca^{2+} transient during SS stimulation at 0.5 Hz in rat ventricular myocytes. The Ca^{2+} transients and contractions were obtained during control conditions and 8 min after the application of 1 and 5 μM KBR, respectively. Figures 1B and 1C summarize the changes in SS twitch cell shortening and Ca^{2+} transient amplitude. KBR up to 5 μM did not significantly affect either SS twitch contractions or Ca^{2+} transients.

Figure 2A represents recordings of twitch cell shortening and Ca^{2+} transient during SS stimulation in guinea pig ventricular myocytes. Figures 2B and 2C indicate the changes

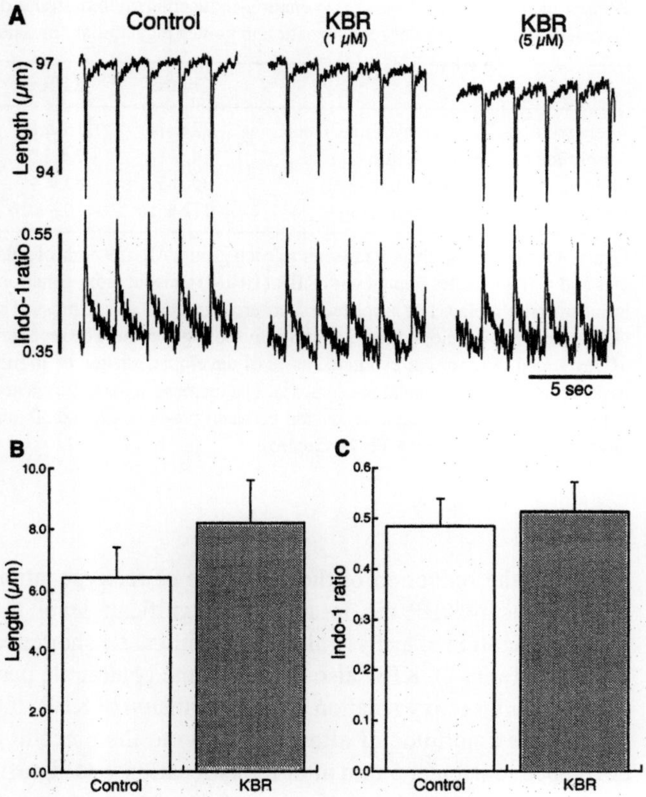

Fig. 1. Effects of KB-R7943 (KBR) on steady-state (SS) twitch contractions and Ca^{2+} transients in rat ventricular myocytes. (A) Representative examples of five SS twitch contractions and Ca^{2+} transients under control condition and during the perfusion of KBR at room temperature. Twitch contractions (upper) and Ca^{2+} transients (lower) were recorded simultaneously in an indo-1-loaded myocyte. (B and C) The summary data of twitch cell shortenings (B) and Ca^{2+} transient amplitudes (C) before (Control) and after application of 5 μM KBR (KBR). KBR up to 5 μM did not affect either cell shortening or Ca^{2+} transient amplitude. Data are means ± S.E. from 12 paired experiments.

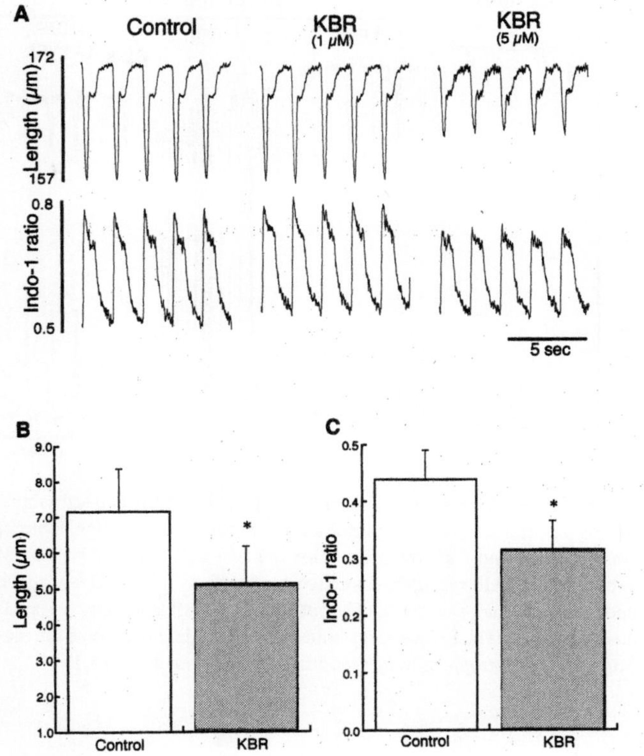

Fig. 2. Effects of KBR on SS twitch contractions and Ca^{2+} transients in guinea pig ventricular myocytes. (A) Representative examples of five SS twitch contractions and Ca^{2+} transients under control condition and during the perfusion of KBR at room temperature. Twitch contractions (upper) and Ca^{2+} transients (lower) were recorded simultaneously in an indo-1-loaded myocyte. (B and C) the summary data of twitch cell shortenings (B) and Ca^{2+} transient amplitudes (C) before (Control) and after application of 5 μM KBR (KBR). In guinea pig, KBR at 1 μM did not change either cell shortening or Ca^{2+} transient amplitude, but at 5 μM it decreased them. Data are means ± S.E. from 5 paired experiments. *$p < 0.05$ vs. Control.

in SS twitch cell shortening and Ca^{2+} transient amplitude. In guinea pig, KBR at 1 μM did not produce significant effects, but at 5 μM, it did reduce both SS twitch contractions and Ca^{2+} transient amplitudes.

Effects of KBR under Na^+-loaded conditions

Inhibition of Na^+/K^+ ATPase by cardiac glycosides leads to an increase in $[Na^+]_i$, which can lead to increased $[Ca^{2+}]_i$ via NCX and thereby can induce a positive inotropic effect or even Ca^{2+} overload and arrhythmias [1]. We previously showed that strophanthidin increased the basal $[Ca^{2+}]_i$ and amplitude of Ca^{2+} transients as $[Na^+]_i$ rose, and when arrhythmias occurred the amplitude of Ca^{2+} transients decreased while $[Na^+]_i$ and basal $[Ca^{2+}]_i$ continued to increase [19]. The increase in Ca^{2+} in response to increased $[Na^+]_i$ can occur by either limiting Ca^{2+} efflux via NCX or if $[Na^+]_i$ is high enough producing net Ca^{2+} influx via NCX. Here we used KBR to examine the role of Ca^{2+} influx via NCX in the positive inotropic and arrhythmogenic effects of strophanthidin in rat ventricular myocytes.

Figure 3 shows representative records of cell shortening and Ca^{2+} transient during SS 0.5 Hz stimulation before and during application of 50 μM strophanthidin. The addition of strophanthidin increased both twitch contraction and Ca^{2+} transient amplitude, but then spontaneous contractile activities and $[Ca^{2+}]_i$ oscillations occurred at 8 min, indicative of Ca^{2+} overload. The application of 5 μM KBR abolished the spontaneous activity and partially restored diastolic $[Ca^{2+}]_i$, but the strophanthidin-induced inotropy remained largely intact. The similar effects were obtained in 3 other cells, which showed spontaneous activities and Ca^{2+} oscillations during strophanthidin application.

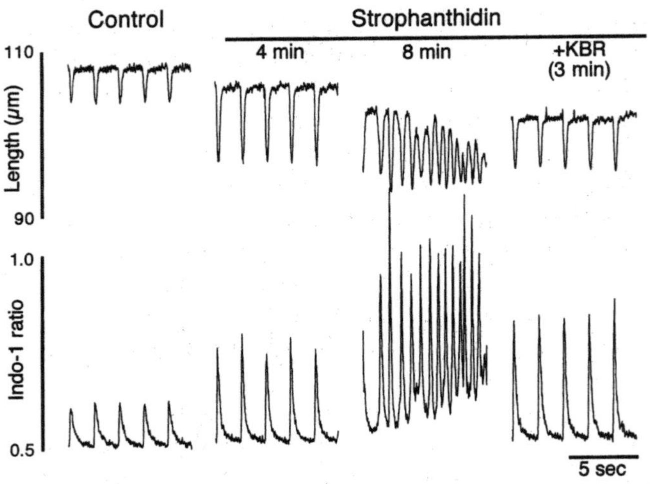

Fig. 3. Effects of KBR on strophanthidin-induced inotropy and arrhythmia in a rat ventricular myocyte. Simultaneous records of SS twitch contractions (upper) and Ca^{2+} transients (lower) in a cell exposed to 50 µM strophanthidin. After spontaneous activities appeared, 5 µM KBR was added under the continuous presence of strophanthidin. KBR abolished the spontaneous activities and reduced diastolic $[Ca^{2+}]_i$, but did not reverse the inotropic effect of strophanthidin. (Modified from Satoh *et al.* (15)).

These results indicate that the Ca^{2+} influx via NCX is not so important for the positive inotropic effect of strophantidin *per se* in rat ventricular myocytes. However, Ca^{2+} influx via NCX may be critical in causing the transition to Ca^{2+} overload in response to elevated $[Na^+]_i$.

Effects of KBR on reoxygenation-induced arrhythmias

Finally, the protective effects of KBR on reoxygenation-induced injuries were examined in guinea pig papillary muscles. In the control experiments, the muscles were perfused for 60 min with the substrate-free hypoxic modified Krebs solution which was achieved by gassing the glucose-free modified Krebs solution with 95% N_2/5% CO_2, and then reoxygenated with the control modified Krebs solution gassed with 95% O_2/5% CO_2. The muscles were stimulated at 1 Hz during the hypoxic period and after reoxygenation. During hypoxia, there was a substantial decline in developed tension and an increase in resting tension. Within minutes of subsequent reoxygenation, arrhythmic activities, which were associated with the occurrence of aftercontractions, were elicited. After the disappearance of arrhythmias, developed tension and resting tension recovered to various degrees.

In the control hypoxia/reoxygenation experiments, all of 9 muscles showed reoxygenation-induced arrhythmias during the 0–5 min reoxygenation period. When KBR (10 µM) was applied from 10 min before reoxygenation, 5 of 9 muscles were exempt from reoxygenation-induced arrhythmias.

Table 1. Effects of KBR on reoxygenation-induced arrhythmias and on contractile parameters after reoxygenation in guinea pig papillary muscles

	Control	KBR
Averaged duration of arrhythmias (sec)	72 ± 14	16 ± 11*
Amplitude of aftercontraction (%)	14 ± 1	7 ± 1*
Recovery of developed tension (%)	69 ± 3	83 ± 4*
Increase in diastolic tension (mg)	112 ± 15	109 ± 16

Data are means ± S.E. from 9 muscles in each group. All of 9 control muscles and 4 of 9 muscles treated with KBR (10 µM) showed reoxygenation-induced arrhythmias. The amplitude of aftercontraction was expressed as % of preceding developed tension at 5 min reoxygenation. The recovery of developed tension was estimated as % of developed tension at 30 min reoxygenation against that at pre-hypoxia. The increase in diastolic tension indicates difference in diastolic tension between pre-hypoxia and 30 min after reoxygenation. *$p < 0.05$ vs. Control.

Although the reduction of the incidence of reoxygenation arrhythmias by KBR did not reach the significant level, the summed duration of arrhythmias was significantly shortened by KBR (Table 1). KBR also improved the contractile performance after reoxygenation. By the perfusion of KBR, the ratio of the amplitude of aftercontraction to the preceding developed tension at 5 min after reoxygenation was significantly decreased, and the recovery of developed tension at 30 min after reoxygenation became greater. However, there was no significant difference in the increase in diastolic tension at 30 min after reoxygenation (Table 1).

Discussion

Ca^{2+} influx via NCX in normal condition; species difference

Thermodynamic calculations indicate that, during the upstroke of the cardiac action potential, membrane potential exceeds the reversal potential of the exchanger such that there is a modest driving force favoring Ca^{2+} entry into the cell. The ability of Ca^{2+} influx mode of NCX to trigger SR Ca^{2+} release was first reported by Berlin *et al.* [20] in a Ca^{2+}-overloaded cardiac preparation. Leblanc and Hume [3] have also presented more direct evidence that Ca^{2+} entry via NCX may induce SR Ca^{2+} release. In this case, they suggested that tetrodotoxin-sensitive Na^+ entry might increase the subsarcolemmal $[Na^+]_i$ thereby activating Ca^{2+} entry via NCX and SR Ca^{2+} release. This hypothesis was supported by other laboratories [4, 21]. In addition, recent works have suggested that voltage-dependent NCX may contribute to E-C coupling even when I_{Na} is inactivated [5–9].

Contrary, Sham *et al.* (10) have suggested that I_{Na}-associated $[Ca^{2+}]_i$ transient results from activation of Ca^{2+} channels during voltage escape. Sipido *et al.* [22] also exhibited that the manipulating Na^+ influx through I_{Na} failed to modulate the SR Ca^{2+} release and suggested that I_{Na} modulates SR Ca^{2+} release by affecting the Ca^{2+} load of the SR. They also assumed in a mathematical model, that Ca^{2+} entering via NCX was only 25% as efficient as that via the L-type channel in promoting Ca^{2+} release from the SR [11]. This may be because NCX does not tightly couple with the SR Ca^{2+} release channel as L-type Ca^{2+} channel does.

This controversy may be ascribed to several factors. First, NCX function is temperature-dependent, such that lower temperature tends to minimize the apparent contribution to E-C coupling [8, 23]. In this regard, we previously indicated that Ca^{2+} entry via NCX has no significant role on E-C coupling at both 23 and 36°C in rat myocytes [15]. Second, there may be species difference in the contribution of NCX to E-C coupling. A recent study using transgenic mice [24] has found that after a treatment with nifedipine, the prominent twitch Ca^{2+} transients were preserved in the transgenic mouse myocyte overexpressing NCX, whereas only small Ca^{2+} transients were seen in the wild type mouse myocyte. In contrast, Adachi-Akahane *et al.* [25] reported that even with the three fold increased density of $I_{Na/Ca}$, the exchanger did not significantly alter the SR Ca^{2+} content and failed to trigger Ca^{2+} release from the SR in the physiological voltage range. In this study, we observed that KBR (5 μM) reduced the cell shortening and Ca^{2+} transient amplitude in guinea pig myocytes, significantly. Previous reports also exhibited negative inotropic effects of KBR in rabbit and canine ventricles [26, 27]. Therefore, the importance of Ca^{2+} influx via NCX may vary because of the difference in the level of expression of the exchanger, action potential configuration and subsarcolemmal $[Na^+]_i$.

Ca^{2+} influx via NCX in Na^+-loaded condition: induction of inotropy and Ca^{2+} overload

It is well known that the inhibition of the Na^+/K^+ pump by cardiac glycosides enhances contractility by increasing $[Ca^{2+}]_i$ through modified NCX activity. The steep relation between intracellular Na^+ activity (aNa_i) and developed tension was first reported in cardiac Purkinje fibers [28]. We also demonstrated in individual guinea pig myocytes that during the inotropic phase of strophanthidin perfusion the increase in the amplitude of Ca^{2+} transient is well associated with an increase in $[Na^+]_i$ [19].

The small increases in $[Na^+]_i$ can have a large impact on the balance of Ca^{2+} fluxes mediated by NCX. For example, an increase of aNa_i from 7–10 mM would shift the reversal potential of NCX by ~ 30 mV in the negative direction. This shift in the reverse potential of NCX ($E_{Na/Ca}$) will shift the competition between SR Ca^{2+} pump and NCX more in favor of the SR Ca^{2+} pump during relaxation [1]. Even though there was no net Ca^{2+} influx via NCX, this may increase both the diastolic $[Ca^{2+}]_i$ and the SR Ca^{2+} load, resulting in greater SR Ca^{2+} release during twitch. Our result that KBR did not prevent the strophanthidin-induced increase in twitch cell-shortening or the amplitude of Ca^{2+} transient well agrees with this idea. However, we have to remind other possible mechanisms of the inotropic action of strophanthidin, such as an increase in the activity of SR Ca^{2+} release channels [29] or an increase in Ca^{2+} influx through Na^+ channels [30].

It has been shown that at high concentrations of cardioactive steroids the positive inotropic action gives way to a negative inotropic effect [1]. We previously showed that when arrhythmic activities were induced by strophanthidin, the Ca^{2+} transient amplitude decreased in spite of the continuous increase in diastolic $[Ca^{2+}]_i$ and $[Na^+]_i$ [19]. When $[Na^+]_i$ becomes high enough (~ 16 mM) by the perfusion of strophanthidin, $E_{Na/Ca}$ goes below resting membrane potential causing net Ca^{2+} influx via NCX even at rest [1]. This influx can induce SR Ca^{2+} release by overloading SR with Ca^{2+} (spontaneous Ca^{2+} release) or by triggering Ca^{2+} release [1]. This SR Ca^{2+} release raises $[Ca^{2+}]_i$ thereby activating net Ca^{2+} extrusion via NCX and unloading the SR [1, 31]. The finding that KBR abolished the spontaneous contractile activities and reduced diastolic $[Ca^{2+}]_i$ well predicts the Ca^{2+} gain via NCX at these high $[Na^+]_i$ levels. Thus, KBR could be useful in limiting dysfunction and arrhythmias induced by Ca^{2+} overload, especially in those where NCX is implicated.

Ca^{2+} influx via NCX in reoxygenation-induced arrhythmias and in recovery of developed tension after reoxygenation

There was a high incidence of arrhythmias on reoxygenation after 60 min of substrate-free hypoxia. It is well known that the SR overloaded with Ca^{2+} causes the oscillatory Ca^{2+} release and the resulting increase in $[Ca^{2+}]_i$ triggers transient inward currents and delayed after depolarizations (DADs) [18, 32, 33]. Although the detailed mechanism of Ca^{2+} overload on reoxygenation remains undefined, the massive and rapid Ca^{2+} influx via NCX is a candidate [2, 34, 35]. In this study, KBR inhibited the reoxygenation-induced arrhythmias and aftercontractions. The reduction of aftercontractions could be due to the amelioration of SR Ca^{2+} overload [36].

We have shown that reoxygenation-induced arrhythmias can be inhibited by various agents such as the Na^+ channel blocker (tetrodotoxin), the L-type Ca^{2+} channel blocker (D-600), the K^+ channel opener (nicorandil), and the SR Ca^{2+} release channel inhibitor (ryanodine) [37]. However, the mode of actions of KBR was obviously different from these agents. For example, tetrodotoxin and nicorandil did not inhibit the amplitude of aftercontractions. Although D-600 suppressed the amplitude of aftercontractions, the ratio to developed tension remained unaffected. Ryanodine reduced both, but also decreased the developed tension under normal condition. The difference in the mode of action by KBR is consistent with the suggested mechanism of the drug, that is, the selective inhibition of the Ca^{2+} influx via NCX. However, KBR at 10 μM did not inhibit the arrhythmic activity completely. This finding raises the possibility that other routes of Ca^{2+} influx during reoxygenation such as the L-type Ca^{2+} channel and/or non-specific cation channel current might contribute to Ca^{2+} overload [38].

We also found that KBR significantly enhanced the post-hypoxic recovery of developed tension. The lower degree of the recovery of developed tension and diastolic tension seems to relate to the cellular Ca^{2+} load on reoxygenation [39]. Therefore, reduced Ca^{2+} influx via NCX by KBR could lead to the amelioration of Ca^{2+} overload and better contractile recovery. However, there were no significant differences in the increase in diastolic tension at 30 min after reoxygenation between the control muscles and those treated with KBR. It could be possible that other factors than $[Ca^{2+}]_i$, (e.g. depletion of ATP) determine the recovery of diastolic tension [40].

In conclusion, the importance of Ca^{2+} influx via NCX for normal E-C coupling is species-dependent, and is involved in potentially arrythmogenic Ca^{2+} overload at digitalis toxicity or hypoxia/reoxygenation. More detailed studies should be addressed to clarify the functional feature of NCX.

Acknowledgments

This work was supported by Japan Heart Foundation & Pfizer Pharmaceuticals Grant for Research on Coronary Artery Disease (H. Satoh). The authors wish to thank New Drug Research Laboratories, Organon Ltd. for supplying KBR.

References

1. Bers DM: Excitation-Contraction Coupling and Cardiac Contractile Force. 2nd edn. Kluwer Academic Press, Dordrecht, Netherlands, 2001
2. Satoh H, Hayashi H, Katoh H, Terada H, Kobayashi A: Na^+/H^+ and Na^+/Ca^{2+} exchange in regulation of $[Na^+]_i$ and $[Ca^{2+}]_i$ during metabolic inhibition. Am J Physiol 268: H1239–H1248, 1995
3. Leblanc N, Hume JR: Sodium current-induced release of calcium from cardiac sarcoplasmic reticulum. Science 248: 372–376, 1990
4. Lipp P, Niggli E: Sodium current induced calcium signals in isolated guinea-pig ventricular myocyte. J Physiol 474: 439–446, 1994
5. Kohmoto O, Levi AJ, Bridge JHB: The relationship between reverse Na-Ca exchange and SR Ca^{2+} release in guinea-pig ventricular myocytes. Circ Res 74: 550–554, 1994
6. Levi AJ, Spitzer KW, Kohmoto O, Bridge JHB: Depolarization-induced Ca entry via Na-Ca exchange triggers SR release in guinea pig cardiac myocytes. Am J Physiol 266: H1422–H1433, 1994
7. Hancox JC, Evans SJ, Levi AJ: The fura-2 transient can show two types of voltage dependence at 36 degrees C in ventricular myocytes isolated from the rat heart. Pflügers Arch 432: 215–224, 1996
8. Wasserstrom JA, Vites AM: The role of Na^+-Ca^{2+} exchange in activation of excitation-contraction coupling in rat ventricular myocytes. J Physiol 493: 529–542, 1996
9. Litwin SE, Li J, Bridge JH: Na-Ca exchange and the trigger for sarcoplasmic reticulum Ca release: Studies in adult rabbit ventricular myocytes. Biophys J 75: 359–371, 1998
10. Sham JK, Cleeman L, Morad M: Gating of the cardiac Ca^{2+} release channel: The role of Na^+ current and Na^+/Ca^{2+} exchange. Science 255: 850–853, 1992
11. Sipido KR, Maes MM, Van de Werf: Low efficiency of Ca^{2+} entry through the Na/Ca exchanger as trigger for Ca^{2+} release from the sarcoplasmic reticulum. Circ Res 81: 1034–1044, 1997
12. Watano T, Kimura J, Morita T, Nakanishi H: A novel antagonist, No. 7943 of the Na^+/Ca^{2+} exchange current in guinea-pig cardiac ventricular cells. Br J Pharmacol 119: 555–563, 1996
13. Iwamoto T, Watano T, Shigekawa M: A novel isothiourea derivative selectively inhibits the reverse mode of Na^+/Ca^{2+} exchange in cells expressing NCX1. J Biol Chem 271: 22391–22397, 1996
14. Elias CL, Lukas A, Shurraw S, Scott J, Omelchenko A, Gross GJ, Hnatowich M, Hryshko LV: Inhibition of Na^+/Ca^{2+} exchange by KB-R7943: Transport mode selectivity and antiarrhthmic consequences. Am J Physiol Heart Circ Physiol 281: H1334–H1345, 2001
15. Satoh H, Ginsburg KS, Qing K, Terada H, Hayashi H, Bers DM: KB-R7943 block of Ca^{2+} influx via Na^+/Ca^{2+} exchange does not alter twitches or glycoside inotropy, but prevents Ca^{2+} overload in rat ventricular myocytes. Circulation 101: 1441–1446, 2000
16. Mukai M, Terada H, Sugiyama S, Satoh H, Hayashi H, Ohno R: Effects of a selective inhibitor of Na^+/Ca^{2+} exchange, KB-R7943, on reoxygenation-induced injuries in guinea pig papillary muscles. J Cardiovasc Pharmacol 35: 121–128, 2000
17. Hayashi H, Miyata H, Noda N, Kobayashi A, Hirano M, Kawai T, Yamazaki N: Intracellular Ca^{2+} concentration and pH_i during metabolic inhibition. Am J Physiol 262: C628–C634, 1992
18. Hayashi H, Ponnambalam C, McDonald TF: Arrhythmic activity in reoxygenated guinea pig papillary muscles and ventricular cells. Circ Res 61: 124–133, 1987
19. Terada H, Hayashi H, Satoh H, Katoh H, Yamazaki N: Simultaneous measurement of $[Na^+]i$ and Ca^{2+} transients in an isolated myocyte: Effects of strophanthidin. Biochem Biophys Res Commun 203: 1050–1056, 1994
20. Berlin JR, Cannel MB, Lederer WJ: Regulation of twitch tension in sheep cardiac Purkinje fibers during calcium overload. Am J Physiol 253: H1540–H1547, 1987
21. Levesque PC, Leblanc N, Hume JR: release of calcium from guinea pig cardiac sarcoplasmic reticulum induced by sodium-calcium exchange. Cardiovasc Res 28: 370–378, 1994
22. Sipido KR, Carmeliet E, Pappano A: Na^+ current and Ca^{2+} release from the sarcoplasmic reticulum during action potentials in guinea-pig ventricular myocytes. J Physiol 489: 1–17, 1995
23. Vornanen M, Shephed N, Isenburg G: Tension-voltage relations of single myocytes reflect Ca release triggered by Na/Ca exchange at 35°C but not 23°C. Am J Physiol 267: C623–C632, 1994
24. Yao A, Nonaka A, Zubair I, Lu L, Philipson KD, Bridge JHB, Barry H: Effects of overexpression of the Na^+-Ca^{2+} exchanger on $[Ca^{2+}]_i$ transients in murine ventricular myocytes. Circ Res 82: 657–665, 1998
25. Adachi-Akahane S, Lu L, Li Z, Frank JS, Philipson KD: Calcium signalling in transgenic mice overexpressing cardiac Na^+-Ca^{2+} exchanger. J Gen Physiol 109: 717–729, 1997
26. Yang HT, Sakurai K, Sugawara H, Watanabe T, Norota I, Endoh M: Role of Na^+/Ca^{2+} exchange in endothelin-1-induced increases in Ca^{2+}

transient and contractility in rabbit ventricular myocytes: pharmacological analysis with KB-R7943. Br J Pharmacol 126: 1785–1795, 1999

27. Kurogouchi F, Furukawa Y, Zhao D, Hirose M, Nakajima K, Tsuboi M, Chiba S: A Na^+/Ca^{2+} exchanger inhibitor, KB-R7943, caused negative inotropic responses and negative followed by positive chronotropic responses in isolated, blood-perfused dog heart preparations. Jpn J Pharmacol 82:155–163, 2000
28. Lee CO, Dagostino M: Effect of strophanthidin on intracellular Na ion activity and twitch tension of constantly driven canine cardiac Purkinje fibers. Biophys J 40:185–198, 1982
29. Morgan JP: The effects of digitalis on intracellular calcium transients in mammalian working myocardium as detected with aequorin. J Mol Cell Cardiol 17: 1065–1075, 1985
30. Santana LF, Gomez AM, Lederer WJ: Ca^{2+} flux through promiscuous cardiac Na^+ channels: slip-mode conductance. Science 279: 1027–1033, 1998
31. Díaz ME, Trafford AW, O'Neill SC, Eisner DA: A measurable reduction of s.r. Ca content follows spontaneous Ca release in rat ventricular myocytes. Pflügers Arch 434: 852–854, 1997
32. Allen DG, Eisner DA, Orchard CH: Characterization of oscillation of intracellular calcium concentration in ferret ventricular muscle. J Physiol 352: 113–128, 1984
33. Ferrier GR, Moffat MP, Lukas A: Possible mechanisms of ventricular arrhythmias elicited by ischemia followed by reperfusion. Studies on isolated canine ventricular tissues. Circ Res 56: 184–194, 1985
34. Poole-Wilson PA, Harding DP, Bourdillon PDV, Tones MA: Calcium out of control. J Mol Cell Cardiol 16: 175–187, 1984
35. Tani M, Neely JR: Role of intracellular Na in Ca overload and depressed recovery of ventricular function of reperfused ischemic rat heart. Circ Res 65: 1045–1056, 1989
36. Nuss HB, Houser SR: Sodium-calcium exchange-mediated contractions in feline ventricular myocytes. Am J Physiol 263: H1161–H1169, 1992
37. Hayashi H, Terada H, Katoh H, McDonald TF: Prevention of reoxygenation-induced arrhythmias in guinea pig papillary muscles. J Cardiovasc Pharmacol 27: 816–824, 1996
38. Colquhoun D, Neher E, Reuter H, Stevens CF: Inward current channels activated by intracellular Ca^{2+} in cultured cardiac cells. Nature 294: 752–754, 1981
39. Harding DP, Poole-Wilson PA: Calcium exchange in rabbit myocardium during and after hypoxia: Effect of temperature and substrate. Cardiovasc Res 14: 435–445, 1980
40. Allen DG, Orchard CH: Intracellular calcium concentration during hypoxia and metabolic inhibition in mammalian ventricular muscle. J Physiol 339: 107–122, 1983

Molecular and Cellular Biochemistry **242**: 19–25, 2003.

Compensated hypertrophy of cardiac ventricles in aged transgenic FVB/N mice overexpressing calsequestrin

Yoji Sato,[1] Albrecht G. Schmidt,[2] Helen Kiriazis,[2] Brian D. Hoit[3] and Evangelia G. Kranias[2]

[1]*Division of Xenobiotics, Metabolism and Disposition, National Institute of Health Sciences, Setagaya, Tokyo, Japan;* [2]*Department of Pharmacology and Cell Biophysics, University of Cincinnati College of Medicine, Cincinnati, OH;* [3]*Division of Cardiology, University Hospitals of Cleveland and Case Western Reserve University, Cleveland, OH, USA*

Abstract

Cardiac-specific overexpression of murine cardiac calsequestrin results in depressed contractile parameters and hypertrophy in transgenic mice. To determine the long-term consequences of calsequestrin overexpression, the cardiac phenotype of young (2–3-months old) and aged (17 months old) transgenic FVB/N mice was characterized. Ventricular/body weight ratios, which were increased in young transgenics compared with wild-types, were unaltered with age. Left atria of aged transgenics exhibited enlargement and mineralization, but their ventricles did not display fibrosis, mineralization and other injuries. Although echocardiography suggested a time-dependent change in ventricular geometry and loading conditions *in vivo*, as well as an age-dependent reduction of left ventricular fractional shortening in transgenic mice, Langendorff-perfused hearts of young and aged transgenics indicated that there were no age-related reductions of contractile parameters (±dP/dt). Furthermore, neither genotype nor age altered lung/body weight ratios. Thus, our findings suggest that left ventricular performance in calsequestrin overexpressing mice becomes apparently depressed with age, but this depression is not associated with progressive reduction of left ventricular contractility and heart failure. (Mol Cell Biochem **242**: 19–25, 2003)

Key words: Sarcoplasmic reticulum, echocardiography, myocardial contractility, cardiac hypertrophy, cardiomyopathy

Abbreviations: SR – sarcoplasmic reticulum; $dCSQ_{OE}$ – transgenic mouse overexpressing canine cardiac calsequestrin; $mCSQ_{OE}$ – transgenic mouse overexpressing murine cardiac calsequestrin; WT – wild-type; SERCA – sarco/endoplasmic reticulum Ca^{2+}-ATPase; h/r – end-diastolic left-ventricular wall thickness/cavity radius ratio; LVFS – left-ventricular percent fractional shortening; Vcf_c – velocity of circumferential fiber shortening corrected for heart rate

Introduction

Calsequestrin is a luminal sarcoplasmic reticulum (SR) protein, which binds Ca^{2+} with low-affinity and high-capacity, and has been suggested to form a functional complex with the ryanodine receptor through interactions with triadin or junctin at the cardiac junctional SR [1]. Recently, two mouse models with cardiac-specific overexpression of calsequestrin were generated, providing valuable insights the physiological role of calsequestrin: (i) mice with 10-fold overexpression of canine cardiac calsequestrin ($dCSQ_{OE}$) [2]; and (ii) mice with 20-fold overexpression of murine cardiac calsequestrin ($mCSQ_{OE}$) [3]. With the latter, mouse lines overexpressing 43- and 56-fold murine calsequestrin were also

Present addresses: A.G. Schmidt, Zentrum Innere Medizin, Abteilung Kardiologie und Pneumologie, Georg-August-Universität Göttingen, Robert-Koch-Strasse 40, 37075 Göttingen, Germany; H. Kiriazis, Department of Physiology, Monash University, Clayton, Victoria 3168, Australia
Address for offprints: E.G. Kranias, Department of Pharmacology and Cell Biophysics, University of Cincinnati College of Medicine, 231 Albert Sabin Way, Cincinnati, OH 45267-0575, USA (E-mail: Litsa.Kranias@uc.edu)

propagated and had similar cardiac function characteristics to the 20-fold overexpression line [3]. Overexpression of canine or murine cardiac calsequestrin resulted in a markedly augmented SR Ca^{2+} storage capacity; however, there was an impairment of SR Ca^{2+} release upon depolarization in transgenic myocytes, as evidenced by depressed Ca^{2+} transients. Consistent with the attenuation in Ca^{2+}-induced Ca^{2+} release from the SR, contractile parameters were depressed for both $dCSQ_{OE}$ and $mCSQ_{OE}$ hearts [2–5]. The depressed cardiac function was associated with the induction of a fetal gene program [3,4], which characterizes pathological cardiac remodeling. Indeed, both $dCSQ_{OE}$ and $mCSQ_{OE}$ models developed cardiac hypertrophy although, surprisingly, the severity and prognosis was quite different between the two.

At 7 weeks of age, $dCSQ_{OE}$ mice were characterized by concentric left ventricular hypertrophy and mild reduction in systolic function [4]. This phenotype rapidly deteriorated to one of dilated cardiomyopathy at 14 weeks of age and led to premature death by 16 weeks [4]. The appearance of severe dilated cardiomyopathy in this model is believed to be due to the altered expression of cardiac calsequestrin and/or associated myocardial Ca^{2+} cycling defects, and has led to the use of the $dCSQ_{OE}$ mouse as a human heart failure model [6]. However, this is not supported by our findings obtained in adult $mCSQ_{OE}$ mice. In contrast to the $dCSQ_{OE}$ model, 8–15 week old $mCSQ_{OE}$ mice display a concentric cardiac hypertrophic phenotype with the absence of severe cardiac dysfunction [3, 5, 7]. Due to this discrepancy between the two models, the direct effects of increased calsequestrin levels on the myocardial phenotype remain controversial.

It is possible that the time-courses of myocardial pathogenesis between the $dCSQ_{OE}$ and $mCSQ_{OE}$ models are different, and that the eventual prognoses are common. If chronic suppression of Ca^{2+} release, upon overexpression of calsequestrin, initiates a hypertrophic pathway that ultimately results in cardiac decompensation, then the latter should become evident with age. Furthermore, cardiac hypertrophy *per se* may secondarily lead to ventricular wall thinning and cavity dilation with time [8, 9]. In this study, to better understand the primary effects of calsequestrin overexpression on the myocardium and its long-term consequences, and to evaluate the significance of SR Ca^{2+} cycling defects in the pathogenesis of myocardial diseases, we examined cardiac hypertrophy and contractile parameters in young (2–3 months old) and aged (17 months old) $mCSQ_{OE}$ transgenic mice.

Materials and methods

Generation of transgenic mice

Transgenic mice (strain: FVB/N) overexpressing murine cardiac calsequestrin ($mCSQ_{OE}$) were generated and maintained, as previously described (Line #418) [3]. Male mice were used for this study. All animals were maintained in pathogen-free facilities with food and water supplied *ad libitum*. The handling and maintenance of the animals in this study were approved by the ethics committee of the University of Cincinnati, and followed the guidelines of the National Institute of Health Sciences in Tokyo.

Immunoblotting

Left ventricles were homogenized at 4°C in a buffer containing 10 mM imidazole (pH 7.0), 300 mM sucrose, 1 mM dithiothreitol, 1 mM sodium metabisulfite, 0.3 mM phenymethylsulfonyl fluoride, 5 μg/ml leupeptin, 2 mM EDTA, 10 μg/ml soybean trypsin inhibitor type II-S, and 7 μg/ml pepstatin A. After solubilization of homogenates, SDS-polyacrylamide electrophoresis and quantitative immunoblotting for calsequestrin, sarco/endoplasmic reticulum Ca^{2+}-ATPase (SERCA) and phospholamban were performed as described previously [3].

Morphological analysis

Histological evaluation of the hearts from 17-month old $mCSQ_{OE}$ and WT mice was performed as described previously [3]. Briefly, the tissues were fixed in 10% formalin, dehydrated through graded alcohols, and embedded in paraffin. Longitudinal sections (5 μm) of the heart (cut at 50-μm intervals) were stained with Masson's trichrome method. Hearts were examined for hypertrophy, mineralization, fibrosis, thrombogenesis and milder changes. Wet tissue/body weight ratios of cardiac chambers and lungs from 3- and 17-month old mice were also measured.

Echocardiography

M-mode and Doppler echocardiography were carried out to non-invasively assess left-ventricular performance and dimensions, as described previously [5]. End-diastolic left-ventricular wall thickness/cavity radius ratio (h/r), left-ventricular percent fractional shortening (LVFS) and velocity of circumferential fiber shortening corrected for heart rate (Vcf_c) were calculated as previously described [5].

Langendorff perfusion

Contractile parameters of isolated hearts from 2–3- and 17-month old mice were determined in Langendorff mode with a constant perfusion pressure of 50 mmHg, as described previously [3, 7]. The frequency-response relation of the pressure monitoring system (including the catheter) was flat up to 46 and 52 Hz within 10 and 20% amplitude distortion levels, respectively, as determined by 'pop tests' [10]. These frequencies were approximately the 7th or 8th harmonics of heart rates of the preparations, indicating adequate fidelity [11].

Statistics

Data are expressed as means ± S.E. Comparisons across groups were evaluated using two-way analyses of variance (ANOVA). When the p-value was less than 0.05, Bonferroni's multiple *t*-test was employed as a *post hoc* test to determine any significance between pairs of means.

Results

Aged calsequestrin overexpressing mice

In contrast to findings in the $dCSQ_{OE}$ model [4, 6], $mCSQ_{OE}$ transgenic mice (Line #418) exhibited no gross abnormality up to 17 months of age with 100% survival (9 out of 9 mice in each group). As shown in Fig. 1, immunoblot analysis of cardiac ventricular homogenates revealed that aging did not alter the relative calsequestrin levels in $mCSQ_{OE}$ hearts in comparison to WTs. Protein level of cardiac calsequestrin in ventricles of 17-month old $mCSQ_{OE}$ mice was 20.2 ± 6.3-fold (n = 5) over the age-matched WT littermates, which was comparable to that of young $mCSQ_{OE}$ mice [3]. Protein levels of SERCA and phospholamban in the aged $mCSQ_{OE}$ hearts were 0.96 ± 0.14-fold and 1.03 ± 0.18-fold (n = 4), respectively, compared to their WT littermates.

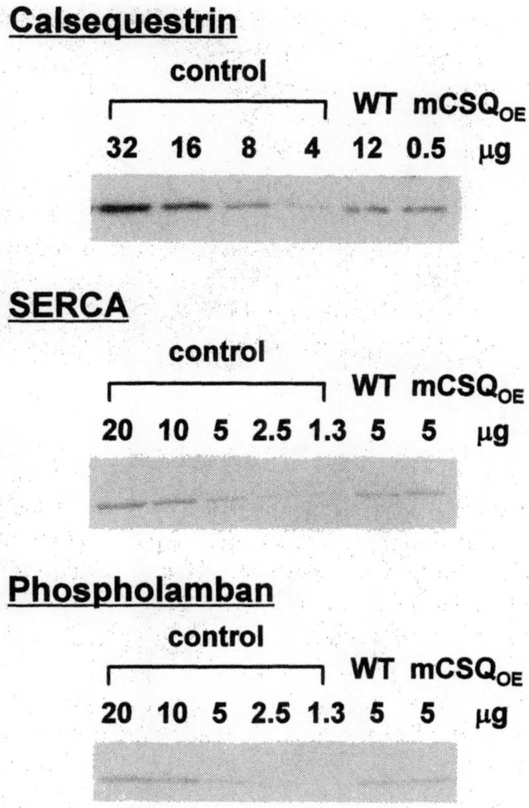

Fig. 1. Immunoblotting for SR Ca^{2+} handling proteins in aged transgenic mouse hearts. Dilution series of a control homogenate from pooled wild-type hearts were used to generate standard lines for quantitation of calsequestrin, sarco/endoplasmic reticulum Ca^{2+}-ATPase (SERCA) and phospholamban levels in cardiac ventricular homogenates from 17-month old calsequestrin-overexpressing mice ($mCSQ_{OE}$) and their age-matched wild-type (WT) littermates. For the quantitation of phospholamban, the solubilized samples were boiled for 5 min to fully dissociate the pentameric form of phospholamban into monomers. Calsequestrin levels in $mCSQ_{OE}$ were approximately 20-fold above WT, and SERCA and phospholamban levels were not different between the two groups.

Morphological analysis

Examination of gross cardiac morphology revealed severe left atrial enlargement with thrombus and mineralization in aged $mCSQ_{OE}$ hearts (Figs 2 and 3). In spite of the severe atrial morbidity, no other obvious pathological changes, except mild ventricular hypertrophy, were observed in histologic sections of aged $mCSQ_{OE}$ hearts (Fig. 2). In young $mCSQ_{OE}$ hearts, mild increases in both left and right ventricular/body weight ratios were observed, indicating biventricular hypertrophy. The increases in the ventricular/body weight ratios by overexpression of murine calsequestrin were not significantly altered with age (Fig. 3). Neither genotype nor age altered lung/body weight ratios. In addition, the body weights of the $mCSQ_{OE}$ mice were not different from those of the age-matched WT controls.

Echocardiography

Mean echocardiographic data are indicated in Fig. 4. As observed previously [5], overexpression of murine cardiac calsequestrin was associated with a 43% increase in the h/r ratio in young $mCSQ_{OE}$ mice, compared to the age-matched WT controls. This observation, combined with the gravimetric data described above, implies a concentric nature of left ventricular hypertrophy. LVFS and Vcf_c were slightly increased in the young $mCSQ_{OE}$ mice, compared to the age-matched WTs, though the differences did not reach statistical significance. There was no difference in the h/r ratios between aged $mCSQ_{OE}$ mice and their age-matched WT controls. LVFS and Vcf_c were significantly decreased in aged $mCSQ_{OE}$ hearts, compared to young $mCSQ_{OE}$ mice. Neither genotype nor age altered intrinsic heart rates.

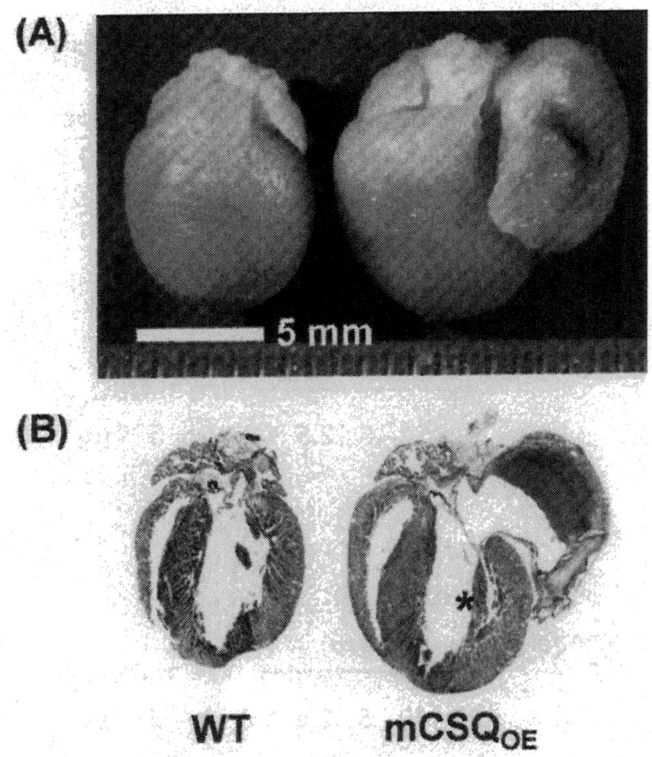

Fig. 2. Representative hearts from aged calsequestrin overexpressing (right) and wild-type (left) mice (anterior view). (A) Gross examination revealed mild ventricular hypertrophy in a 17-month old calsequestrin-overexpressing mouse (mCSQ$_{OE}$), in comparison to its wild-type littermate (WT). Note the massive left atrial enlargement and mineralization in the mCSQ$_{OE}$ heart. (B) A four-chamber section of the mCSQ$_{OE}$ heart demonstrates mild increases in right and left ventricular chamber size and wall thickness (in the mCSQ$_{OE}$ section, a papillary muscle is visible in the left ventricular cavity (*)). The mCSQ$_{OE}$ heart also exhibited thrombus in the left atrial chamber, and enlargement of the left atrium with mineralization and fibrosis.

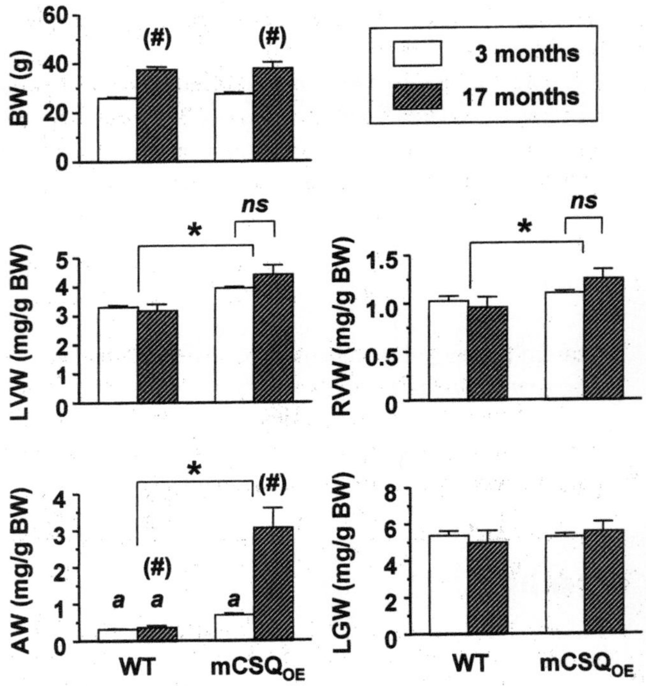

Fig. 3. Cardiopulmonary weights in young and aged mice. Body weight (BW), and blotted left ventricular (LVW), right ventricular (RVW), atrial (AW) and lung (LGW) weights normalized to body weight for 3- and 17-month old wild-type and calsequestrin-overexpressing (mCSQ$_{OE}$) mice. There was no difference in body weight between mCSQ$_{OE}$ and wild-type groups at the same age. Note the mild and stable biventricular hypertrophy and the marked increase in atrial weight in the 17-month old mCSQ$_{OE}$s. Genotype and age did not alter lung weight. Values are means ± S.E. (n = 3–5). $^{\#}p < 0.05$ association with age, $^{*}p < 0.05$ association with genotype, $^{a}p < 0.05$ vs. 17-month old mCSQ$_{OE}$s, ns – not significant.

Langendorff perfusion

To examine whether the isovolumic parameters further deteriorate with aging in the calsequestrin overexpressing myocardium, isolated hearts from young and aged mCSQ$_{OE}$ mice were perfused in parallel with age-matched WT controls (Fig. 5). Age had no effect on end-diastolic pressure in either WT or mCSQ$_{OE}$ hearts (WT: 2.6 ± 0.6 mmHg (young) and 2.5 ± 0.8 mmHg (aged); mCSQ$_{OE}$: 5.4 ± 0.5 mmHg (young) and 7.2 ± 0.9 mmHg (aged), n = 3–7). In addition, the maximal rates of pressure development (+dP/dt) and decline (–dP/dt), which were significantly depressed in mCSQ$_{OE}$ hearts compared to WTs, did not further deteriorate with age. Neither genotype nor age altered intrinsic heart rates (Fig. 5).

Discussion

In the present study, we demonstrate that left ventricular function was sufficiently compensated in transgenic animals overexpressing mouse cardiac calsequestrin up to 17 months of age. The mild increases in ventricular/body weight ratios in young mCSQ$_{OE}$ hearts were not progressive with age, and echocardiography indicated that the increases in the h/r ratios, which were observed in young mCSQ$_{OE}$ mice, were not present in the aged transgenics, when compared to their respective WT controls. These data suggest time-dependent remodeling of the ventricle.

We have also previously shown that overexpression of mouse cardiac calsequestrin leads to low intrinsic myocardial contractility, as evidenced by decreases in isovolumic left ventricular contractile parameters in Langendorff preparations [3]. In the present study, echocardiography revealed age-

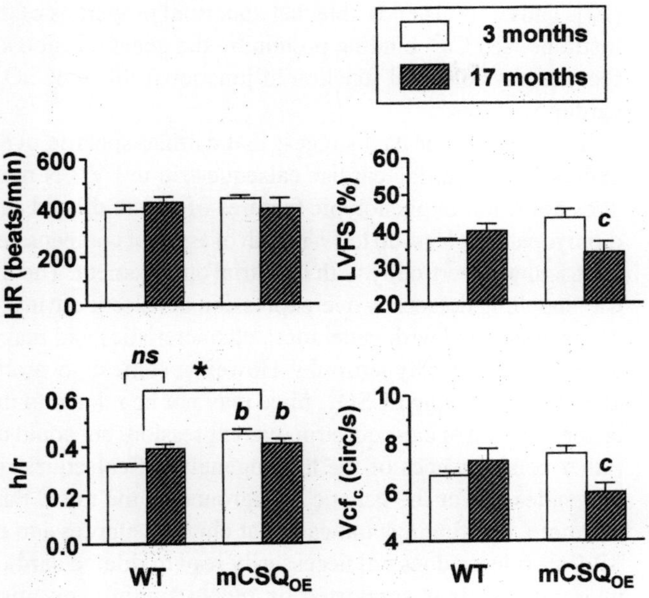

Fig. 4. Echocardiography of anesthetized intact mice. M-mode and Doppler echocardiography on left ventricles of young (3-month old) and aged (17-month old) mice demonstrated a concentric profile of left ventricular hypertrophy in young mice overexpressing calsequestrin ($mCSQ_{OE}$). Left ventricular fractional shortening (LVFS) and velocity of circumferential fiber shortening corrected for heart rate (Vcf_c) were decreased in $mCSQ_{OE}$ mice with age. WT: wild-type, HR: heart rate, h/r: end-diastolic wall thickness to cavity radius ratio. Values are mean ± S.E. (n = 6–8). *$p < 0.05$ association with genotype, $^{b}p < 0.05$ vs. 3-month old WTs , $^{c}p < 0.05$ vs. 3-month old $mCSQ_{OE}$s, ns – not significant.

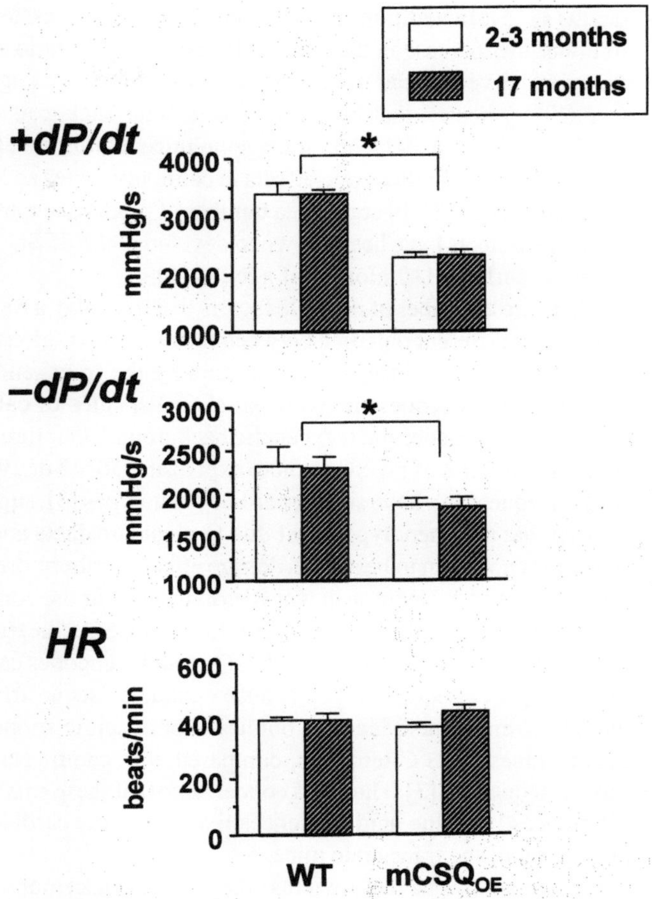

Fig. 5. Left ventricular contractile parameters of Langendorff perfused hearts. Langendorff perfused hearts from young (2–3-month old) and aged (17-month old) mice demonstrated depressed maximal rates of left ventricular pressure development (+dP/dt) and decline (–dP/dt) in mice overexpressing calsequestrin ($mCSQ_{OE}$). No change was observed in the intrinsic heart rate (HR), compared to wild-types (WT). The depressed contractile parameters in the $mCSQ_{OE}$s were not further deteriorated with age. Values are the mean ± S.E. (n = 3–7). *$p < 0.05$ association with genotype.

dependent decreases in both LVFS and Vcf_c in the $mCSQ_{OE}$s, implying attenuation of *in vivo* cardiac performance with age. However, these changes in ejection phase indices do not necessarily reflect similar temporal attenuation of myocardial contractility, as both remodeling and loading conditions importantly influence LVFS and Vcf_c. Indeed, in contrast to the *in vivo* auxotonic indices, *ex vivo* isovolumic contractile parameters of $mCSQ_{OE}$ ventricles were not further depressed with age, suggesting that the depression of cardiac contractility by calsequestrin overexpression is not progressive. The discrepancy in left ventricular function between *in vivo* and *ex vivo* measurements could be due, at least partly, to altered external loads by the enlarged and mineralized left atrium in aged $mCSQ_{OE}$ mice. Nonetheless, the aged $mCSQ_{OE}$ mice did not have congested lungs that characterize congestive heart failure. Furthermore, the observed atrial morbidity in the $mCSQ_{OE}$ hearts appears to be due to sustained atrial wall stress by reduced left ventricular function, though we cannot exclude the possibility that overexpression of calsequestrin in the atria contributed to generation of the atrial abnormalities.

The phospholamban/SERCA protein ratio has been postulated to be a major determinant of cardiac contractile parameters [12], and an increase in this ratio, which is mainly due to downregulation of SERCA, has been reported to be associated with morbidity in failing hearts [13, 14]. We have previously shown parallel increases in protein levels of SERCA and phospholamban in young $mCSQ_{OE}$ hearts, compared to those of their WT littermates, without alteration in the ratio of phospholamban to SERCA [3]. Although increases in these proteins were not observed in 17-month old $mCSQ_{OE}$ mice, the phospholamban/SERCA ratios remained comparable between $mCSQ_{OE}$ and WT. This finding is consistent with the non-progressive feature of the reduced contractility in $mCSQ_{OE}$ hearts.

The findings of the present study are clearly different than those from the $dCSQ_{OE}$ model, which displays severe and progressive dilation of ventricular cavity and premature death [4, 15]. In addition, we have observed no progression of either the increase in the ventricular/body weight ratio or the reduced ventricular contractility up to 7 months of age in $mCSQ_{OE}$ hearts, that have a mixed genetic background (FVB/N+129SvJ/CF-1), and transgenic mice exhibit no attenuation of the β-adrenergic effects in cardiomyocytes at 2–3 months of age [3]. In contrast, a number of β-adrenoceptor signaling defects have been shown to develop in the $dCSQ_{OE}$ hearts as early as 1–2 months of age [4, 15].

There are a number of possible factors that may play a role in the very different outcome between the models: (i) clonal artifact, such as an insertional mutation by the transgene; (ii) levels of calsequestrin expression; (iii) species of calsequestrin transgene; and (iv) genetic background. Our three independent lines of $mCSQ_{OE}$ mice (expressing 56, 43 or 20-fold calsequestrin), have similar cardiac phenotypes [3], suggesting that the phenotype is not due to clonal artifacts and there does not seem to be a gene-dosage effect. Furthermore, murine cardiac calsequestrin was overexpressed in the context of the same endogenous protein in the $mCSQ_{OE}$ hearts [3], whereas the transgene of the $dCSQ_{OE}$ model encodes canine cardiac calsequestrin [2]. Mouse cardiac calsequestrin has a 90% amino acid identity, but has 5 more acidic amino acid residues in its C-terminus, compared with canine cardiac calsequestrin [3]. Thus, it is conceivable that these small differences in amino acid sequences may affect the cardiac phenotype of the transgenic mice.

Several studies have shown that the genetic backgrounds of the mouse strains, used in transgenesis studies, may considerably influence the observed phenotype (for review, see ref. [16]). The $dCSQ_{OE}$ has a genetic background of an inbred strain DBA/2 (or C3H/DBA) [15, 17], whereas we used FVB/N for generation of the $mCSQ_{OE}$ model [3]. DBA/2 and C3H mouse strains have been known to develop spontaneous dystrophic cardiac calcinosis as early as 1–2 months of age [18–20], characterized with cardiomyocyte injury, inflammation, necrosis, and calcium deposition, which may lead to cardiac hypertrophy and/or congestive heart failure [19]. In fact, $dCSQ_{OE}$ mice develop dystrophic ventricular calcifications [15]. Therefore, overexpression of calsequestrin in the DBA/2 (or C3H/DBA) background may influence the onset or severity of the spontaneous dystrophic cardiac calcinosis, resulting in severe dilated cardiomyopathy. In addition, recent genetic analyses revealed significant co-segregation of the spontaneous dystrophic cardiac calcinosis in both DBA/2 and C3H strains with a single gene located on Chromosome 7. This region contains *Hrc*, a candidate gene responsible for the spontaneous calcinosis, which encodes the histidine-rich Ca^{2+} binding protein [20, 21]. The histidine-rich Ca^{2+} binding protein is a low-affinity-high-capacity Ca^{2+} binding protein in skeletal and cardiac muscle, and is associated with the ryanodine receptors through binding to triadin at the junctional SR of skeletal muscle, reminiscent of calsequestrin [22]. Thus, it is also possible that abnormal properties of the histidine-rich Ca^{2+} binding protein by the genetic lesion affect the physiological function of junctional SR in $dCSQ_{OE}$ cardiomyocytes.

In summary, our data suggest that cardiac-specific overexpression of murine cardiac calsequestrin in FVB/N mice does not result in phenotypic features of severe dilated cardiomyopathy at least up to 17-month of age, but compensated ventricular hypertrophy with left atrial enlargement. The SR Ca^{2+} handling defects by overexpression of calsequestrin *per se* are associated with generation, characteristics and maintenance of cardiac hypertrophy. However, some myocardial alterations in young $dCSQ_{OE}$ mice may not be related to the primary action of calsequestrin overexpression, but could be rather consequences of the heterogeneity of calsequestrin molecules and/or the genetic background of the mice. Furthermore, our findings indicate that chronic suppression of SR Ca^{2+} release does not necessarily lead to dilated cardiomyopathy. This is supported by recent findings on mice chronically treated with ryanodine [23]. Cardiac hypertrophy and cardiomyopathy are multifactorial, and diverse etiologies are associated with their onset or progression. Therefore, further examination of endogenous traits of genetic backgrounds and materials for transgenesis should help to elucidate the etiological roles of intracellular Ca^{2+}-cycling defects for myocardial pathogenesis *in vivo*.

Acknowledgements

This work was supported in part by the US National Institute of Health Grants HL26057, HL52318, P400RR12358 and HL64018 (to E.G.K.), and Grants from Japan Ministry of Health, Labour and Welfare, the Japan Health Sciences Foundation (#23004 and KH23106), and Uehara Memorial Foundation (to Y.S.).

References

1. Zhang L, Kelley J, Schmeisser G, Kobayashi YM, Jones LR: Complex formation between junctin, triadin, calsequestrin, and the ryanodine receptor. Proteins of the cardiac junctional sarcoplasmic reticulum membrane. J Biol Chem 272: 23389–23397, 1997
2. Jones LR, Suzuki YJ, Wang W, Kobayashi YM, Ramesh V, Franzini-Armstrong C, Cleemann L, Morad M: Regulation of Ca^{2+} signaling in transgenic mouse cardiac myocytes overexpressing calsequestrin. J Clin Invest 101: 1385–1393, 1998
3. Sato Y, Ferguson DG, Sako H, Dorn GW II, Kadambi VJ, Yatani A, Hoit BD, Walsh RA, Kranias EG: Cardiac-specific overexpression of mouse cardiac calsequestrin is associated with depressed cardiovascular function and hypertrophy in transgenic mice. J Biol Chem 273: 28470–28477, 1998
4. Cho MC, Rapacciuolo A, Koch WJ, Kobayashi Y, Jones LR, Rockman HA: Defective beta-adrenergic receptor signaling precedes the de-

velopment of dilated cardiomyopathy in transgenic mice with calsequestrin overexpression. J Biol Chem 274: 22251–22256, 1999
5. Schmidt AG, Kadambi VJ, Ball N, Sato Y, Kranias EG, Hoit BD: Cardiac specific overexpression of calsequestrin results in left ventricular hypertrophy, depressed force-frequency relation and pulsus alternans *in vivo*. J Mol Cell Cardiol 32: 1735–1744, 2000
6. Harding VB, Jones LR, Lefkowitz RJ, Koch WJ, Rockman HA: Cardiac β ARK1 inhibition prolongs survival and augments β blocker therapy in a mouse model of severe heart failure. Proc Natl Acad Sci USA 98: 5809–5814, 2001
7. Sato Y, Kiriazis H, Yatani A, Schmidt AG, Hahn H, Ferguson DG, Sako H, Mitarai S, Honda R, Mesnard-Rouiller L, Frank KF, Beyermann B, Wu G, Fujimori K, Dorn GW II, Kranias EG: Rescue of contractile parameters and myocyte hypertrophy in calsequestrin overexpressing myocardium by phospholamban ablation. J Biol Chem 2001 276: 9392–9399
8. Seiler C, Jenni R, Vassalli G, Turina M, Hess OM: Left ventricular chamber dilatation in hypertrophic cardiomyopathy: Related variables and prognosis in patients with medical and surgical therapy. Br Heart J 74: 508–516, 1995
9. Levy D, Larson MG, Vasan RS, Kannel WB, Ho KK: The progression from hypertension to congestive heart failure. JAMA 275: 1557–1562, 1996
10. Nichols WW, O'Rourke MF: In: D.A. McDonald (ed). Blood Flow in Arteries: Theoretic, Experimental and Clinical Principles, 3rd edn. Lea & Febiger, Philadelphia, 1990, pp 143–195
11. McDonald DA (ed): Blood Flow in Arteries, 2nd edn. Williams & Wilkins, Baltimore, 1974, pp 162–164
12. Luo W, Wolska BM, Grupp IL, Harrer JM, Haghighi K, Ferguson DG, Slack JP, Grupp G, Doetschman T, Solaro RJ, Kranias EG: Phospholamban gene dosage effects in the mammalian heart. Circ Res 78: 839–847, 1996
13. Meyer M, Schillinger W, Pieske B, Holubarsch C, Heilmann C, Posival H, Kuwajima G, Mikoshiba K, Just H, Hasenfuss G: Alterations of sarcoplasmic reticulum proteins in failing human dilated cardiomyopathy. Circulation 92: 778–784, 1995
14. Ito K, Yan X, Tajima M, Su Z, Barry WH, Lorell BH: Contractile reserve and intracellular calcium regulation in mouse myocytes from normal and hypertrophied failing hearts. Circ Res 87: 588–595, 2000
15. Knollmann BC, Knollmann-Ritschel BE, Weissman NJ, Jones LR, Morad M: Remodelling of ionic currents in hypertrophied and failing hearts of transgenic mice overexpressing calsequestrin. J Physiol 525: 483–498, 2000
16. Sigmund CD: Viewpoint: Are studies in genetically altered mice out of control? Arterioscler Thromb Vasc Biol 20: 1425–1429, 2000
17. Suzuki YJ, Ikeda T, Shi SS, Kitta K, Kobayashi YM, Morad M, Jones LR, Blumberg JB: Regulation of GATA-4 and AP-1 in transgenic mice overexpressing cardiac calsequestrin. Cell Calcium 25: 401–407, 1999
18. Nabors CE, Ball CR: Spontaneous calcification in hearts of DBA mice. Anat Rec 164: 153–161, 1969
19. Eaton GJ, Custer RP, Johnson FN, Stabenow KT: Dystrophic cardiac calcinosis in mice: Genetic, hormonal, and dietary influences. Am J Pathol 90: 173–186, 1978
20. Ivandic BT, Qiao JH, Machleder D, Liao F, Drake TA, Lusis AJ: A locus on chromosome 7 determines myocardial cell necrosis and calcification (dystrophic cardiac calcinosis) in mice. Proc Natl Acad Sci USA 93: 5483–5488, 1996
21. van den Broek FAR, Bakker R, den Bieman M, Fielmich-Bouwman AXM, Lemmens AG, van Lith HA, Nissen I, Ritskes-Hoitinga JM, van Tintelen G, van Zutphen LFM: Genetic analysis of dystrophic cardiac calcification in DBA/2 mice. Biochem Biophys Res Commun 253: 204–208, 1998
22. Sacchetto R, Turcato F, Damiani E, Margreth A: Interaction of triadin with histidine-rich Ca^{2+}-binding protein at the triadic junction in skeletal muscle fibers. J Muscle Res Cell Motil 20: 403–415, 1999
23. Meyer M, Trost SU, Bluhm WF, Knot HJ, Swanson E, Dillmann WH: Impaired sarcoplasmic reticulum function leads to contractile dysfunction and cardiac hypertrophy. Am J Physiol 280: H2046–H2052, 2001

Molecular and Cellular Biochemistry **242:** 27–33, 2003.

The myocardial protein S100A1 plays a role in the maintenance of normal gene expression in the adult heart

James N. Tsoporis,[1] Alexander Marks,[2] Danna B. Zimmer,[3] Chris McMahon[1] and Thomas G. Parker[1]

[1]*The Centre for Cardiovascular Research, Division of Cardiology, Department of Medicine, The Toronto Hospital, University of Toronto, Toronto, ON;* [2]*Banting and Best Department of Medical Research, University of Toronto, Toronto, ON, Canada;* [3]*Department of Pharmacology, College of Medicine, University of South Alabama, Mobile, AL, USA*

Abstract

S100A1 and S100B are members of a family of 20 kDa Ca^{2+}-binding homodimers that play a role in signal transduction in mammalian cells. S100A1 is the major isoform in normal heart and S100B, normally a brain protein, is induced in hypertrophic myocardium and functions as an intrinsic negative modulator of the hypertrophic response. In order to examine the function of S100A1, we first showed that, in contrast to S100B, S100A1 was downregulated in rat experimental models of myocardial hypertrophy following myocardial infarction or pressure overload. Second, in co-transfection experiments in cultured neonatal rat cardiac myocytes, S100A1 inhibited the α_1-adrenergic activation of promoters of genes induced during the hypertrophic response including the fetal genes skeletal α actin (skACT), and β-myosin heavy chain (MHC) and S100B, but not the triiodothyronine (T3) activation of the promoter of the α-MHC gene, that is normally expressed in adult myocardium. These results suggest that S100A1 is involved in the maintenance of the genetic program that defines normal myocardial function and that its downregulation is permissive for the induction of genes that underlie myocardial hypertrophy. (Mol Cell Biochem **242**: 27–33, 2003)

Key words: S100B, S100A1, myocardial infarction, pressure overload, gene expression

Introduction

The transition of undifferentiated and dividing fetal cardiac myocytes to differentiated and non-dividing adult myocytes involves changes in gene expression. These include the replacement of fetal isoforms of contractile muscle proteins such as skeletal α-actin (skACT) and β-myosin heavy chain (MHC) by their counterparts cardiac α-actin and α-MHC. In disease states, such as pressure overload due to congestive heart failure or massive necrosis due to myocardial infarction, viable adult myocytes hypertrophy as an adaptive response in an attempt to preserve cardiac function [1, 2]. Myocyte hypertrophy is accompanied by a program of fetal gene re-expression that includes skACT and β-MHC [3–5].

The hypertrophic response can be reproduced in rat and mouse experimental models and cultured neonatal cardiac myocytes from these species. Studies in these models, have implicated hormonal stimulation including α_1-adrenergic agents, angiotensin II, peptide growth factors, and mechanical stretch as instigators of the hypertrophic response [3–9]. Some of these, including α_1-adrenergic agents, activate the PKC signaling pathway [8–11]. The mechanisms that maintain the expression of adult-type genes in the normal myocardium and contribute to a switch to a program of fetal gene re-expression in the hypertrophic myocardium are likely multifactorial and not completely elucidated.

S100A1 and S100B are 20-kDa Ca^{2+}-binding homodimers that belong to a 19-member family of related intracellular pro-

Address for offprints: T.G. Parker, The Centre for Cardiovascular Research, Division of Cardiology, The Toronto Hospital, 200 Elizabeth Street, EN12-208, Toronto, Ontario M5G 2C4, Canada (E-mail: tom.parker@utoronto.ca)

teins that are believed to play a signal transduction role in the regulation of growth and differentiation of mammalian cells [12]. These proteins are expressed selectively in normal adult tissues. S100B is an astrocyte protein and is absent from cardiac myocytes, whereas S100A1 is highly expressed in skeletal and cardiac muscle. The changes in expression of these proteins may be related to their function in normal and disease states. We have previously reported that S100B was induced in hypertrophic myocardium following myocardial infarction or α_1-adrenergic stimulation and inhibited the hypertrophic response including the re-expression of fetal genes [13, 14]. This suggests that S100B plays a role as a negative modulator to mitigate the potential maladaptive consequences of unchecked hypertrophy which, in its chronic state, can lead to irreversible heart failure. S100A1 has been implicated in the control of Ca^{2+} homeostasis in cardiac myocytes but its functional role in the myocardium has not been defined [12].

In the present communication, we report that S100A1 inhibits the α_1-adrenergic-mediated induction of the S100B, skACT, and β-MHC promoters, but not the induction of the α-MHC promoter by triiodothyronine (T3) in cultured neonatal rat cardiac myocytes. This suggests that S100A1 plays a role in the maintenance of the program of normal adult myocardial gene expression. Interestingly, we also show that the expression of S100A1 is markedly downregulated in rat myocardium following experimental left ventricular myocardial infarction or pressure overload, while conversely S100B is induced in these states. This downregulation of S100A1 may be required for the unhindered induction of the genetic program that underlies the hypertrophic response of the damaged myocardium, including S100B, the intrinsic negative modulator of this response.

Materials and methods

Human S100B promoter driven reporter plasmids

A luciferase reporter gene system (Promega, Madison, WI, USA) was used to study the relative ability of different 5′ DNA regions of the human S100B gene to promote transcription in transient transfection assays, as previously described [15]. In this system, a maximal and a minimal human S100B promoter were cloned upstream of the luciferase gene into the plasmid pGL3 that lacks eukaryotic enhancer and promoter sequences. Initially, a 9.8 kb *HindIII* fragment of human genomic DNA [16], spanning exons 1 and 2 (and the intervening intron) of the human S100B gene, and 6689 bp of upstream and 320 bp of downstream DNA sequence, respectively, was subcloned into a Bluescript (pBS) plasmid (Stratagene, La Jolla, CA, USA) to generate pBS100 9.1. A *HindIII/BamHI* fragment spanning the DNA nucleotide sequence –6689/+698 relative to the transcription initiation site was excised from pBS100 9.1, and designated as the maximal S100B promoter. This fragment includes 6689 bp of DNA sequence upstream of exon 1, exon 1 (71 bp) and 627 bp of downstream sequence and was subcloned into pGL3 to generate pGBS–6689/+698. The minimal promoter *SmaI/BamHI* fragment which includes 162 bp of DNA sequence upstream of the transcription initiation site was subcloned into pGL3 to generate pGBS -6689/+698.

Expression plasmids

The S100A1 expression vector pCMVA1 containing the rat S100A1 coding sequence driven by the cytomegalovirus enhancer was constructed by subcloning the human S100A1 coding sequences from a bacterial expression plasmid pKKA1 [17] into the *EcoR1* and *HindIII* sites of the pCMV5 vector. The following reported plasmids were obtained from the indicated sources: skACT-CAT (containing the mouse skACT promoter driving the chloramphenicol (CAT) reporter gene) [9]; β-MHC-CAT (containing the rat β-MHC promoter driving the CAT reporter gene) [8]; α-MHC (containing the human α-MHC promoter driving the CAT reporter gene) [18]; Rous sarcoma virus (RSV)-CAT (containing a RSV long terminal repeat promoter driving the CAT reporter gene) from Stratagene (La Jolla, CA, USA); RSV-Luciferase (LUX) (containing a RSV long terminal repeat promoter driving the LUX reporter gene) [8].

Left coronary artery ligation

Left coronary artery ligation was performed in the rat as previously described [19]. Briefly, 12–14 week-old male Sprague Dawley rats were anaesthetized with ketamine (90 mg/kg) and xylazine (10 mg/kg) i.p. A left thoracotomy was performed in the fifth intercostal space and the pericardium was opened. The proximal left coronary artery under the tip of the left atrial was encircled and ligated using a 6-0 silk suture. Proximal left anterior descending artery ligation in a rat model creates a reproducibly large lateral wall infarction. In the sham-operated animals, the left coronary artery was encircled but not ligated and the muscle layers and skin were closed similarly.

Aortic banding

Aortic banding was performed in the rat as previously described [20], with modification. Briefly, 12–14 week-old male Sprague Dawley rats were anaesthetized with ketamine (90 mg/

kg) and xylazine (10 mg/kg) intraperitoneally. A left thoracotomy was performed in the fifth intercostal space and the pericardium was opened. A 21-gauge needle was placed along the aortic arch, distal to the subclavian artery. A 6-0 silk suture was tightened against the needle. The needle was then removed to produce a predefined constriction. Sham operation was performed by isolation of aorta without ligation.

Cell cultures

Neonatal cardiac myocytes were isolated from ventricles of 2-day-old Sprague Dawley rats and established in culture as previously described [21].

Cell transfection

Transfection was carried out by calcium phosphate precipitation [8], using specific quantities of the following plasmids: pCMVA1 (1 μg); skACT-CAT, β-MHC-CAT, α-MHC-CAT (9 μg); pGBS–162/+698, pGBS–6689/+698 (5 μg). RSV-LUX (0.1 μg) was included as an internal control for transfection efficiency for co-transfections with skACT-CAT, β-MHC-CAT, and α-MHC-CATRSV-CAT, whereas RSV-CAT (0.1 μg) was included for pGBS–162/+698 and pGBS–6689/+698. Myocyte cultures were maintained in medium supplemented with 5% fetal bovine serum for 18 h following transfection, prior to transfer to serum-free medium and treatment with norepinephrine (NE, 20 μM), phenylephrine (PE, 20 μM), isoproterenol (ISO, 20 μM), triiodothyronine T3 (20 ng/mL), phorbol ester phorbol 12-myristate 13-acetate (PMA) (10 nM) or vehicle diluent (100 μM ascorbic acid for NE, PHE, ISO, T3 or 0.01% DMSO for PMA) for 72 h. The cell lysates were assayed for LUX and CAT activity using published techniques [22, 23]. Co-transfection with any of the plasmids or treatment of cultures as described above did not affect the respective internal control CAT or LUX activity. The activity of the reporter genes was normalized based on the activity of the respective controls in the same dish. The differences of the respective LUX and CAT activities between duplicate dishes were < 10% of their mean. Treated/control ratios were tested from deviation from unity by calculation of confidence limits.

Ribonuclease protection assay

RNA was isolated from (i) normal rat tissues, including fetal, neonatal, and adult left ventricular myocardium, adult brain (ii) residual non-infarcted rat left ventricular myocardium outside the territory supplied by the ligated coronary artery [13] and (iii) left ventricular myocardium following aortic banding by the one-step acid guanidium phenol method [24]. Antisense riboprobes for S100B and glyceraldehyde-3 phosphate dehydrogenase (GAPDH) were derived as described [13]. The antisense riboprobe for S100A1 was derived by *in vitro* transcription with T7 RNA polymerase of a plasmid containing 352 bp of S100A1 conserved coding sequence generated by RT-PCR using S100A1 primer sequences based on DNA sequences in the National Center for Biotechnology data base. RNAse protection assays to determine steady-state levels of rodent S100B, S100A1 and GAPDH mRNAs were performed as previously described [13].

Western blotting

Recombinant human S100A1 and goat antiserum to the protein [25], were obtained from Dr. Claus Heinzmann (Zurich, Switzerland). Rat heart lysates were prepared from frozen powder and Western blotting was performed as described previously [14]. Briefly, aliquots of extracts containing 25 μg of total protein or recombinant S100A1 (10 ng) were dissociated and subjected to electrophoresis in 16% sodium dodecyl sulfate-polyacrylamide slab gels under reducing conditions. The proteins were transferred electrophoretically to membranes (PVDF; Milipore Corp., Milford, MA, USA) and the blots were incubated with goat anti-S100A1 antiserum at a 1:10000 dilution followed by peroxidase-conjugated rabbit anti-goat Ig antibody (Sigma, St. Louis, MO, USA) at a 1:1000 dilution. The blots were developed using enhanced chemiluminescence (Pierce, Rockford, IL, USA) and exposed to X-ray film.

Statistical analysis

Treated/control ratios were tested for deviation from unity by calculation of confidence limits. Mean values were compared by analysis of variance by Student-Newman-Keuls test, with significance defined as $p < 0.05$.

Results

S100A1 mRNA and protein is reduced in adult rat myocardium following coronary artery ligation or aortic banding

Steady-state levels of S100A1, S100B, and GAPDH mRNAs were determined using an RNAse protection assay in fetal, neonatal, and adult rat heart, adult rat brain, non-infarcted adult myocardium adjacent to an infarct 35 days post-coronary artery ligation and adult myocardium 28 days post-aortic banding (Fig. 1A). S100A1 mRNA was detected in

A.

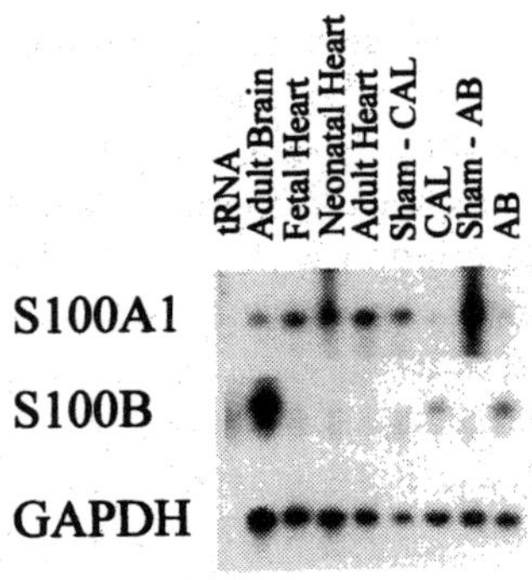

B.

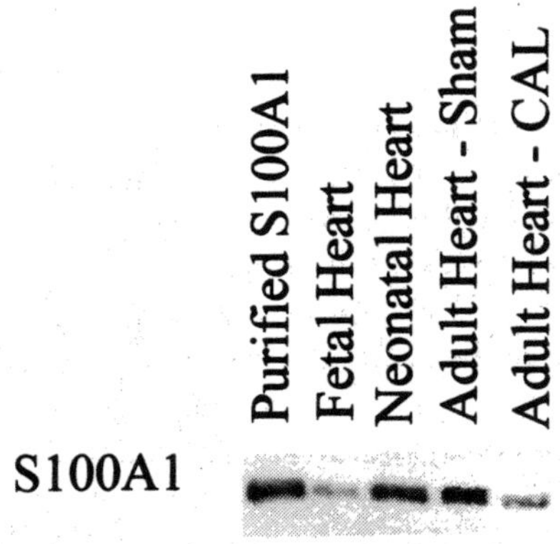

Fig. 1. S100B is induced and S100A1 is downregulated in the left ventricular myocardium following experimental myocardial infarction or aortic banding. Adult rats were sacrificed at 35 days after coronary artery ligation (CAL) or 28 days after aortic banding (AB). (A) Steady-state levels of S100B, S100A1 and GAPDH mRNAs were determined by RNAse protection. Protected fragments specific for rat S100A1 (293 bp), S100B (264 bp) and GAPDH (413 bp) mRNAs are indicated. The results of a representative experiment include RNA from fetal, neonatal, and adult heart and adult brain from normal animals, and heart from ligated, banded or sham-operated experimental animals as indicated. Rat tRNA was used as a control. (B) The expression of S100A1 was analyzed by Western blotting in protein extracts from fetal and neonatal heart, and from adult heart 35 days after sham operation or coronary artery ligation (CAL).

fetal, neonatal, and adult rat heart, and adult rat brain. Following coronary artery ligation or aortic banding S100A1 mRNA was markedly diminished in adult heart. S100B mRNA was not detected in fetal, neonatal or adult rat myocardium, but was highly expressed in adult rat brain. Following coronary artery ligation or aortic banding S100B mRNA was expressed in adult myocardium. GAPDH steady state mRNA levels served as a control for the quality and loading of the mRNA. S100A1 protein levels were determined by Western analysis in fetal, neonatal, and adult rat heart, and non-infarcted adult myocardium adjacent to an infarct 35 days post-coronary artery ligation (Fig. 1B). S100A1 protein was present in neonatal and adult heart but was markedly diminished in adult peri-infarct myocardium.

S100A1 inhibits transcription from the human S100B promoter through elements located upstream of the minimal promoter

The β-agonist ISO did not activate the maximal or minimal S100B promoters (Figs 2A and 2B). Co-transfection with S100A1 inhibited transcription from the maximal but not the minimal S100B promoter, implicating elements located upstream of the minimal promoter (Figs 2A and 2B). S100A1 inhibited basal expression and decreased the PE, NE, and PMA-induced activation of the maximal S100B promoter.

S100A1 inhibits the α_1-adrenergic induction of the fetal genes β-MHC and skACT but not the induction of the α-MHC promoter by T3

The β-agonist ISO did not induce the β-MHC or skACT promoters (Figs 3A and 3B). Co-transfection with the S100A1 expression plasmid had no effect on basal transcription from the β-MHC or skACT promoters, but inhibited the approximately 2–4 fold induction of these promoters by the α_1-adrenergic agonists NE and PHE and the PKC activator PMA (Figs 3A and 3B). Co-transfection with S100A1 did not decrease the induction of the α-MHC promoter by T3 (Fig. 3C).

Discussion

We have previously inferred an induction of S100B in the hypertrophic myocardium based on the detection of the protein or its mRNA in the peri-infarct region of the human heart and in the rat or mouse heart following experimental myocardial infarction or α_1-adrenergic infusion, respectively [13, 14]. In the present study, we have extended these results by demonstrating first, the presence of S100B mRNA in rat heart following left ventricular experimental pressure overload due to aortic banding, and second, the activation of the S100B promoter by α_1-adrenergic agonists or the phorbol ester PMA (an activator of PKC) in cultured neonatal rat cardiac myocytes. This suggests that the S100B gene is induced in cardiac myocytes by the α_1-adrenergic pathway involving PKC. The same pathway plays a major role in initiating and sustaining the hypertrophic response and, as we have reported previously, is subject to negative modulation by S100B [13]. Taken together, these results raise the possibility that S100B is a negative feedback regulator of this pathway during myocardial hypertrophy.

In contrast to S100B, which was induced during hypertrophy, we show that the expression of S100A1, the major S100 protein isoform in normal adult myocardium, was downregulated in left ventricular myocardium following experimental infarction or pressure overload. Furthermore, in cultured

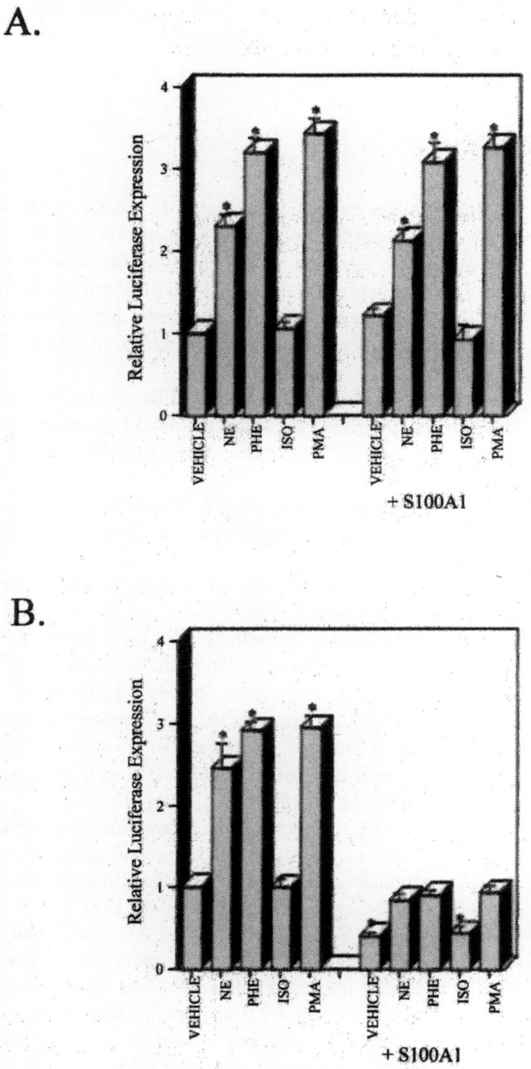

Fig. 2. S100A1 inhibits transcription from the human S100B promoter through elements located upstream of the minimal promoter. Neonatal rat cardiac myocytes were co-transfected with RSV-CAT and pGBS –162/+698 (minimal S100B promoter) (A) or pGBS –6698/+698 (maximal S100B promoter) (B) or co-transfected with the above and S100A1 as indicated. Myocytes treated with vehicle were co-transfected with the plasmid vector. The transfected or co-transfected myocytes were treated with NE, PHE, ISO, or PMA and assayed for luciferase activity which is expressed relative to myocytes transfected with the plasmid vector and treated with vehicle as described in 'Materials and methods'. Bars represent mean ± S.E. of normalized luciferase expression (i.e. LUX/CAT activities); *$p < 0.05$ vs. vehicle-treated myocytes assigned a value of 1.

neonatal rat cardiac myocytes, co-transfection with S100A1 inhibited the activation of the promoters of S100B and the fetal genes skACT and β-MHC by α_1-adrenergic agonists or PMA, but not that of the adult-type isoform α-MHC by T3. These results suggest that the downregulation of S100A1 is permissive for the induction of the genetic program that un-

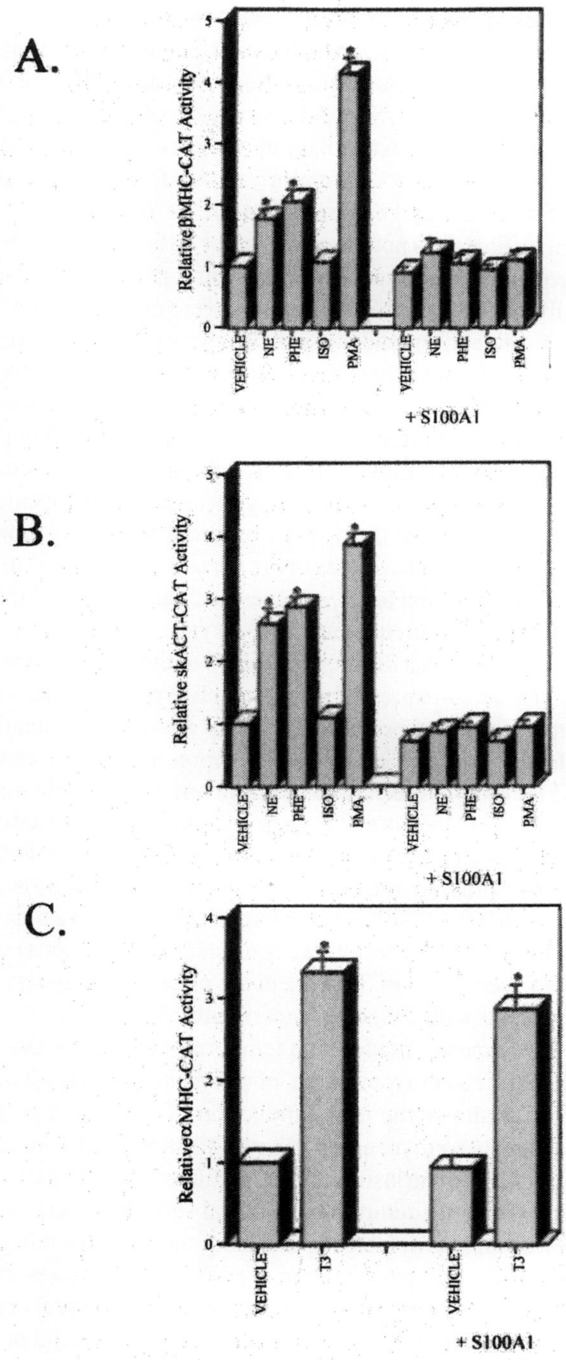

Fig. 3. S100A1 inhibits the a_1-adrenergic activation of the β-MHC and skACT promoters but not the T3 activation of the α-MHC promoter. Neonatal rat cardiac myocytes were transfected with RSV-LUX and β-MHC-CAT (A), skACT-CAT (B), or α-MHC-CAT (C) or co-transfected with the above and S100A1 as indicated. The transfected or co-transfected myocytes were treated with NE, PHE, ISO or PMA and assayed for CAT activity which is expressed relative to myocytes transfected with the plasmid vector and treated with vehicle as described in 'Materials and methods'. Bars represent mean ± S.E. of normalized CAT expression (i.e. CAT/LUX activities); *$p < 0.05$ vs. vehicle-treated myocytes assigned a value of 1.

derlies the hypertrophic response, including the re-expression of the fetal genes and the expression of S100B, the intrinsic negative modulator of this response. In the normal adult myocardium, S100A1 may play a role in silencing these genes in the context of the maintenance of the genetic program that underlies normal myocardial function. A similar role has recently been proposed for the transcription factor TEF-1 [26]. It is not known whether S100A1 and TEF-1 act through related or unrelated signaling pathways. Paradoxically, S100A1 mRNA and protein were detected in fetal and neonatal rat heart and it is not clear why the basal expression of the protein in these tissues does not silence the expression of the fetal genes or their induction by α_1-adrenergic agonists in cultured neonatal cardiac myocytes. One possibility is that S100A1 interacts with specific co-factors in adult-myocardium in repressing these genes. It appears that overexpression of the protein, as in the case of co-transfection experiments, bypasses this requirement and allows S100A1 to inhibit the α_1-adrenergic induction of skACT and β-MHC in cultured neonatal myocytes.

It is known that different members of the S100 protein family can be co-expressed in some cell types and that heterodimers of these proteins may play specialized functional roles [12, 27]. However, our work represents the first example of a transition from the expression of one S100 protein isoform (S100A1) in the normal state to another (S100B) in a disease state. At some point in this transition, S100A1 and S100B may be co-expressed in cardiac myocytes raising the possibility that an S100A1/S100B heterodimer may also have a specialized function. Our results are consistent with previous reports of reduced levels of S100A1 in the plasma and myocardium of patients with congestive heart failure [28, 29].

We propose a model for a temporal sequence of events in left ventricular myocardium, in which the downregulation of S100A1 allows the initial induction of hypertrophic genes subsequent to myocardial injury, followed by the induction of S100B, the intrinsic negative regulator of the hypertrophic response. In the return to the normal state, S100A1 is re-expressed and participates in silencing the hypertrophic genes and S100B. The participation of S100A1 in the trans-repression of the S100B promoter is an example of a novel mechanistic interaction between two members of the S100 protein family.

Acknowledgements

All animal experiments conformed with protocols approved by the respective host institutions. We thank D. O'Hanlon and F. Dawood for excellent technical assistance, Drs. L.R. Karns, P.C. Simpson, C.C. Liew and L.J. Van Eldik for the gift of essential plasmids and helpful discussion, and Dr. C. Heizmann for recombinant human S100A1 and goat antiserum to the protein. This work was supported in part by the Canadian Institute of Health Research.

References

1. Schwartz K, Boheler KR, de la Bastie D, Lompre AM, Mercadier JJ: Switches in cardiac muscle gene expression as a result of pressure or volume overload. Am J Physiol. 262: R364–R369, 1992
2. Gidh-Jain M, Huang B, Jain P, Gick G, El-Sherif N: Alterations in cardiac gene expression during ventricular remodeling following experimental myocardial infarction. J Mol Cell Cardiol 30: 627–637, 1998
3. Chien KR, Zhu H, Knowlton KU, Miller-Hance W, van Bilsen M, O'Brien TX, Evans S: Transcriptional regulation during cardiac growth and development. Annu Rev Physiol 55: 77–95, 1993
4. Parker TG, Schneider MD: Growth factors, proto-oncogenes, and plasticity of the cardiac phenotype. Ann Rev Physiol 53: 1 79–200, 1991
5. Parker TG: Molecular biology of cardiac growth and hypertrophy. Hertz 18: 245–255, 1993
6. Bishopric NH, Simpson PC, Ordahl CP: Induction of the skeletal α-actin gene in α_1-adrenoceptor-mediated hypertrophy of rat cardiac myocytes. J Clin Invest 80: 1194–1199, 1987
7. Waspe LE, Ordahl CP, Simpson PC: The cardiac β-myosin heavy chain isogene is induced selectively in α_1-adrenergic receptor-stimulated hypertrophy of cultured rat heart myocytes. J Clin Invest 85: 1206–1214, 1990
8. Kariya K, Karns LR, Simpson PC: Expression of a constitutively activated mutant of the β-isozyme of protein kinase C in cardiac myocytes stimulates the promoter of the β-myosin heavy chain isogene. J Biol Chem 266: 10023–10026, 1991
9. Karns LR, Kariya K, Simpson PC: M-CAT, CArG, and Sp1 elements are required for α_1-adrenergic induction of the skeletal α-actin promoter during cardiac myocyte hypertrophy. Transcriptional enhancer factor-1 and protein kinase C as conserved transducers of the fetal program in cardiac growth. J Biol Chem 270: 410–417, 1995
10. Kariya K, Farrance IKG, Simpson PC: Transcription enhancer factor-1 in cardiac myocytes interacts with an α_1-adrenergic- and protein kinase C- inducible element in the rat β-myosin heavy chain promoter. J Biol Chem 268: 26658–26662, 1993
11. Kariya K, Karns LR, Simpson PC: An enhancer core element mediates stimulation of the rat β-myosin heavy chain promoter by an α_1-adrenergic agonist and activated β-protein kinase C in hypertrophy of cardiac myocytes. J Biol Chem 269: 3775–3782, 1994
12. Donato R: S100: A multigene family of calcium-modulated proteins of the EF-hand type with intracellular and extracellular functional roles. Intern J Biochem Cell Biol. 3: 637–668, 2001
13. Tsoporis JN, Marks A, Kahn HJ, Butany JW, Liu PP, O'Hanlon D, Parker TG: S100β inhibits α_1-adrenergic induction of the hypertrophic phenotype in cardiac myocytes. J Biol Chem 272: 31915–31921, 1997
14. Tsoporis JN, Marks A, Kahn HJ, Butany JW, Liu PP, O'Hanlon D, Parker TG: Inhibition of norepinephrine-induced cardiac hypertrophy in S100β transgenic mice. J Clin Invest 102: 1609–1616, 1998
15. Castets F, Griffin W ST, Marks A, Van Eldik LJ: Transcriptional regulation of the human S100B gene. Mol Brain Res 46: 208–216, 1997

16. Allore RJ, Friend WC, O'Hanlon D, Neilson KM, Baumal R, Dunn RJ, Marks A: Cloning and expression of the human S100B gene. J Biol Chem 265: 15537–15543, 1990
17. Landar A, Rustandi RR, Weber DJ, Zimmer DB: S100A1 utilizes different mechanisms for interacting with calcium-dependent and calcium-independent proteins. Biochem 37: 17429–17438, 1998
18. Yamauchi-Takihara K, Sole MJ, Liew J, Ing D, Liew CC: Characterization of the human cardiac myosin heavy chain genes. Proc Natl Acad Sci USA 86: 3504–3508, 1989
19. Orenstein TL, Parker TG, Butany JW, Goodman JM, Dawood F, Wen W-H, Wee L, Martino T, McLaughlin PR, Liu PP: Favorable left ventricular remodeling following large myocardial infarction by exercise training. Effect on left ventricular morphology and gene expression. J Clin Invest 96: 858–866, 1995
20. Anderson PG, Allard MF, Thomas GD, Bishop SP, Digerness SB: Increased ischemic injury but decreased hypoxic injury in hypertrophied rat hearts. Circ Res 67: 948–959, 1990
21. Simpson PC: Stimulation of hypertrophy of cultured neonatal rat heart cells through an α_1-adrenergic receptor and induction of beating through an α_1- and β_1-adrenergic receptor interaction. Circ Res 56: 884–894, 1985
22. De Wet JR, Wood KV, De Luca M, Helinski DR, Subramani S: Firefly luciferase gene: Structure and expression in mammalian cells. Mol Cell Biol 7: 725–737, 1987
23. Gorman CM, Moffat LF, Howard BH: Recombinant genomes which express chloramphenicol acetyltransferase in mammalian cells. Mol Cell Biol 2: 1044–1051, 1982
24. Chomczynski P, Sacchi N: Single-step method of RNA isolation by acid guanidium thiocyanate-phenol-chloroform extraction. Anal Biochem 162: 156–159, 1987
25. Ilg EC, Schafer BW, Heizmann CW: Expression pattern of S100 calcium-binding proteins in human tumors. Int J Cancer 68: 325–332, 1996
26. Stewart AFR, Suzow J, Kubota T, Ueyama T, Hsiao-Huei C: Transcription factor RTEF-1 mediates α_1-adrenergic reactivation of the fetal gene program in cardiac myocytes. Circ Res 83: 43–49, 1998
27. Yang Q, O'Hanlon D, Heizman CW, Marks A: Demonstration of heterodimer formation between S100B and S100A6 in the yeast two-hybrid system and human melanoma. Exp Cell Res 246: 501–509, 1999
28. Kiewitz R, Acklin C, Minder E, Huber PR, Schafer BW, Heizmann CW: S100A1, a new marker for acute myocardial ischemia. Biochem Biophys Res Commun 274: 865–871, 2000
29. Remppis A, Greten T, Schafer BW, Hunziker P, Erne P, Katus HA, Heizman CW: Altered expression of the Ca^{2+}-binding protein S100A1 in human cardiomyopathy. Biochim Biophys Acta. 1313: 253–257, 1996

Molecular and Cellular Biochemistry **242:** 35–38, 2003.

The carboxy-tail of connexin-43 localizes to the nucleus and inhibits cell growth

Xitong Dang, Bradley W. Doble and Elissavet Kardami

Institute of Cardiovascular Sciences, St. Boniface General Hospital Research Centre, and Department of Physiology, University of Manitoba, Winnipeg, Manitoba, Canada

Abstract

Gap junctions are plasma membrane intercellular communication channels that in addition to ensuring electrical coupling and coordinated mechanical activity, can act as growth suppressors. To define the role of a non-channel forming domain of connexin-43 (Cx43), the main constituent of cardiomyocyte gap junctions, on growth regulation, we expressed its C-terminal portion (CT-Cx43) in cardiomyocytes and HeLa cells. In addition to broad cytoplasmic localization, CT-Cx43 was also localized to the nucleus of both cell types, detected by immunofluorescence as well as immunoblotting of subcellular fractions. Furthermore, stable expression of CT-Cx43 in HeLa cells induced a significant decrease in proliferation. It is therefore suggested that plasma membrane localization and formation of channels are not required for growth inhibition by Cx43, and that nuclear localization of CT-Cx43 may exert effects on gene expression and growth. (Mol Cell Biochem **242**: 35–38, 2003)

Key words: structure-function, cell proliferation, gap junctions, nucleus

Introduction

The ability of cardiac myocytes for timely cell division and regeneration after injury is greatly reduced in the post-natal heart [1]. Understanding and interfering with the mechanism(s) limiting cardiomyocyte regeneration may provide a way to improve the repair response of the myocardium. One as yet unexplored mechanism for cardiomyocyte growth inhibition may be exerted at the level of intercellular communication channels, termed gap junctions (GJ) [2]. In addition to ensuring electrical coupling and coordinated mechanical activity, GJ are considered to act as growth (tumor) suppressors [3]. Adult cardiomyocytes are extensively connected via gap junctions at the intercalated disk region [4]. While this is essential, ensuring the coordinated action of the heart, it may also contribute to suppression of regeneration.

GJ are composed of connexins (Cx) [5]. The mechanism of growth inhibition by connexins has not been defined. Forced expression of connexin genes in tumor cells increases cell-cell communication, and inhibits cell growth [6]; this effect appears to be connexin-species specific and does not always correlate with the level of GJ mediated intercellular communication (GJIC) [7].

One of our goals is to define the role of the different domains of Cx43 (the main connexin of working cardiomyocytes) on growth regulation. As a first step, we have transiently expressed the COOH-terminal portion of Cx43 (CT-Cx43) in cardiomyocytes and HeLa cells and examined its subcellular localization in comparison to the NH2-terminal portion of Cx43 (NT-Cx43) and wild type (WT) Cx43. We have also used HeLa cells to achieve stable expression of CT-Cx43 and WT Cx43, and examined effects on cell proliferation. In this communication we show that CT-Cx43 localizes to the cytosol and nucleus and that it decreases cell proliferation.

Materials and methods

Materials

Rat Cx43 full-length cDNA [8] was used as described previously [9]. Cell culture media, G418, fetal bovine serum

Address for offprints: E. Kardami, Institute of Cardiovascular Sciences, SBGH Research Center, 351 Tache Avenue, Winnipig, Manitoba, Canada R2H 2A6 (E-mail: ekardami@sbrc.ca)

(FBS), and bovine pancreatic trypsin were purchased from Gibco/BRL (Rockville, MD, USA). Polyclonal rabbit anti-Cx43 antibodies have been described previously [10].Monoclonal anti-FLAG antibodies and all other chemicals were purchased from Sigma (Oakville, ON, Canada).

Plasmids

All the Cx43 constructs were obtained by PCR amplification of full-length or part of coding region of rat Cx43. For PCR, the primers were designed to include restriction sites, start code on sense-primer and stop code on anti-sense primer respectively. Amplification reactions were carried out in a Thermal Cycler using parameters as 2 min at 95°C; 30 cycles of 1 min at 94°C, 1 min at 56°C and 1 min at 72°C; followed by 10 min at 72°C. To generate wild type (wt)-Cx43, the coding region of Cx43 was amplified with sense primer 5′-CCGAATTCAGACATGGGTGACTGGAGT-3′ and antisense primer 5′- CGCTCTAGATTAAATCTCCAG-3′. PCR products were gel purified, digested with EcoRI and XbaI, and then cloned into pcDNA3.1. For the N-portion (NT) of CX43, the coding region from amino acids 1–242 (NT242) were amplified with sense primer 5′-CGCGAATTCAGACATGGGTGAC-3′, and anti-sense primer 5′-CGCGGATCCTCTTCCCTTCACGCGAT-3′. PCR products were gel purified, digested with EcoRI and BamHI, and cloned into pFLAG-CMV2. For construction of C-terminal Cx43 (CT-Cx43), the coding region from amino acids 243–382 (CT243) were amplified with sense primer 5′-CCGGAATTCAGCATGGATCTCTTCTACGTC-3′, and anti-sense primer 5′-CGCTCTAGATTAAATCTCCAG-3′. PCR products were gel purified, digested with EcoRI and XbaI, and cloned into pcDNA3.1. All constructs were fully sequenced in both directions (Cortec DNA service Laboratories Inc., Kingston, ON, Canada).

Transient transfection and stable expression of Cx43 constructs

Cardiac myocytes were cultured on collagen-coated coverslips or plastic, in 35 mm dishes, while HeLa cells were grown directly on coverslips or plastic. Transient transfections were done using the calcium phosphate precipitation method with modifications, as described previously [10]. At 48 h after transfection, coverslips were processed for immunofluorescence, while cells grown on plastic were used for Western blotting analysis. For stable transfections, HeLa cells were first transiently transfected with plasmids expressing cDNAs for WT-Cx43, and CT-Cx43. At 48 h after transfection, cells were split at a ratio of 1:10 into fresh medium containing 700 μg/ml G418. Selective medium was replaced every three days until G418 resistant colonies were detected. Each selected colony was further expanded and analyzed by Western blot and Immunofluorence.

Immunofluorence

After transfection, cells on coverslips were rinsed with phosphate-buffered saline (PBS) three times, fixed for 15 min in 4% paraformaldehyde in PBS and permeabilized in 0.1% Triton X-100 in PBS for 15 min. Cells were processed for detection of Cx43, using enhanced chemiluminescence detection, exactly as described previously [9, 10].

Subcellular fractionation and analysis by Western blotting

The Nuclei-EZ kit (SIGMA) was used to obtain nuclei-enriched and cytosolic fractions from cardiomyocyte or HeLa cell homogenates, according to manufacturer's instructions. Fractions (20 μg protein/lane) were analysed by SDS-PAGE, using 12% polyacrylamide gels, transferred subsequently to PVDF membranes and probed with anti-Cx43 polyclonal antibodies as described previously [9].

Cell growth

Stable cell lines were grown in 10% FBS in DMEM-hi-glucose medium with G418 until they reached 80% confluence. They were then lifted by trypsinization and seeded at 20,000/well in 24-well plates in DMEM supplemented with 1% FBS. The medium was changed every 48 h. On days 3, 4, 5, 6, 7 from plating, cells were trypsinized and counted using a hemocytometer.

Results and discussion

Connexin-43 has four transmembrane, three cytosolic and two extracellular domains [11]. We constructed cDNAs coding for residues 243–382, containing the third, C-terminal cytosolic 'tail' of Cx43 (CT-Cx43), and residues 1–242 , containing the plasma membrane, channel forming portion of the molecule (NT-Cx43). NT-Cx43 was 'tagged' with FLAG epitopes, so that it could be detected by anti-FLAG monoclonal antibodies. Our rabbit polyclonal antibodies, raised against a 17-residue peptide from the C-tail of Cx43 [10] were used to detect expression of WT- and CT-Cx43.

Immunofluorescent staining for FLAG (Fig. 1A) confirmed expression of the cDNA coding for NT-Cx43 and showed the expected localization to sites of intercellular contact between cardiomyocytes; it also showed that our transient transfection method and 'tagging' did not interfere with correct subcellular targeting. Anti-FLAG staining was completely absent in cultures that had not been transfected with a FLAG-tagged constructs (data not shown). Myocyte cultures transfected with cDNA coding for CT-Cx43 displayed two cellular patterns of anti-Cx43 staining. One pattern, identical to that of myocytes from non-transfected cultures, consisted of strong punctate staining along sites of intercellular contact as well as absence of cytosolic staining; it is presumed to represent untransfected myocytes. The other pattern, encountered in approximately 15% of myocytes (consistent with the expected efficiency of transfection), was characterized by strong intracellular staining (Fig. 1B), and is presumed to reflect expression of CT-Cx43, since it was seen exclusively in cultures transfected with the corresponding cDNA. Interestingly, an apparent association of CT-Cx43 with myofibrils could be discerned. Unexpectedly, CT-Cx43 was also localized to the nuclei of cardiomyocytes. Myofibrillar and nuclear localization pointed to a potential for the CT-Cx43 fragment to exert plasma-membrane independent function, and perhaps to affect gene expression directly.

To examine whether the localization pattern of CT-Cx43 was similar in another cell type, one that expresses undetectable levels of endogenous Cx43, we used transient transfections to express the cDNA for CT-Cx43 in Hela cells. Parallel cultures were transfected with the cDNA for WT-Cx43, as a positive control for correct targeting of the molecule. As shown in Fig. 1C, anti-Cx43 antibodies detected Cx43 in transfected HeLa cells, and showed plasma membrane localization, at both cell-cell contact as well as cell-substrate attachment sites (Fig. 1C); non-transfected HeLa cultures did not stain for Cx43 (data not shown). HeLa cultures, transiently transfected with the cDNA coding for CT-Cx43 also contained anti-Cx43 immunoreactive cells (Fig. 1D). In contrast to the membrane staining of cells expressing WT-Cx43, CT-Cx43 localized predominantly to the nucleus of HeLa cells. It would thus appear that the ability of CT-Cx43 to enter the nucleus is not cell-type specific, and can occur independently of whether the cells express or not endogenous Cx43 channels, in myocytes and HeLa cells, respectively.

To obtain independent evidence for nuclear localization of CT-Cx43 we prepared subcellular (cytosolic and nuclear) fractions from transfected myocytes or HeLa cells, and analyzed them for Cx43 by Western blotting. As shown in Fig. 2, immunoreactive band(s) displaying apparent electrophoretic motility corresponding to approximately 20 kDa, could be detected in both cytosolic and nuclear fractions, from either myocytes or Hela cells. Cytosolic fractions had more than one immunoreactive bands, while the nuclear fractions consisted of a single band of ~ 20 kDa. Non-transfected HeLa cells or myocytes displayed no immunoreactive bands or only the ~43–45 kDa intact Cx43 bands, respectively (data not shown).

The apparent size of immunoreactive bands at ~ 20 kDa, detected only in CT-Cx43-transfected cultures, is higher than

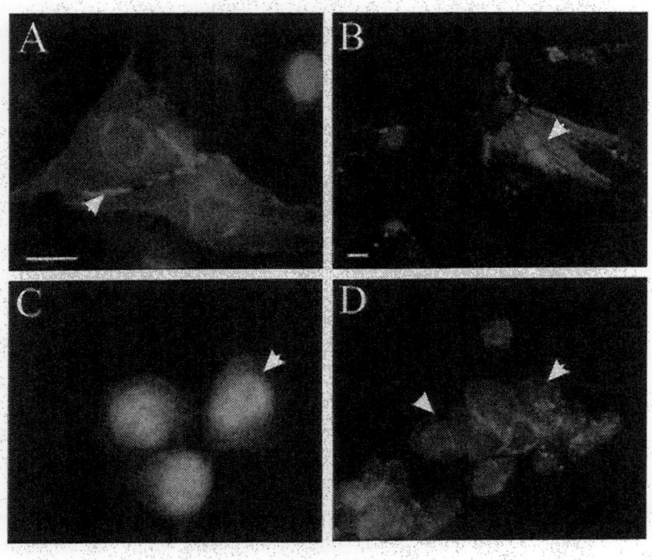

Fig. 1. Expression and localization of Cx43 constructs in cardiomyocytes and HeLa cells, detected by immunofluorescence. (A and B) Cardiomyocyte cultures transfected with cDNAs expressing NT-Cx43 (FLAG-tagged) or CT-Cx43, respectively and stained for FLAG (A) or Cx43 (B). (C and D) HeLa cells transfected with cDNAs expressing WT-Cx43 or CT-Cx43, respectively, and stained with anti-Cx43 antibodies. A, C and D are of the same magnification. Bar = 20 μM.

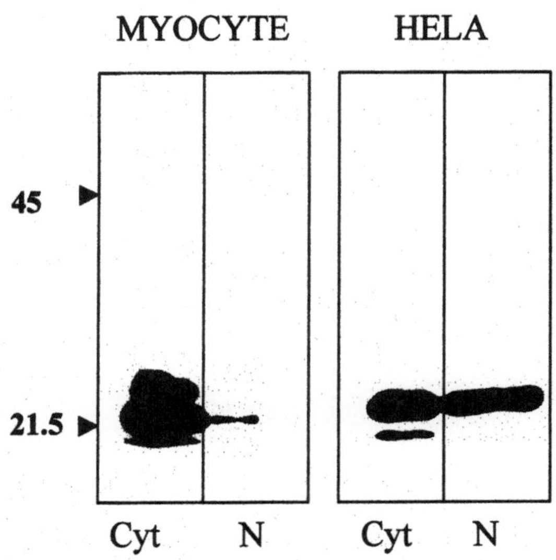

Fig. 2. Western blotting of subcellular fractions from cardiomyocyte or HeLa cell cultures transfected with cDNA coding for CT-Cx43. Cytosolic or nuclear fractions (20 μg protein per lane) from cardiomyocytes or HeLa cells (as indicated) were analyzed in 12% SDS/PAGE gels, transferred to PVDF membranes and probed with anti-Cx43 polyclonal antibodies. Migration of electrophoretic motility markers is indicated.

the predicted molecular weight for CT-Cx43, at ~ 15 kDa, suggesting presence of post-translational modifications.

Evidence therefore from both approaches, immunolocalization and subcellular fractionation indicates that, in addition to the cytosol, CT-Cx43 can localize to the nucleus. Since no conventional nuclear localization sequence (NLS) is present in CT-Cx43, one can speculate that it may enter the nucleus by interacting, directly or indirectly, with other protein(s) possessing NLS signals. CT-Cx43 is known to interact with proteins such as src [13], epsilon PKC [5] and β-catenin [13], all of which can translocate to the nucleus.

To determine whether CT-Cx43 can affect cell proliferation, we obtained HeLa cell lines stably overexpressing CT-Cx43, or WT-Cx43, for comparative purposes. These cell lines were maintained in culture for up to 7 days, in the presence of low (1%) FBS. Figure 3 shows representative data, cell number counts at 3–7 days of culture, from one set of stable cell lines. All cell lines expressing the same construct presented similar growth characteristics. While expression of WT-Cx43 caused only a small decrease in cell number (compared to untransfected HeLa cells), becoming significant only at 7 days ($p < 0.05$), expression of CT-Cx43 resulted in significant inhibition of growth, compared to either control HeLa cells or WT-Cx43 expressing cells ($p < 0.01$), clearly detectable at 4–7 days (Fig. 3). Daily morphological examination of the various cultures did not provide evidence for presence of above background cell death/ cell detachment or nuclear compaction, suggesting that the observed differences in cell numbers were not due to effects on cell survival. It is thus concluded that the 'free' C-tail of Cx43 can have significant biological effects on cells such as inhibition of proliferation, and that these effects do not require plasma membrane-dependent, channel-forming functions.

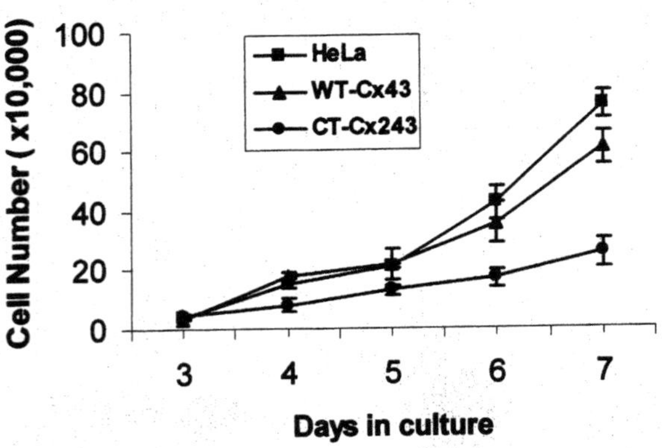

Fig. 3. Effect of WT-Cx43 or Cx43-CT on HeLa cell proliferation. HeLa cells stably expressing WT- or CT-Cx43 (as indicated) were plated at 20,000 cell per well (24-well plates) and maintained in the presence of 1% serum. Cell numbers were determined at 3–7 days from plating and shown as ± S.D. (n = 4). Cell numbers at different time points were compared between the different cell lines using one way ANOVA statistical analysis, where $p < 0.05$ was considered significant.

The mechanism of growth inhibition by CT-Cx43 remains to be established. Since CT-Cx43 was apparently more effective as a growth suppressor compared to WT-Cx43, it is possible that its properties are dependent on its localization away from the plasma membrane. Localization data pointed to a possible link between nuclear localization and growth inhibition by CT-Cx43. Another important question that remains to be addressed is whether 'free' endogenous CT-Cx43 can be present in cells, under physiological or pathophysiological conditions, and if so, whether it exerts effects on cell proliferation or other functions. This may be of significance in the presence of increased protein degradation, as would be encountered during ischemia.

References

1. Schneider MD: Myocardial infarction as a problem of growth control: Cell cycle therapy for cardiac myocytes? J Card Fail 2: 259–263, 1996
2. Lampe PD, Lau AF: Regulation of gap junctions by phosphorylation of connexins. Arch Biochem Biophys 384: 205–215, 2000
3. Yamasaki H, Naus CC: Role of connexin genes in growth control. Carcinogenesis 17: 1199–1213, 1996
4. Gourdie RG, Harfst E, Severs NJ, Green CR: Cardiac gap junctions in rat ventricle: Localization using site- directed antibodies and laser scanning confocal microscopy. Cardioscience 1: 75–82, 1990
5. Beyer EC: Gap junctions. Int Rev Cytol: 1–37, 1993
6. Zhang ZQ, Zhang W, Wang NQ, Bani-Yaghoub M, Lin ZX, Naus CC: Suppression of tumorigenicity of human lung carcinoma cells after transfection with connexin43. Carcinogenesis 19: 1889–1894, 1998
7. Mesnil M, Krutovskikh V, Piccoli C, Elfgang C, Traub O, Willecke K, Yamasaki H: Negative growth control of HeLa cells by connexin genes: Connexin species specificity. Cancer Res 55: 629–639, 1995
8. Beyer EC, Paul DL, Goodenough DA: Connexin43: A protein from rat heart homologous to a gap junction protein from liver. J Cell Biol 105: 2621–2629, 1987
9. Doble BW, Chen Y, Bosc DG, Litchfield DW, Kardami E: Fibroblast growth factor-2 decreases metabolic coupling and stimulates phosphorylation as well as masking of connexin43 epitopes in cardiac myocytes. Circ Res 79: 647–658, 1996
10. Doble BW, Ping P, Kardami E: The epsilon subtype of protein kinase C is required for cardiomyocyte connexin-43 phosphorylation. Circ Res 86: 293–301, 2000
11. Yancey SB, John SA, Lal R, Austin BJ, Revel JP: The 43-kD polypeptide of heart gap junctions: Immunolocalization, topology, and functional domains. J Cell Biol 108: 2241–2254, 1989
12. Giepmans BN, Hengeveld T, Postma FR, Moolenaar WH: Interaction of c-Src with gap junction protein connexin-43. Role in the regulation of cell-cell communication. J Biol Chem 276: 8544–8549, 2001
13. Ai Z, Fischer A, Spray DC, Brown AM, Fishman GI: Wnt-1 regulation of connexin43 in cardiac myocytes. J Clin Invest 105: 161–171, 2000

Molecular and Cellular Biochemistry **242**: 39–45, 2003.

Porcine aortic endothelial cells show little effects on smooth muscle cells but are potent stimulators of cardiomyocyte growth

Thomas Kubin,[1] Sabina Vogel,[2] Jutta Wetzel,[1] Stefan Hein,[3] Frederic Pipp,[2] Jörg Herold,[2] Matthias Heil,[2] Andreas Kampmann,[1] Stephanie Hehlgans,[1] Dietmar von der Ahe,[1] Wolfgang Schaper[2] and René Zimmermann[1]

[1]*Research Group Vascular Genomics, Kerckhoff Clinic, Bad Nauheim;* [2]*Department of Experimental Cardiology, Max Planck Institute, Bad Nauheim;* [3]*Department Kerckhoff Clinic, Bad Nauheim, Germany*

Abstract

Smooth muscle cells (SMC) and endothelial cells (EC) play a pivotal role in arteriogenesis and atherosclerosis. We evaluated the role of EC on the growth of SMC and neonatal cardiomyocytes (NEO) by using serum-free EC-supernatant (AoCM). Five percent fetal calf serum was used in order to mimic growth effects of blood. EC and SMC purities were 99% as determined by absence or presence of markers such as CD31, desmin, α-smooth muscle actin and tropomyosin using immunostaining and FACS analysis. AoCM markedly influenced the morphology of NEO as determined by α-actinin staining but showed only little effect on the phenotype of SMC. Protein synthesis after 2 days increased 2.5-fold in SMC and 3.7-fold in NEO as determined by tritium incorporation. The values for serum (2.8 and 2.3-fold, respectively) were comparable. The induction of DNA-synthesis by serum in NEO was twice that of AoCM (3.9-fold). However, proliferative effects of serum and AoCM on SMC differed markedly: Serum induced a 66-fold increase in DNA-synthesis resulting in a 54% higher cell number. DNA-synthesis after AoCM treatment lead to a nonsignificant small increase and no proliferation was detected. Platelet derived growth factor (PDGF-AB), present in blood, induced a 47-fold increase in DNA-synthesis and a 38% increase in cell number. Our data suggest that EC in the absence of physical forces exert strong morphogenic effects on cardiomyocytes but they lack specific effects on smooth muscle cells. In vessels EC might function as a border to isolate SMC from key regulators in blood such as PDGFs. (Mol Cell Biochem **242**: 39–45, 2003)

Key words: smooth muscle cells, cardiomyocytes, endothelial cells, proliferation, remodeling, growth

Introduction

It is well-known that smooth muscle cells (SMC) and endothelial cells (EC) play important roles in atherosclerosis and arteriogenesis. Atherosclerosis is the primary cause of heart disease and is the underlying cause of about 50% of all deaths in westernized countries. Interestingly, while atherosclerosis finally leads to the occlusion of the artery there are also reports on patients with symptomatic ischemic heart disease with occlusions of all three major coronary arteries or even the left main coronary artery but did not suffer from infarction. These patients profited from arteriogenetic processes where growth of preexisting arterioles leads to the development of an extensive collateral circulation [1]. Obviously smooth muscle cells and endothelial cells play in both processes, arteriogenesis and atherosclerosis, a pivotal role despite possible convergent and divergent mechanisms of arteriogenesis and atherosclerosis. Endothelial cells function

Address for offprints: T. Kubin, Research Group Vascular Genomics, Kerckhoff Clinic, Benekestrasse 2, D-61231 Bad Nauheim, Germany (E-mail: t.kubin@vascular-genomics.de)

as an active growth regulatory element as well as a potential selective barrier between blood and muscle cells. Furthermore, EC are well-known potent producers of a variety of growth factors/cytokines [2]. We have recently demonstrated that serum-free culture medium conditioned by endothelial cells of high cell purity (99%) [3] was highly potent in the induction of remodeling processes in cultured adult cardiac myocytes resulting in marked increases in protein synthesis, concomitant enhanced proteolysis, increases in surface area, *de novo* synthesis of sarcomeres and degradation of preexisting ones. Furthermore, EC induced a reexpression of the fetal α-smooth muscle actin in cardiomyocytes reminiscent of the effects of *in vivo* pressure overload [2]. On this basis we proposed that endothelial cells stimulate proliferation and growth of cultured adult aortic smooth muscle by secretion of trophic factors. Therefore, the effect of conditioned medium on the growth of SMC was investigated and compared with the effect of 5% fetal calf serum (a very strong cell culture stimulant for muscle cells) and platelet derived growth factor, a potent well-known growth factor for smooth muscle cells secreted by EC. Cultured neonatal cardiomyocytes which retain potency for DNA-synthesis in contrast to terminally differentiated adult cardiomyocytes were used to show the potency of conditioned medium and for better evaluation of its activity on smooth muscle cells.

Materials and methods

Materials

Basic medium consists of medium 199 with Earle's balanced salts (medium 199, Life Technologies) including 25 mM HEPES, 25 mM $NaHCO_3$, 100 IU/ml penicillin (Life Technologies), 100 mg/ml streptomycin (Life Technologies), 0.5 UG/ml amphotericin B (Life Technologies), supplemented with 2 mM L-carnitine, 5 mM creatine, 5 mM taurine (all Sigma). Fetal calf serum (FCS), newborn calf serum (NCS) were obtained from Sigma, HBSS⁻ (Hanks balanced salt solution, Ca^{2+}-free, Mg^{2+}-free) was from C.C.Pro (Neustadt, Germany) and trypsin from Life Technologies. All antibodies used for FACS (fluorescence activated cell sorter) and immunostaining (confocal microscopy) were obtained and diluted for studies as previously described in detail [3]. Pictures with a phase-contrast microscope (Leica, Germany) were taken with a digital camera (Leica) using its supplementary software.

Isolation and culture of porcine aortic smooth muscle cells

Porcine aorta were obtained from the local slaughterhouse. A part of the descending aorta close to the arch was removed, cut into one half and washed in $HBSS^-$. After incubation in 70% ethanol for 30 sec in order to get rid off endothelial cells the aorta was washed in $HBSS^-$ and the intima was removed. One gram media was after predigestion for 1 h in $HBSS^-$ containing 0.05% collagenase plus 0.025% elastase (CellSystems, St. Katharinen, Germany) digested in a new aliquot of the same solution for further 3 h. The digest was filtered through a mesh to remove debris and centrifuged for 10 min at 1200 rpm. The cell pellet was suspended in medium 199 containing 20% FCS, plated on a 148 cm^2 dish (Sigma) and incubated in 5% CO_2-95% O_2 at 37°C for approximately 2 weeks until confluency. Four medium changes were performed. Cells of fourth to sixth passage (split ratio 1:3) were used for the assays. For growth assay determinations smooth muscle cells were seeded at a density of 6.5×10^4 cells/cm^2. Twelve-well dishes (Falcon) were used for protein synthesis, DNA-synthesis determination and phase-contrast microscopy. Thirty-five millimeter dishes (Falcon) were used for FACS analysis and chamber slides (Nunc) for fluorescence microscopy.

Isolation and culture of rat neonatal cardiac myocytes

Ventricular myocytes were isolated from 1–2 day old Wistar rats. After removing atria, ventricles from 20 hearts were dissected and dissociated in $HBSS^-$ containing 0.018% collagenase (the same collagenase as for smooth muscle cell isolation) with shaking at 37°C for 30–45 min. Dissociated myocytes of digestion cycle 2–4 were collected in basic medium containing 1% NCS. Cells were preplated on an uncoated 148 cm^2 dish (Falcon) to remove non-myocytes and incubated in 5% CO_2-95% O_2 at 37°C for 1.5 h. Non-attached neonatal myocytes were removed from the dish and centrifuged at 1200 rpm for 10 min. Cell pellet was suspended in 1% FCS and cells were plated at a density of 6.5×10^4 cells/cm^2 on fibronectin (10 µg/ml; PromoCell, Heidelberg, Germany) coated culture dishes for 1 day. Before stimulation cells were washed twice with basic medium. Twelve-well dishes (Falcon) were used for protein synthesis, DNA-synthesis determination and light microscopy. Chamber slides (Nunc) were used for fluorescence microscopy.

Preparation of conditioned medium (AoCM)

Porcine aortic endothelial cells were isolated and cultured as described previously [3]. Cultures of second to fourth passage were grown to confluence in a multitray cell factory (Nunc), washed three times with serum-free medium 199 with Earle's balanced salts and then cultured for 2 days in the same medium to produce conditioned medium (AoCM). Prior to use ingredients were added as mentioned before to produce the same basal composition as basic medium leading to a final proportion of AoCM of 4:1 (vol/vol) because this dilution gave highest activity. AoCM was not supplemented with growth promoting agents.

Measurement of DNA synthesis, protein synthesis, DNA content and cell number

To investigate stimulatory effects on protein synthesis and DNA synthesis cells were stimulated with conditioned medium (AoCM), 5% fetal calf serum (5%) or 10 ng/ml platelet-derived growth factor AB (PDGF-AB). Medium change was performed daily. Cells were radiolabelled during the last 12 h of the culture period with 2.5 μCi/ml of [^{3}H]-phenylalanine or [^{3}H]-thymidine (Amersham). As baseline control served untreated cells. At the end of cultivation cells were washed 3 times with HBSS$^-$, fixed for 1 h in 10% trichloracetic acid (TCA), washed twice with 10% TCA, 3 times with 95% ethanol, air-dried and extracted in 0.3 M NaOH. Aliquots were used for counting of tritium incorporation. For DNA-synthesis we determined total [^{3}H]-thymidine incorporation per well. Cell numbers were determined with a haemocytometer for smooth muscle cells after trypsinization and trypan-blue staining. DNA content was determined as previously described [2]. Non-replicative DNA is considered to be a reliable measure for the relative cell number [2]. For measurement of protein synthesis smooth muscle cells (SMC) and neonatal cardiomyocyte cultures (NEO) were constantly treated with 10 μM 1-(β-D-arabinofuranosyl)cytosine to prevent SMC proliferation and proliferation of non-myocytes in NEO cultures. For determination of protein synthesis we calculated the ratio of [^{3}H]-phenylalanine incorporation-to-DNA.

Statistics

Statistical comparisons were performed with ANOVA and Bonferroni tests. P values < 0.05 were taken as statistical significant.

Results

Isolation and characterization of porcine aortic smooth muscle cells (AoSMC)

Treatment of the aorta with 70% ethanol and selective dissection of the media enabled us to isolate defined regions of AoSMC and to minimize contaminations with other cell layers and cell types. Primary isolates from 1 g of tissue attached within 1–2 days, start to flatten within 5 days similar to human AoSMC [4] and became confluent within 2 weeks corresponding to approximately 15×10^6 primary cells. Since AoCM in contrast to many other cell types show overgrowth, cultures which almost completely cover the culture dish were regarded as confluent as depicted in Fig. 1A. Cultures were then split at a ratio of 1:3. Passaged cells become confluent within 3–4 days which corresponds to a general doubling rate of approximately 36–48 h. Figure 1A shows a phase-contrast micrograph of the typical confluent phenotype of AoSMC in 20% FCS just before passaging. The phenotype, growth behaviour and speed can be maintained at least until passage 6. Cells show the typical elongated shape. Multilayers but also monolayers are visible (Fig. 1A). Note the low number of light cell spots indicating only a few dying or dead cells. Increasing the number of passages higher than 10 lead to cultures of minor quality visible in slower growth rate, clustering during trypsinization, increased cell death, some formation of growing islands without covering the culture dish (not shown). For that reason we terminated cultures at passage six or earlier.

Figures 1B–1D show intense immunostainings of cultured AoSMC (70% confluence) for the muscle specific marker desmin (B), muscle-specific tropomyosin (C) and α-smooth muscle actin (D). However, data obtained with α-smooth muscle actin as a typical muscle marker has to be interpreted with some care and will be reasoned in the Discussion section. The use of antibodies and the performance of immunostaining and FACS analysis were exactly the same as previously described in detail [3]. Figures 1B–1D show that almost all cells are positive (green colour) for the markers mentioned above. Blue colour indicates nuclei. In addition, Figs 1B and 1D show AoSMC areas of monolayers and multilayers. Figure 1E depicts phalloidin staining of actin and Fig. 1F staining for CD31 (PECAM-1). Since the instrumental settings of the confocal microscope are the same in the figures the complete absence of CD31 positive cells in AoSMC cultures is clearly demonstrated.

In order to determine the purity of AoSMC flow cytometric (Fluorescence-activated cell sorter; FACS) analysis was performed by loading the cells with anti-desmin (Fig. 2A), anti-tropomyosin (Fig. 2B) and anti-α-smooth muscle actin (Fig. 2C) antibodies. For exclusion of non-specific signals, AoSMC were also loaded with isotype matching antibodies. In comparison to isotype loaded cells, there was a distinct rightward shift in the FACS profile for all three muscle markers (Figs 2A–2C) revealing 99% positive staining of AoSMC. In contrast, this shift was absent in the presence of anti-CD31 indicating the absence of endothelial cells (Fig. 2D).

Fluorescence and FACS data obtained with endothelial cells are not shown here because they have been extensively presented previously [3].

Effect of AoCM and FCS on the phenotype of aortic smooth muscle cells (AoSMC) and neonatal cardiac myocytes (NEO)

We examined the effect of AoCM and 5% fetal calf serum on the phenotype of smooth muscle cells and neonatal cardiomyocytes. Before stimulation cultured smooth muscle cells were kept for 2–3 days in pure basic medium to induce

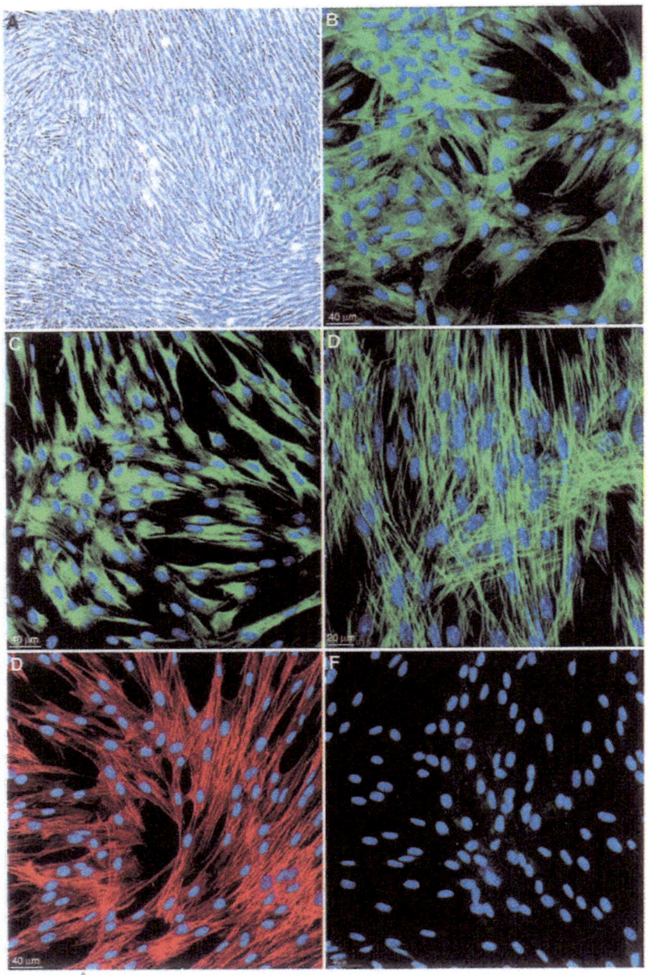

Fig. 1. Phase contrast micrograph (10× lens) of smooth muscle cells at confluence (A). SMC (all green) were immunostained for desmin (B), tropomyosin (C), α-smooth muscle actin (D), actin (E, red)) and CD-31 (F).

quiescence. Quiescent AoSMC were then treated with various agents for 3 days (Fig. 3). Control cells kept in serum-free basic medium did not change significantly but showed a slightly spread phenotype with the tendency to cluster in cell groups with typically interspersed cell-free areas (Fig. 3A). AoCM- treated cells showed a similar phenotype but cell areas appeared to be slightly more densely packed (Fig. 3B). In contrast proliferating AoSMC in 5% FCS showed a diminution of interspersed areas and cells started partly to overgrowth (Fig. 3C) by building extensions. They will finally reach a confluent phenotype similar to 20% FCS treated cells after 1 week as seen in Fig. 1A. The slower growth speed of cells in 5% is due to a lower serum concentration and the period of quiescence which is absent during the passages of AoSMC in 20% serum. PDGF-AB shows a comparable growth but AoSMC appeared to be thinner with more pronounced long and thin extensions (Fig. 3D).

Figure 4 shows NEO 2 days after treatment with basic medium (A) or with AoCM (B). Fluorescence labeling (green) of sarcomeric α-actinin clearly indicates cross-striated pattern typical for cardiomyocytes. Pure phalloidin (actin; red) staining indicates also the presence of some fibroblasts (Figs 4A and 4B). Purity of neonatal cardiomyocytes is usually about 95%. After AoCM stimulation of cardiomyocytes there are clearly visible increases in surface area, length, diameter and large nuclei. A more intense staining of α-actinin indicates synthesis of new filaments in contrast to cultures kept in basic medium. In the control a visible increase is obvious in the number of dying or dead cells (light green colour in round shaped cells, right area of Fig. 4A) which accounts for approximately 20–30% cell loss until 48 h with the number of dead cells increasing markedly at later time points. In AoCM-stimulated cultures cardiomyocytes start to form a confluent monolayer and develop contractile activities. The number of dying cells is approximately 10% and does not increase markedly within a week. Five percent FCS-stimulation shows similar but less pronounced effects on cardiomyocytes (not shown) as described previously [2].

Effect of AoCM and FCS on DNA- and protein-synthesis of neonatal cardiac myocytes and on proliferation of aortic smooth muscle cells

We examined the effect of AoCM, 5% fetal calf serum, and PDGF-AB on protein synthesis and DNA synthesis of AoSMC by determination of total [^{3}H]-phenylalanine and [^{3}H]-thymidine incorporation after 2 days. Cell numbers were counted after trypsinization. The same parameters were determined in neonatal cardiac myocytes. However, cell numbers were not counted since cultured NEO are generally non-proliferative under the conditions presented here. Values of un-

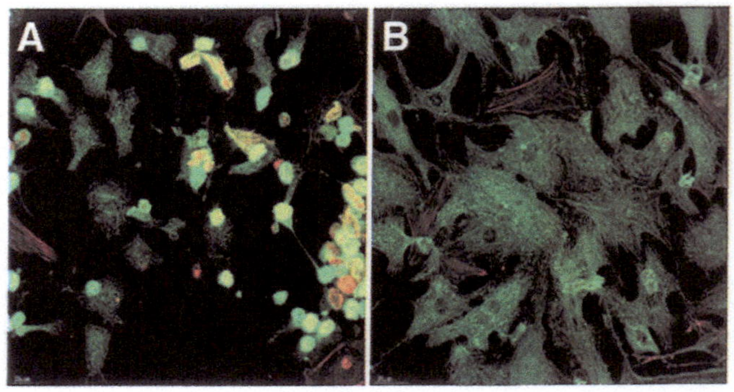

Fig. 4. Photomicrographs of α-actinin (green) and actin (red) immunostained neonatal cardiomyocytes. Untreated control cells (A) or cells stimulated with conditioned medium (B) after 2 days. Note large nuclei and broad cross-striated distribution of filaments on large surface areas (B) which are little developed in control cells (A).

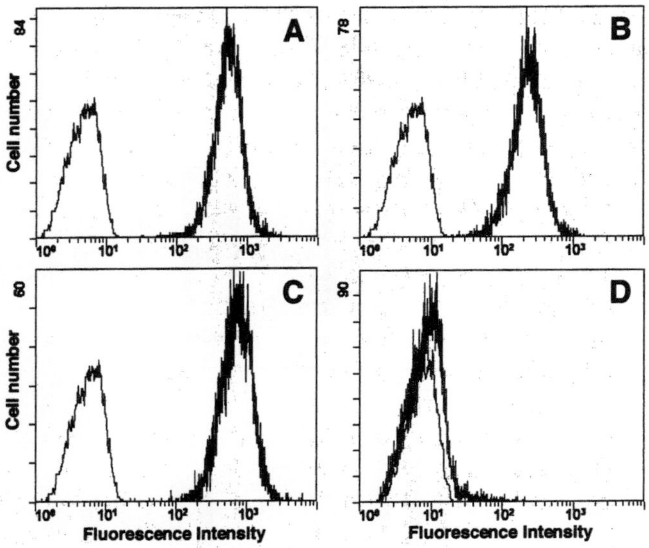

Fig. 2. Flow cytometric analysis of AoSMC loaded with desmin, tropomyosin, α-smooth muscle actin and CD-31. Fluorescence was compared with isotype labeled cells as described in the 'Materials and methods' section. More than 99% SMC were positively labeled and show a rightward shift for desmin (A), tropomyosin (B) and α-smooth muscle actin (C). AoSMC were negative for CD-31 (D).

stimulated cells served as control for both cell types and were set 100%.

In neonatal cardiomyocytes the effect of AoCM on protein-synthesis was stronger than that of FCS while FCS acted more potent on DNA-synthesis. AoCM caused significant increases of 3.7-fold in protein synthesis (Fig. 5A) and 3.9-fold in DNA-synthesis (Fig. 5B) in NEO. FCS induced increases of 2.3-fold in protein synthesis and 7.3-fold in DNA-synthesis (Figs 5A and 5B). In smooth muscle cells activities of AoCM and serum were contrasting. The effect of AoCM and FCS on protein-synthesis of AoCM (Fig. 5C) were comparable to 2.5- and 2.8-fold, respectively. However, FCS lead (66- vs. 6.6-fold) significantly to a 10 times stronger stimulation than AoCM (Fig. 5D) treatment. The proliferative effect of FCS is seen in increases in the total cell number of 54% in comparison to the cell number at the time of stimulation (100%) (Fig. 5E). Ten percent of cells were lost in the control group and almost no change in cell number is seen in AoCM treated cultures. PDGF-AB also significantly induced a 47-fold increase in DNA synthesis resulting in a 38% increase in cell number (Figs 5C and 5E).

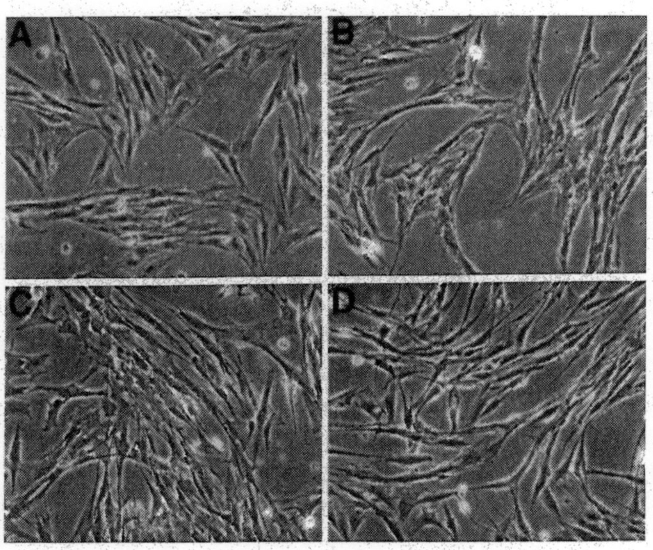

Fig. 3. Phase contrast micrographs of aortic smooth muscle cells (20× lens). Control cells kept in basic medium (A). Cells were stimulated with conditioned medium (B), 5% fetal calf serum (C) or 10 ng/ml platelet derived growth factor-AB (D) for 3 days.

Discussion

We have previously demonstrated that serum-free conditioned medium produced by endothelial cells is able to induce remarkable effects on the morphology and dedifferentiation of cultured primary adult rat cardiomyocytes resulting in an extensive remodeling of myocyte structure [2]. The species independent activity of endothelial conditioned medium was stronger and in its morphological effects different from that of serum. Therefore, our results fitted to an accumulating body of experimental data supporting the presence of a paracrine pathway for the modulation of myocardial function by endothelial cells [5]. We now tested whether endothelial cells had similar effects on smooth muscle cells derived from the aorta. We expected that aortic endothelial cells as a potent source of cardiovascular growth factors/cytokines will be effective in inducing proliferative and growth responses of smooth muscle cells.

Immunofluorescence and FACS analysis showed a 99% cell purity of AoSMC. In addition to α-smooth muscle actin we used desmin and muscle-specific tropomyosin antibodies for determination of cell purity because we have observed that α-smooth muscle actin is not necessarily a marker for smooth muscle cells. We have previously described that microvascular endothelial cells express α-smooth muscle actin *in vitro* and *in vivo* in pigs [3] and in human myocardium [6]. Its expression has also been reported [7] in other cell types. Thus, α-smooth muscle actin expression alone does not provide definitive evidence for α-smooth muscle cell lineage [8]. The stringent isolation procedure from the media lead to homogeneous cell cultures with constant growth rates and similar morphological appearance. Cultures until passage 6 guaranteed stable results. This is in contrast to our experience with smaller animals. For three reasons we preferred porcine AoSMC to small standard laboratory animals such as rat or mouse. First, human coronary morphology as well as

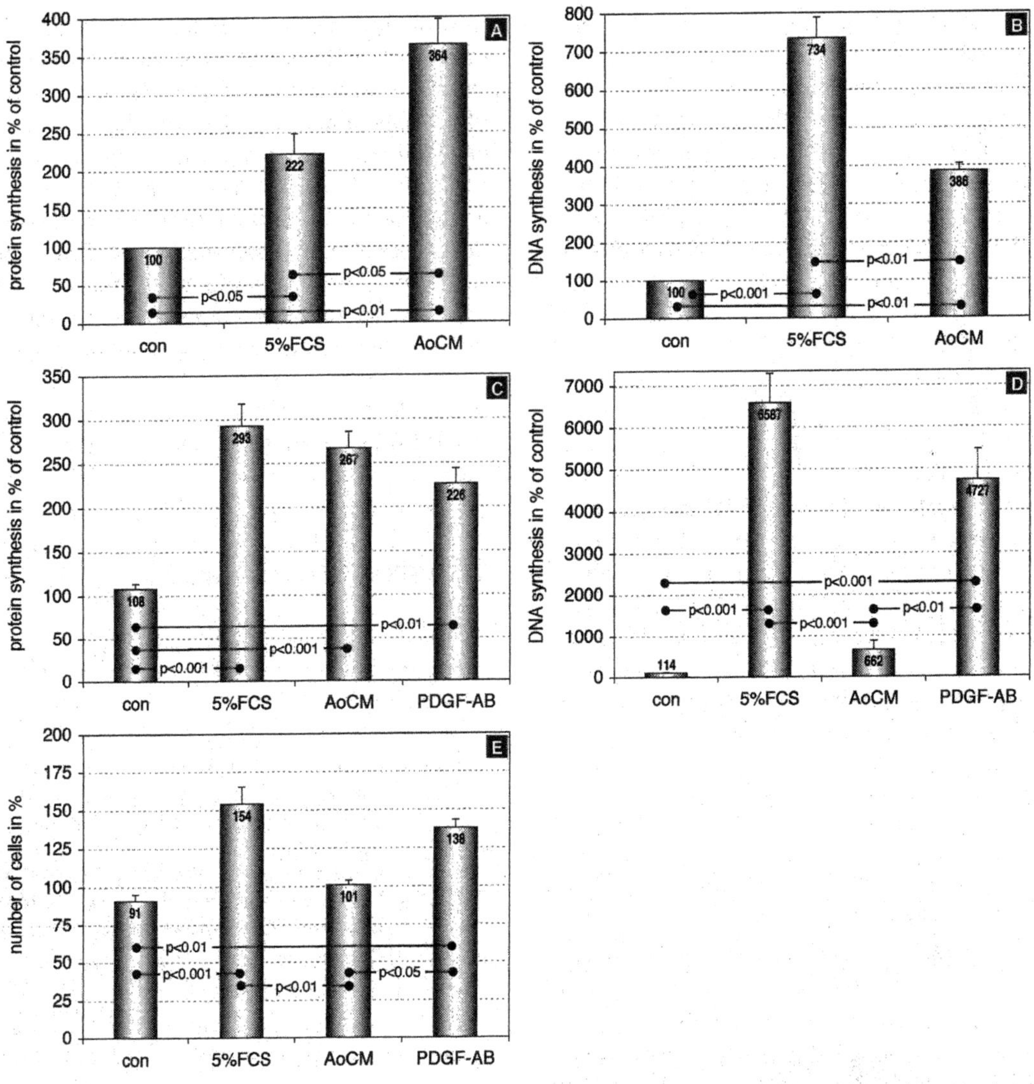

Fig. 5. The effect of conditioned medium (AoCM). Five percent fetal calf serum (5%) and platelet derived growth factor-AB (PDGF-AB) on protein- and DNA-synthesis of smooth muscle cells (C, D) and neonatal cardiomyocytes (A, B) after 2 days. Protein synthesis (A, C). DNA-synthesis (B, D). Rate of proliferation only shown for SMC after 3 days (E). Data (A–E) are the average of 4 independent experiments performed in duplicate.

hemodynamic parameters are similar to that of pigs [3, 9, 10]. For example heart beating frequencies of pigs with 60–90 beats per min (bpm) is close to humans while beating frequencies of rats and mice are between 300–600 and 300–800 bpm, respectively. Therefore functional and morphological parameters are different between large and small animals. Second, in our hands cell cultures from mice or rats gave variabilities in their results. Thus, constant growth rates and comparable morphology between subcultures are difficult to obtain. Third, since we are presently investigating signal transduction pathways by performing proteomic approaches it is necessary to obtain large-scale cultures of AoSMC. High passage numbers and doubling rates of AoCM from small animals might bring the Hayflick-limit (the limited capacity of cells to divide; extensively reviewed in [11]) into play. This might result in increased numbers of senescent cells and pronounced inconsistencies in the results. The most appropriate system would be the culture of human aortic smooth muscle cells. However, these frequently do not provide an appropriate *in vitro* model to investigate biologically relevant functions because cells are either only available from diseased hearts in patients treated with various drugs before transplantation or from commercial sources which do not correspond to the quality criteria described here.

Our results showed that the effects of AoCM and serum on smooth muscle cells were different. While both media induced significant and comparable increases in protein synthesis, in comparison to AoCM DNA-synthesis was 10-fold higher after serum stimulation. AoCM did not result in an increased cell number in contrast to a 54% increase in cell number after serum stimulation. Furthermore, AoCM did not induce observable morphological changes in smooth muscle cells when compared to untreated controls. This is in contrast to data obtained with AoCM-stimulated cardiomyocytes where a dramatic remodelling was quite obvious. Therefore under the culture conditions presented here it appears unlikely that endothelial cells are active in a way that they stimulate smooth muscle cells as they influence cardiomyocytes. Reasons for the differences in the activity of endothelial cells on AoSMC and NEO may lie in differential responsiveness of these cell types but also in the potential secretion of smooth muscle cell specific inhibitors by endothelial cells. Other parameters such as physical forces in the activation of EC (which are absent in our culture system) have also taken into consideration.

Platelet derived growth factor-AB (PDGF) one of the most potent growth factors [12] for AoSMC showed an 8-fold stronger effect on DNA-synthesis than AoCM and smooth muscle cell number increased 38%. PDGF occurs in blood in potent mitogenic concentrations [13] and is produced and secreted by blood derived cells such as platelets and monocytes [12]. Furthermore, it is a well-known fact that blood (cells) contain large amounts of various growth factors such as transforming growth factor-β and insulin-like growth factors which show potent activities on AoSMC. Therefore, serum is the strongest growth promoting medium *in vitro* for smooth muscle cells presently known.

In order to demonstrate that the absence of proliferative and morphogenic activity of AoCM on smooth muscle cell is not due to a lack of agonists we investigated the effect of AoCM on neonatal cardiomyocytes. NEO retain the capacity of DNA-synthesis in contrast to terminally differentiated adult cardiac myocytes. It has to be noted that rat neonatal cardiomyocytes are not in our *in vitro* system proliferating to a measurable degree. However, a capacity for DNA-synthesis can be seen. They have usually one nucleus but many of them show two nuclei per cell after 1 week of serum treatment (not shown). This is reflected in a significant 7.3-fold increase of DNA-synthesis after serum stimulation. AoCM were less effective on DNA-synthesis but surprisingly more potent on protein synthesis. The reason lies probably in different growth factor compositions of these growth promoting media. The profound influence of AoCM on NEO is visible in the change of the morphological appearance. Large nuclei indicate growth activities resulting in myofibrillogenesis visible in α-actinin staining and enlarged surface areas. These processes are hardly visible in the control group. Furthermore, the establishment of cell-cell contact with concomitant contractile activity indicates the development of a 'tissue-like' sheet and show the great potential of endothelial cells to participate in regenerative processes of the myocardium.

Our experimental data provide evidence for the existence of paracrine pathways between endothelial cells and cardiomyocytes characterized by the bioactivities of trophic factors in conditioned medium. The lack of significant effects on smooth muscle cells by endothelial cells leads to the following conclusions: First, key regulators such as PDGF may originate from blood. Second, disruption of endothelial cells as a selective borderline could lead to the availability of growth factors/cytokines or mitogens produced by blood cells or the loss of endothelial-derived AoSMC inhibitors. Third, endothelial cells might be activated by other mechanisms such as physical forces in order to transmit endothelial-derived mitogenic signals for smooth muscle cells.

Acknowledgements

The authors thank Brigitte Matzke for excellent technical assistance, Gunther Schuster for computer reproduction of the immunohistochemical illustrations, and Gerhard Stämmler for preparation of the manuscript.

References

1. Schaper W, Piek JJ, Munoz-Chapuli R, Wolf C, Ito W: Collateral circulation of the heart. In: J.A. Ware, M. Simons (eds). Angiogenesis and Cardiovascular Disease. Oxford University Press, New York, Oxford, 1999, pp 159–198
2. Kubin T, Ando H, Scholz D, Bramlage P, Kostin S, van Veen A, Hein S, Fischer S, Breier A, Schaper J, Schaper W: Microvascular endothelial cells remodel cultured adult cardiomyocytes and increase survival. Am J Physiol 45: H2179–H2187, 1999
3. Ando H, Kubin T, Schaper W, Schaper J: Porcine coronary microvascular endothelial cells express α-smooth muscle actin and show low NOS III activity. Am J Physiol 45: H1755–H1768, 1999
4. Smirnov VN, Orekhov AN: Smooth muscle cells from adult human aorta. In: H.M. Piper (ed). Cell Culture Techniques in Heart and Vessel Research. Springer, Berlin, 1990, pp 271–289
5. Shah AM: Paracrine modulation of heart cell function by endothelial cells. Cardiovasc Res 31: 847–867, 1996
6. Elsässer A, Schlepper M, Klövekorn WP, Cai WJ, Zimmermann R, Müller KD, Strasser R, Kostin S, Gagel C, Münkel B, Schaper W, Schaper J: Hibernating myocardium. Circulation 96: 2920–2931, 1997
7. Schwartz SM, deBlois D, O'Brien ERM: The intima-soil for atherosclerosis and restenosis. Circ Res 77: 445–465, 1995
8. Owens GK: Regulation of differentiation of vascular smooth muscle cells. Physiol Rev 75: 487–517, 1995
9. Kassab GS, Rider CA, Tang NJ, Fung YCB: Morphometry of pig coronary arterial trees. Am J Physiol 265: H350–H365, 1993
10. Kassab GS, Fung YCB: Topology and dimensions of pig coronary capillary network. Am J Physiol 267: H319–H325, 1994
11. Shay JW, Wright WE: Hayflick, his limit, and cellular aging. Nature Rev 1: 72–76, 2000
12. Hughes AD, Clunn GF, Refson J, Demoliou-Mason C: Platelet-derived growth factor (PDGF): Actions and mechanisms in vascular smooth muscle. Gen Pharmac 27: 1079–1089, 1996
13. Bowen-Pope DF, Hart CE, Seifert RA: Sera and conditioned media contain different isoforms of platelet-derived growth factor (PDGF) which bind to different classes of PDGF receptor. J Biol Chem 264: 2502–2508, 1989

Molecular and Cellular Biochemistry **242:** 47–55, 2003.

Differential cytokine expression in myocytes and non-myocytes after myocardial infarction in rats

Alexander Deten, Hans Christian Volz, Wilfried Briest and Heinz-Gerd Zimmer

Carl-Ludwig-Institute of Physiology, University of Leipzig, Leipzig, Germany

Abstract

The proinflammatory cytokines interleukin (IL)-1β and IL-6 are increased after acute myocardial infarction (MI). Moreover, serum IL-6 level is elevated after MI, but has also been associated with heart failure. In the present study, heart function was monitored in a rat model of chronic MI. Cytokine expression in the infarcted and non-infarcted myocardium as well as in hearts of sham-operated controls was measured by the ribonuclease-protection assay. To identify the cells contributing to the increased cytokine expression, we further analyzed myocytes and non-myocytes isolated in the acute phase as well as during congestive heart failure (CHF) after MI. There was a strong induction in cytokine expression in the myocytes of the infarct area 6 h after MI. In the non-infarcted myocardium, cytokine expression increased only slightly in the non-myocytes after 6 h. This was not different from sham-operated controls and may, therefore, be induced by stress and catecholamines. In CHF, however, cytokine expression level in myocytes was normal. It increased slightly but significantly in the non-myocytes 4 and 8 weeks after MI. In conclusion, we suggest that pro-inflammatory cytokines, produced by the ischemic myocytes may be involved in the initiation of wound healing of the necrotic area, whereas the effect of pro-inflammatory cytokines in CHF, if any, seems not to be crucial. (Mol Cell Biochem **242**: 47–55, 2003)

Key words: myocardial infarction, chronic heart failure, cytokine expression

Introduction

Substantial cardiac remodeling is initiated in response to large myocardial infarction (MI). These changes include replacement of the necrotic area by scar as well as compensatory hypertrophy and adverse fibrosis of the surviving myocardium [1]. Although believed to serve initially as a compensatory mechanism to maintain cardiac function, this remodeling is also supposed to contribute to the development of congestive heart failure (CHF). This is associated with a local and systemic elevation of both norepinephrine (NE) and pro-inflammatory cytokines, such as interleukin (IL)-1β, IL-6, and tumor necrosis factor (TNF)-α [2–6]. Therefore, a growing number of studies investigated the cardiovascular actions of cytokines. It is known that IL-6 plasma levels are elevated in acute MI [7]. Recently, we have shown that pro-inflammatory cytokine expression is strongly induced in the acute state of myocardial infarctions in rats [8]. Similar results were obtained in the reperfused myocardium after short periods of coronary artery occlusion in dogs [9, 10]. In addition, Ono *et al.* [11] and Prabhu *et al.* [12] showed elevated myocardial gene expression of IL-1β, IL-6 and TNF-α up to 20 weeks after coronary artery ligation when cellular inflammatory infiltration had abated. Although cytokines like IL-1β or TNFα have negative inotropic effects [13–15], transgenic mice overexpressing IL-6 and the soluble IL-6 receptor developed cardiac hypertrophy [16].

In the present study, heart function was monitored in a rat model of chronic MI. In addition, cytokine expression in the infarcted and non-infarcted myocardium as well as in hearts of sham-operated controls was measured by ribonuclease-protection assay. To identify the cells contributing to increased cytokine expression, we further analyzed myocytes and non-myocytes isolated in the acute phase as well as during congestive heart failure after MI.

Address for offprints: A. Deten, Carl-Ludwig-Institute of Physiology, University of Leipzig, Leibigstrasse 27, D-04103 Leipzig, Germany (E-mail: deta@medizin.uni-leipzig.de)

Materials and methods

Animal model

Myocardial infarctions were induced in female Sprague-Dawley rats (240–260 g) by ligation of the left anterior descending coronary artery (LAD) under ether anesthesia essentially as previously described [17, 18]. In brief, the fourth intercostal space was opened, the heart was exteriorized and the pericardium was incised. Thereafter, the hearts were held with forceps and a needle was passed around the LAD near its origin between the left auricle and the pulmonary outflow tract. Next, the ligation (6/0 prolene, Johnson & Johnson) was tightened, the thorax was closed, and the air removed by gently compressing the thorax. Intubation was not necessary since the entire procedure took no longer than 60 sec. In case spontaneously breathing had stopped, the rats were artificially ventilated trough the nose with room air for a few seconds. The rats were allowed to recover and to move freely in their cages with access to tap water and standard pellet diet (Altromin) *ad libitum*. Sham-operated animals underwent the same procedure except that no ligation was performed.

Heart and circulatory function was measured with ultra-miniature catheter pressure transducers (model SPR-249 and model SPR-291 for the left and right heart, respectively; Millar Instruments) in closed-chest spontaneously breathing rats anesthetized with thiopental sodium (Trapanal® 60 mg/kg i.p., Byk Gulden) as previously described [17, 18]. Heart rate (HR), right (RV) and left ventricular (LV) pressure and the maximal rate in rise of ventricular pressure (LV and RV dp/dt_{max}) were recorded continuously for 15 min on a Brush 2600 recorder (Gould Instruments Systems). At the end of hemodynamic characterization, a thermodilution microprobe (1,5 French, Columbus Instruments) was placed in the ascending aorta and cardiac output was determined by the thermodilution method by a computer (Cardiomax IIR, Columbus Instruments). Total peripheral resistance (TPR) was calculated by dividing mean arterial pressure by cardiac output, normalized to body weight. After the hemodynamic measurements had been obtained, the hearts were rapidly excised. The infarct area was cut from the non-infarcted LV and RV. The hearts of sham-operated animals were also cut into three parts: the RV, the LV and the anterior region of the LV free wall. The latter corresponded to the infarcted area in MI animals and served as control. The tissue pieces were snap frozen in liquid nitrogen for RNA isolation. All experiments were approved by the appropriate Federal State Agency.

Cell isolation

Cardiac myocytes and non-myocytes were isolated as previously described with minor modifications [19]. Four rats per isolation procedure were anaesthetized by an i.p. injection of 100 mg/kg thiopental sodium and 6.000 IU/kg heparin (Liquemin®, Hoffmann-La Roche) at different times after induction of MI or sham operation. Rats without any surgery served as additional controls. The hearts were excised, mounted on a modified Langendorff perfusion system and rinsed with Joklik's minimal essential medium (JMEM, F. Messi, Cell Culture Technologies) at 37°C continuously gassed with 95% O_2/5% CO_2. Perfusion was continued by recirculating JMEM, containing 29 U/ml collagenase (Worthington), for 35 min. Thereafter, the hearts were removed, the ventricles dissected, and the infarct area cut from the non-infarcted LV. Due to the small tissue samples the sections of two hearts were pooled. The samples were then cut with tissue scissors in about 1 × 1 mm pieces and incubated for another 15 min in 20 ml collagenase buffer (KB buffer: 70 mmol/L KCl; 30 mmol/L K_2PO_4; 22 mmol/L glucose; 5 mmol/L $MgSO_4$; 0.5 mmol/L EGTA; 20 mmol/L taurine; 5 mmol/L creatine; 10 mmol/L succinic acid; 2 mmol/L pyruvic acid; 5 mmol/L ATP; 2 mmol/L butyric acid; pH 7.4 and 115 U/ml collagenase. The cell suspension was filtered through a 250 µm nylon mesh and spun down at 25 × g for 3 min. The supernatant was centrifuged at 250 × g for 10 min. Both cell pellets were used for RNA preparation without further manipulations.

RNase protection assay (RPA)

The rat cytokine template set (rCK1) was obtained from PharMingen and labeled with [α-^{32}P]-UTP (3000 Ci/mmol, Amersham) by means of RiboQuant® *In Vitro* Transcription Kit (PharMingen) as described by the manufacturer. Total RNA was isolated using the Trizol®-Reagent (Gibco BRL) according to the protocol supplied by the manufacturer. 7.5 µg of total RNA were used in the RNase protection assay (RiboQuant® RPA Kit, PharMingen) as earlier described [18]. Protected probes were quantified using the Molecular Imager (BioRad). The signals of specific mRNAs were normalized to those of L32 mRNA. The rCK1 template sets contained the following cDNA probes (probe length in bp/protected): IL-1α (432/403), IL-1β (390/361), TNF-β (351/322), IL-3 (315/286), IL-4 (285/256), IL-5 (255/226), IL-6 (231/202), IL-10 (210/181), TNF-α (189/160), IL-2 (171/142), Interferon γ (156/127), L32 (141/112) and GAPDH (126/97).

Statistical analysis

The data are expressed as mean ± S.E.M. Multiple range test and Kruskall-Wallis test on ranks was used for multigroup comparison (STATGRAPHICS Plus 4.1, Statistical Graphics Corp.) utilizing multiple comparison procedure according to Tukey's HSD method. The Mann-Whitney *U*-test was used for two-group comparison. A value of $p < 0.05$ was considered significant.

Results

Hemodynamic measurements

Large myocardial infarctions were confirmed in all experimental animals by inspection of the LV myocardium and by ECG. As in our previous studies, all animals had severe depression of LV function. This was characterized by a LV developed pressure below 110 mmHg and a marked increase in LV end-diastolic pressure (LVEDP, Fig. 1). Also, cardiac output was reduced, while total peripheral resistance was only slightly reduced. Between 3 and 9 days after MI, circulatory function had recovered partially as indicated by an increase in cardiac output, but myocardial contractile dysfunction was still evident, since LV developed pressure was reduced and LVEDP was increased. Thereafter, cardiac output declined, while LVEDP further increased. In parallel, RVSP progressively increased (Fig. 1). The hemodynamic measurements further showed a strong reduction in the rate of rise and fall in LV pressure (LV dP/dt_{max} and LV dP/dt_{min}, respectively) as well as in LV stroke work (not shown). RV function, on the other hand, showed a progressive increase in RV dP/dt_{max}, RV dP/dt_{min} and RV stroke work and from day 9 onwards (not shown).

Cardiac cytokine expression after myocardial infarction

In the RNA samples, extracted from tissue pieces in the first 12 h after MI or sham operation, cardiac cytokine expression was increased (Fig. 2). This induction was more pronounced in the infarct area compared to the non-infarcted myocardium, but also occurred after sham operation. Moreover, the cytokine expression decreased in the non-infarcted myocardium after 12 h compared to 6 h after MI, while it was even more increased in the infarct area 12 h post-MI. On day 3 after MI, there were no differences in cytokine expression between MI-animals and sham-operated controls (not shown). Also at later times after MI, i.e. after 4 and 8 weeks, cytokine expression was at control level (Fig. 2). These results were obtained in samples of the infarct area as well as in the non-infarcted myocardium of animals with reduced heart function and signs of congestive heart failure, but also in animals with MI that had normal heart function (details not shown).

Cytokine expression in isolated cells after myocardial infarction

The expression of the pro-inflammatory cytokines IL-1, IL-6 and TNFα was induced in the myocyte as well as in the non-

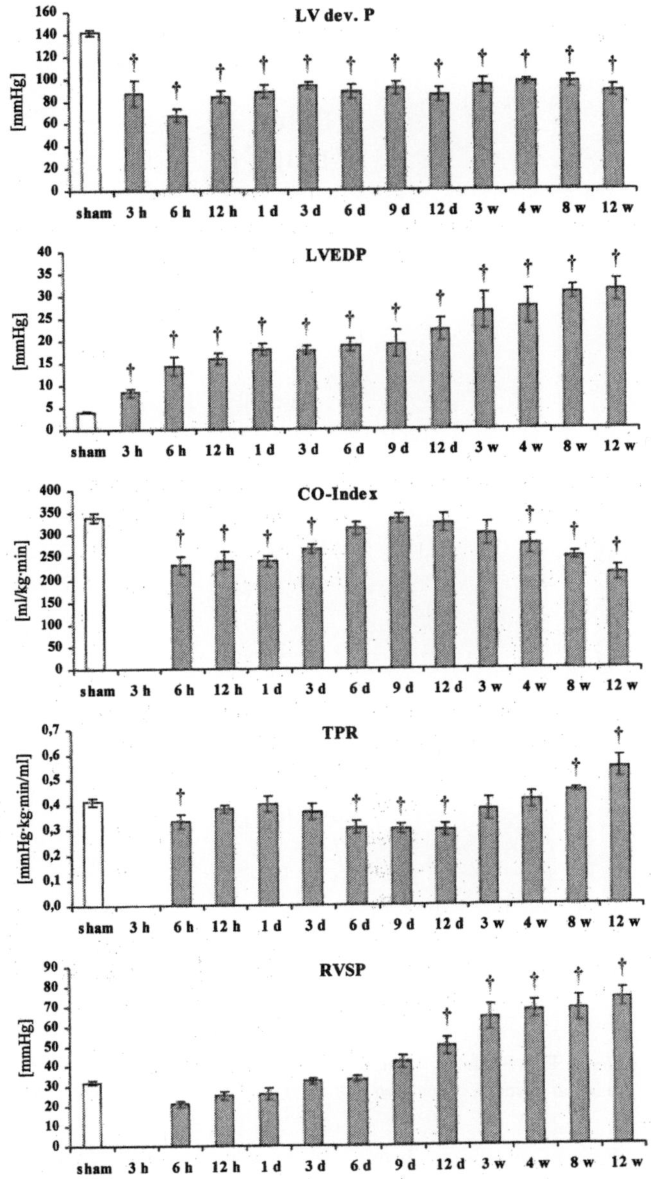

Fig. 1. Time course of parameters of cardiac function after MI. Left ventricular (LV) developed pressure; LV end-diastolic pressure (LVEDP); cardiac output (CO) index; total peripheral resistance index (TPR) and right ventricular systolic pressure (RVSP). Since there were no significant differences between sham-operated controls at different times after surgery, the data of all sham-operated controls (open bars) were combined (n = 43). Data are shown as mean ± S.E.M.; n = 6–12 for each MI-group; †$p < 0.05$ vs. sham-operated controls.

myocyte fraction isolated 6 h after MI (Fig. 3). Compared to untreated animals, also sham operation caused a significant increase in cytokine expression especially in the RV 6 h after surgery. Moreover, the expression level of all investigated cytokines was more pronounced in the RV myocytes compared to the LV myocytes in animals without surgery as well as 6 h after sham operation (Fig. 4, left; data are shown as x-fold differences from LV myocytes of untreated animals). In

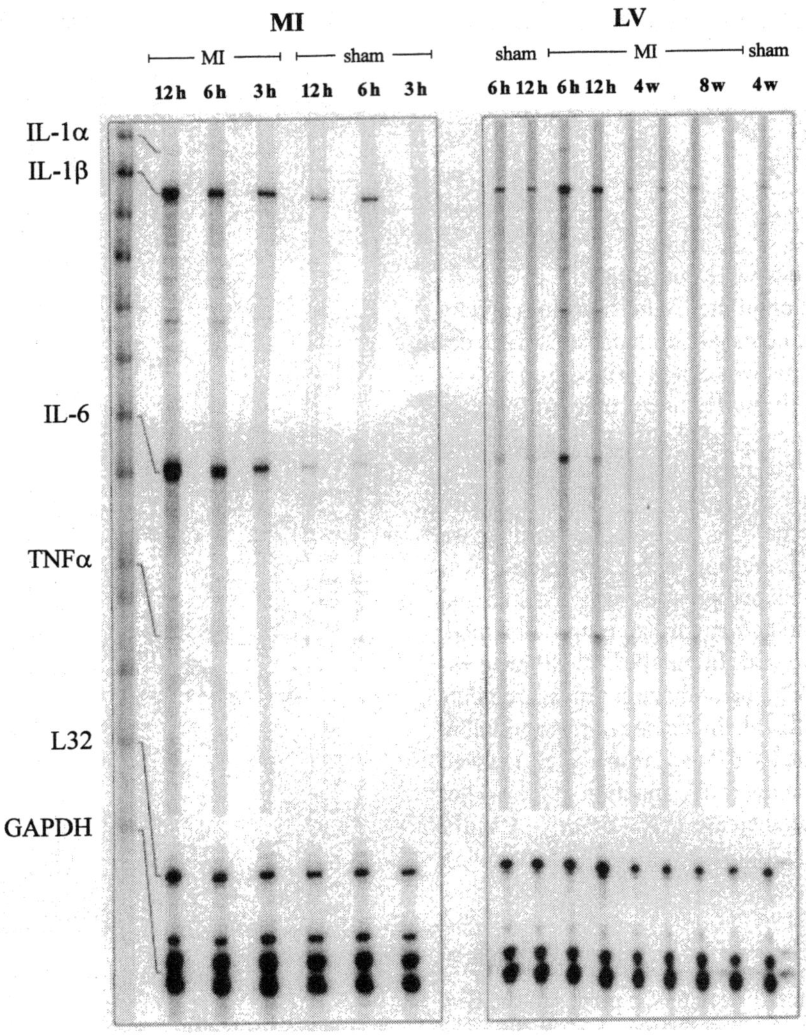

Fig. 2. Representative autoradiographs from RNase protection assays of cytokines in the infarct area (MI, left panel) and in the non-infarcted LV (LV, right panel) at different times after induction of MI (MI) or sham operation (sham) as indicated. Each lane was loaded with 7.5 μg (A) or 5 μg (B) total RNA.

the non-myocytes, however, there were no differences in cytokine expression levels between LV and RV. Therefore, the data of LV and RV non-myocytes were combined in the control groups (Fig. 5, left; data are shown as x-fold differences from LV and RV non-myocytes of untreated animals).

Cytokine expression in isolated myocytes after myocardial infarction

Significant differences were detected for IL-1, IL-6 and TNFα mRNA expression in the myocyte cell fraction isolated from the infarct area 6 and 12 h after MI. Also in the myocytes of the non-infarcted myocardium, cytokine expression increased compared to untreated animals but to a much lower magnitude when compared to the infarct area. Moreover, the only just significant difference in cytokine mRNA expression from the respective and time corresponding sham-operated control was observed for IL-6 after 6 h (Fig. 4, about 6-fold when compared to the time corresponding controls).

Cytokine mRNA expression in isolated non-myocytes after myocardial infarction

The mRNA expression level of IL-1 and IL-6 increased in the non-myocyte cell fraction isolated from the infarcted as well as from the non-infarcted myocardium to a comparable degree (Fig. 5). However, these expression levels were not significantly different from time corresponding controls, since an increase of the same magnitude was also observed for cells isolated 6 h after sham operation. 12 h after MI, the expression level of all cytokines was lower than that after 6 h of sham operation. Moreover, it reached control level of untreated animals or 4 weeks after sham operation. Nevertheless, IL-1 and IL-6 mRNA expression levels were slightly but

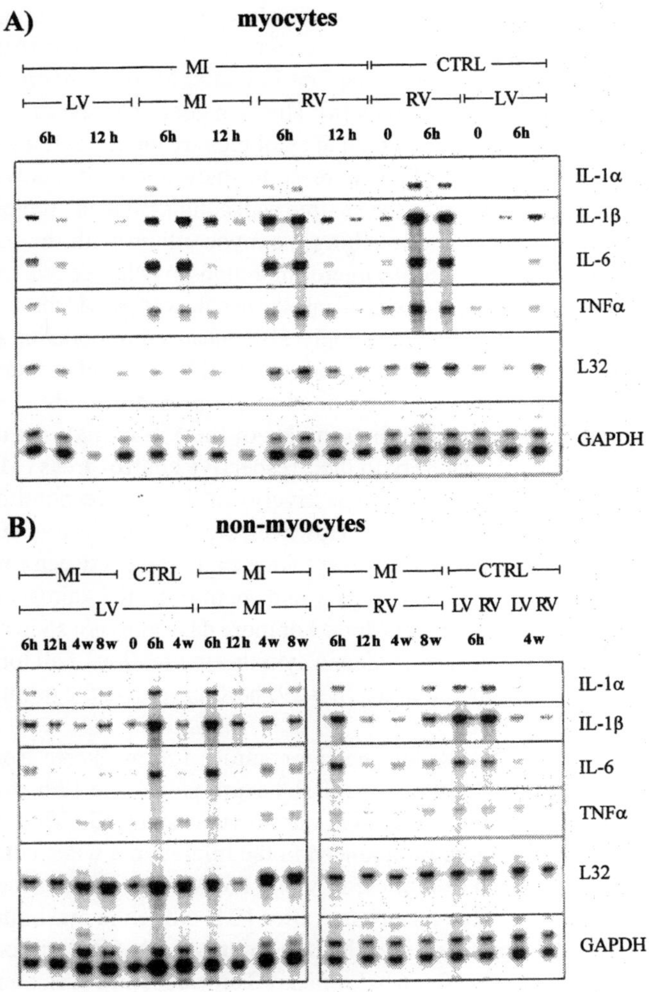

Fig. 3. Representative autoradiographs from RNase protection assays of cytokines in the myocyte (A) and non-myocyte (B) cell fraction isolated at different times after induction of MI (MI) or sham operation (CTRL) as indicated. 0 indicates rats without any surgery. Each lane was loaded with 7.5 μg total RNA.

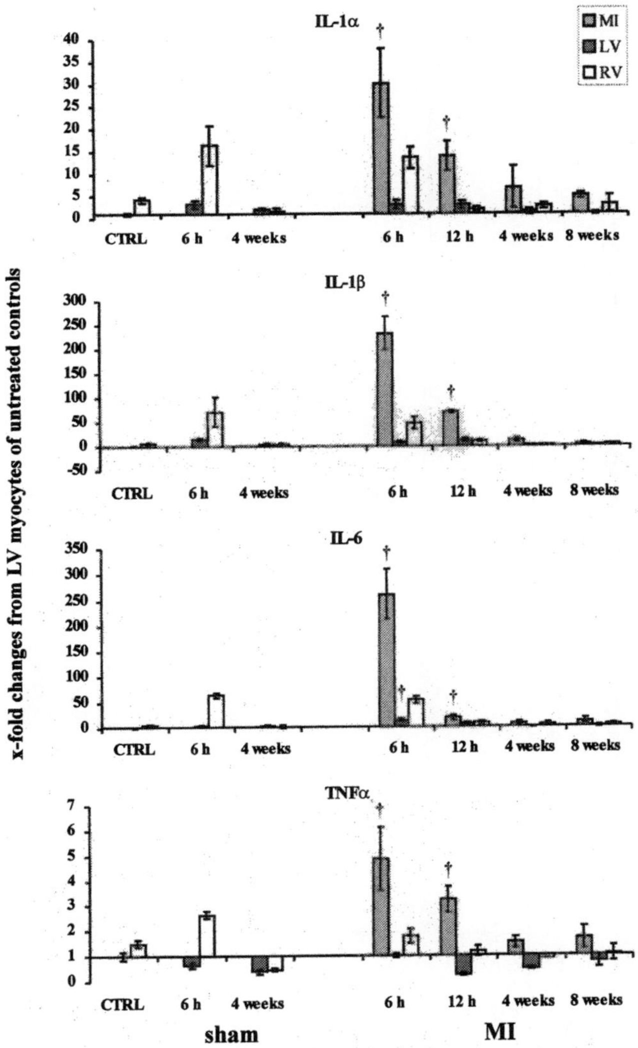

Fig. 4. Relative IL-1α, IL-1β, IL-6 and TNFα mRNA expression in the myocyte cell fraction isolated from the infarct area (MI) and the non-infarcted LV (LV) or RV (RV) at different times after surgery. As indicated on the bottom line, the left hand panel shows the data of control rats without any surgery (CTRL) and rats 6 h and 4 weeks after sham operation. The right hand panel shows the data for myocytes isolated at different times after MI as indicated. Data are expressed as x-fold changes from LV myocytes of untreated controls after normalization to L32 mRNA. Values are mean ± S.E.M.; n = 4 for each group; †$p < 0.05$ vs. respective time corresponding sham-operated controls.

significantly higher in the non-myocytes from the infarct area when compared to the non-infarcted LV 12 h after MI.

At later times after MI, i.e. 8 weeks after MI, the cytokine expression again increased slightly, but significantly in the non-myocytes isolated from the infarct scar (Fig. 5, right). In addition, IL-6 mRNA expression increased in the RV non-myocytes 8 weeks after MI, while the changes in the non-myocytes of the non-infarcted LV were not significant.

Discussion

Functional alterations

The experimental animals in this study were characterized by large myocardial infarctions which invariably extended over almost the entire free wall. The severity of MI was evaluated on the basis of the hemodynamic alterations as in previous studies [17, 18, 20]. All experimental animals had severe LV dysfunction (Fig. 1). Heart function was globally reduced in the acute state after MI, but recovered partially from day 3 onwards as indicated by nearly normal cardiac output. From day 12 onwards, however, LVEDP further increased and, in parallel, also RVSP progressively increased (Fig. 1). Moreover, this was accompanied by a progressive decrease in car-

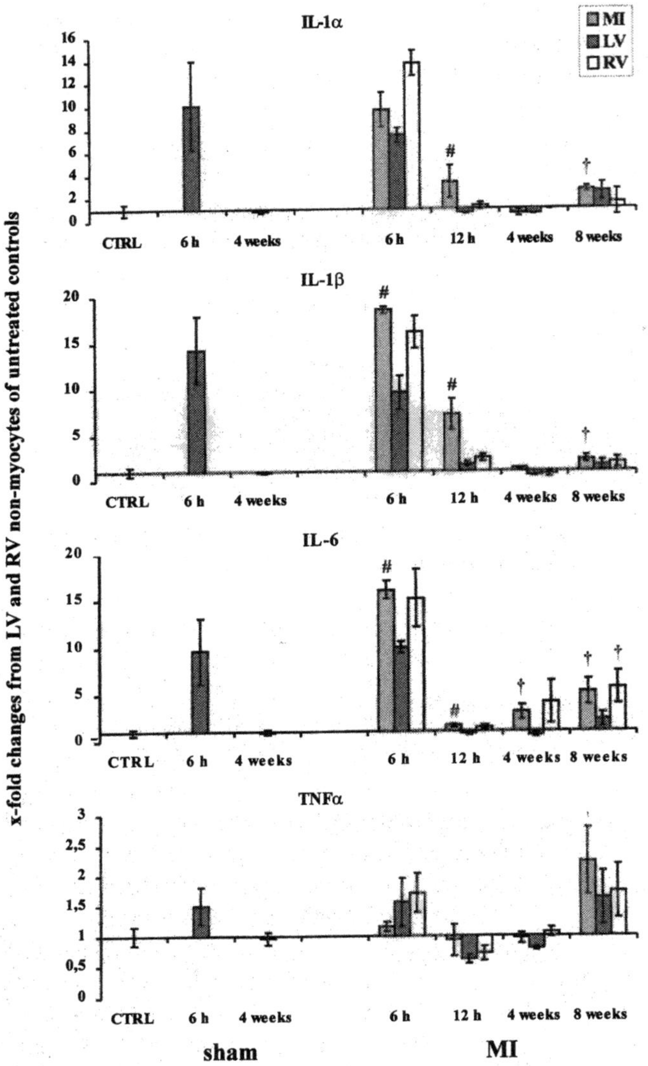

Fig. 5. Relative IL-1α, IL-1β, IL-6 and TNFα mRNA expression in the non-myocyte cell fraction isolated from the infarct area (MI) and the non-infarcted LV (LV) or RV (RV) at different times after surgery. As indicated on the bottom line, the left hand panel shows the data of control rats without any surgery (CTRL) and rats 6 h and 4 weeks after sham operation. The right hand panel shows the data for non-myocytes isolated at different times after MI as indicated. Data are expressed as x-fold changes from mean value of LV and RV non-myocytes of untreated controls after normalization to L32 mRNA, since these data were not significantly different. Values are mean ± S.E.M.; n = 4 for each group; †$p < 0.05$ vs. time corresponding sham-operated controls; #$p < 0.05$ for non-myocytes from the infarcted area vs. time-corresponding non-infarcted LV non-myocytes after MI.

diac output, indicating the development of congestive heart failure.

Cardiac cytokine expression after myocardial infarction

The mRNA expression of the pro-inflammatory cytokines IL-1β and IL-6 was strongly induced in the first 12 h after MI (Fig. 2). Also in sham-operated rats, cytokine expression significantly increased. Moreover, we also observed an increase in serum IL-6 levels in the first 12 h after MI or sham operation in our previous study [8]. This is in good accordance with the well-established fact that cytokines are induced after any stress or surgery. This increase in sham-operated rats was more pronounced in the LV free wall compared to the septum presumably due to touching the myocardium with the forceps. In this region also micro-infarctions may have occurred, since also in sham-operated rats the needle was passed through the myocardium. Interestingly, also enhanced remodeling of the extracellular matrix occurred in the LV free wall of sham-operated animals as described in our previous study [18]. Thus, we compared the samples of the MI-animals for up to 1 day after surgery to their respective control areas (MI-control, LV and RV, respectively) in time corresponding sham-operated controls. However, we did not observe any differences with respect to cytokine expression between control rats without any surgery and sham-operated animals in which the surgery was done 3 or more days ago (not shown). Therefore, the mean value in cytokine expression of all long term sham-operated rats (n = 29) best represents the normal level to us.

Compared to their respective controls, a strong increase occurred in IL-1β and IL-6 mRNA expression 6 and 12 h after MI as discussed in detail in our previous study [8]. This increase was more pronounced in the infarct area when compared to the non-infarcted LV and RV. Moreover, it reached maximum in the infarct area after 12 h, while the maximum cytokine expression in the non-infarcted myocardium occurred already after 6 h. From day 3 onwards, however, there were no more significant differences from sham-operated controls (Fig. 2; right hand panel). Also, mRNA expression of IL-1α was detectable in the myocardium after MI, but not after sham operation. Therefore, quantitative analysis was not done. In addition, TNFα was slightly increased with a maximum on day 1 after MI. This was surprising, especially since also at early times after MI no increase occurred. So, we suggest that preformed TNFα might be liberated from resident cardiac mast cells. For the other investigated cytokines, no mRNA transcripts were detected by RPA method.

Cytokine expression in isolated myocytes after myocardial infarction

Cytokines are expressed in the myocytes of normal myocardium of untreated rats at very low levels (Fig. 3A). However, after normalization to either L32 or GAPDH, cytokine expression level is higher in RV myocytes than in LV myocytes. We suggest that this is due to the isolation method, since the thinner RV is more stressed than the LV during the 35 min perfusion period. As for tissue samples, cytokine expression increased also in the myocytes of sham-operated controls 6 h after surgery. This increase was more pronounced in the RV myocytes compared to the LV myocytes. However, when compared to the higher level in RV myocytes of untreated controls, this increase was of the same magnitude as for the LV myocytes. Therefore, we suppose that these differences are mostly due to the cell isolation procedure. However, we can not completely rule out the possibility that the pneumothorax during surgery induces cytokine expression in the RV. Nevertheless, the increase in cytokine expression after surgery is much less pronounced in RV tissue samples of sham-operated controls (not shown). Therefore, the RV myocytes might also be more sensitive to the additional isolation stress due to surgery.

Myocardial infarction caused a strong induction of cardiac cytokine expression in the myocyte cell fraction isolated after 6 h. As for the tissue samples, this increase was much more pronounced in the infarct area compared to the non-infarcted myocardium. Moreover, when compared to respective and time corresponding controls, this increase was significant only in the infarct area. This is in good agreement with a previous study by Gwechenberger *et al.* [21] who demonstrated increased IL-6 expression in the myocytes of reperfused infarctions of dogs. Intense signals for IL-6 were obtained by immunohistochemical analysis only in the infarct area 6 and 12 h after MI (not shown). This is in accordance with our recent report of increased phosphorylation of the IL-6 downstream signal pathway, the signal transducer and activator of transcription (STAT) 3, in the same timely ordered fashion as for IL-6 protein [8]. The early expression of pro-inflammatory cytokines by ischemic myocytes may be an important signal in initiating wound healing as we have recently suggested [8]. The precise role, however, is still a matter of debate and needs to be elucidated in further studies.

For tissue samples, this increase in cytokine expression after MI is similar to that induced by norepinephrine [22, 23]. Therefore, norepinephrine, which is strongly increased systemically and locally after MI [2], might contribute to the IL-6 expression besides other factors like hypoxia, complement derived factors (C5a) or TNF-α. The further increase in the infarct area is likely due to invading leukocytes since the resident myocardial cells in the necrotic area have died. This is further supported by the lack in high level cytokine expression in the myocytes from the infarct area 12 h after MI. Moreover, this provides evidence that the results obtained for myocytes of the infarct area indeed represent myocytes. This is important, since the cell fractions were not cultured after isolation and, therefore, contamination with leucocytes adherent to myocytes or endothelial cells can not completely ruled out. The evident increase in cytokine expression in the myocytes of the infarct area 12 h after MI might be due to surviving myocytes of the adjacent border zone which was included in the infarct area during isolation. For the same reason, myocyte data are presented for the scar 4 and 8 weeks after MI which normally does not contain any myocytes. Therefore, these data represent myocytes from the adjacent border zone.

The lack in significantly increased expression of pro-inflammatory cytokines in the myocytes of the non-infarcted myocardium is likely due to the cytokine induction during the isolation of the pre-stressed cells. This view is supported by the less pronounced cytokine induction in the tissue samples of sham-operated rats 6 h after surgery. However, one might also speculate that cells other than myocytes, i.e. fibroblasts, leucocytes, resident mast cells or endothelial cells, account for the increased cytokine expression in tissue samples of the non-infarcted myocardium.

Cytokine mRNA expression in isolated non-myocytes after myocardial infarction

The pro-inflammatory cytokines are expressed at relatively low levels also in the non-myocyte cell fraction of control rats without any surgery. A comparison to myocytes, however, is difficult, since the content of L32 as well as of GAPDH varies as indicated by the relation L32/GAPDH. In contrast to the myocytes, we did not observe differences between non-myocytes isolated from the LV and RV. The expression of cytokines increased 6 h after both MI or sham operation (Fig. 3B). However, this was not significantly different after MI compared to sham operation (Fig. 5). This might also be due to the isolation procedure as discussed for myocytes. The induction in IL-1β and IL-6, on the other hand, was more pronounced in the non-myocytes of the infarct area when compared to the non-myocytes of the non-infarcted LV 6 and 12 h after MI. Nevertheless, the magnitude of this increase in the infarct area was much less pronounced in the non-myocytes relative to the myocyte fraction, each compared to non-treated controls. We, therefore, suggest that the observed increase in cytokine expression in the tissue samples 6 h after MI is by far mainly due to the increase in myocytes. The increase in cytokine expression in the non-infarcted myocardium in the acute state after MI might reflect a response to stress.

At later times after MI, i.e. after 4 and 8 weeks, the situation is different. There was a slight but significant increase in cytokine expression in the non-myocyte fraction of the infarct scar as well as in the IL-6 mRNA expression in the RV non-myocytes (Fig. 5). Similar results were recently reported for the non-infarcted LV, although the same problem of increased cytokine expression in sham-operated controls was reported [24]. The increased cytokine expression weeks post-MI, however, is at variance to our results in the tissue samples where significant changes were not detectable. For the RV, this might be explained by the fact that this slight increase is not measurable in the mixture of cells in tissue samples. Nevertheless, we can not rule out the possibility that the measured increase is due to the isolation procedure. Therefore, we would like to conclude that slight increases in cytokine expression long times after MI, as recently shown by Ono *et al.* [11], are presumably due to an increase in non-myocytes but not in myocytes.

There is evidence for a direct influence of pro-inflammatory cytokines on contractile heart function [13–15]. IL-6, on the other hand, might contribute to hypertrophy in a paracrine manner in the chronic state after MI. Hypertrophic effects of the cytokines of the IL-6 family have been demonstrated in a transgenic mouse model in a previous study [16]. However, an important role in developing compensatory hypertrophy is less likely, since there is a gap between the early elevated IL-6 and the late onset of hypertrophy in the RV. Therefore, we rather suggest a role of pro-inflammatory cytokines, if any, in remodeling and fibrosis [25–29]. Serum IL-6 levels, on the other hand, are elevated in patients with CHF [30]. Due to the lack in clearly elevated cytokine expression in the myocardium in the chronic state after MI, we suggest that elevated serum IL-6 level reflects an increased sympathetic drive and is rather of prognostic value.

Acknowledgements

This work was supported by the Deutsche Forschungsgemeinschaft (ZI 199/10-3). The excellent technical assistance of Brigitte Mix and Grit Marx is gratefully appreciated.

References

1. Sutton MG, Sharpe N: Left ventricular remodeling after myocardial infarction. Circulation 101: 2981–2988, 2000
2. Schomig A: Catecholamines in myocardial ischemia. Circulation 82: II13–II22, 1990
3. Thomas JA, Marks BH: Plasma norepinephrine in congestive heart failure. Am J Cardiol 41: 233–243, 1978
4. Long CS: The role of IL-1 in the failing heart. Heart Failure Rev 6: 81–94, 2001
5. Wollert K, Drexler H: The role of IL-6 in the failing heart. Heart Failure Rev 6: 95–103, 2001
6. Mann DL: Recent insights into the role of TNF in the failing heart. Heart Failure Rev 6: 71–80, 2001
7. Cruickshank AM, Oldroyd KG, Cobbe SM: Serum interleukin-6 in suspected myocardial infarction. Lancet 1994 343: 974, 1994
8. Deten A, Volz HC, Briest W, Zimmer H-G: Cardiac cytokine expression is upregulated in the acute phase after myocardial infarction. Experimental studies in rats. Cardiovasc Res 55: 329–340, 2002
9. Kukielka GL, Smith CW, Manning AM, Youker KA, Michael LH, Entman ML: Induction of interleukin-6 synthesis in the myocardium. Potential role in postreperfusion inflammatory injury. Circulation 92: 1866–1875, 1995
10. Frangogiannis NG, Youker KA, Rossen RD, Gwechenberger M, Lindsey MH, Mendoza LH, Michael LH, Ballantyne CM, Smith CW, Entman ML: Cytokines and the microcirculation in ischemia and reperfusion. J Mol Cell Cardiol 30: 2567–2576, 1998
11. Ono K, Matsumori A, Shioi T, Furukawa Y, Sasayama S: Cytokine gene expression after myocardial infarction in rat hearts: Possible implication in left ventricular remodeling. Circulation 98: 149–156, 1998
12. Prabhu SD, Chandrasekar B, Murray DR, Freeman GL: Beta-adrenergic blockade in developing heart failure. Circulation 101: 2103–2109, 2000
13. Hosenpud JD, Campbell SM, Mendelson DJ: Interleukin-1-induced myocardial depression in an isolated beating heart preparation. J Heart Transplant 8: 460–464, 1989
14. Finkel MS, Oddis CV, Jacob TD, Watkins SC, Hattler BG, Simmons RL: Negative inotropic effects of cytokines on the heart mediated by nitric oxide. Science 257: 387–389, 1992
15. Balligand JL, Ungureanu D, Kelly RA, Kobzik L, Pimental D, Michel T, Smith TW: Abnormal contractile function due to induction of nitric oxide synthesis in rat cardiac myocytes follows exposure to activated macrophage-conditioned medium. J Clin Invest 91: 2314–2319, 1993
16. Hirota H, Yoshida K, Kishimoto T, Taga T: Continuous activation of gp130, a signal-transducing receptor component for IL-6-related cytokines, causes myocardial hypertrophy in mice. Proc Natl Acad Sci USA 92: 4862–4866, 1995
17. Zimmer H-G, Gerdes AM, Lortet S, Mall G: Changes in heart function and cardiac cell size in rats with chronic myocardial infarction. J Mol Cell Cardiol 22: 1231–1243, 1990
18. Deten A, Hölzl A, Leicht M, Barth W, Zimmer H-G: Extracellular matrix changes in the infarct area and in the non-infarcted left ventricle after coronary artery ligation in rats. J Mol Cell Cardiol 33: 1191–1206, 2001
19. Leicht M, Greipel N, Zimmer H-G: Comitogenic effect of catecholamines on rat cardiac fibroblasts in culture. Cardiovasc Res 48: 274–284, 2000
20. Pfeffer MA, Pfeffer JM, Fishbein MC, Fletcher PJ, Spadaro J, Kloner RA, Braunwald E: Myocardial infarct size and ventricular function in rats. Circ Res 44: 503–512, 1979
21. Gwechenberger M, Mendoza LH, Youker KA, Frangogiannis NG, Smith CW, Michael LH, Entman ML: Cardiac myocytes produce interleukin-6 in culture and in viable border zone of reperfused infarctions. Circulation 99: 546–551, 1999
22. Barth W, Deten A, Bauer M, Reinohs M, Leicht M, Zimmer H-G: Differential remodeling of the left and right heart after norepinephrine treatment in rats: Studies on cytokines and collagen. J Mol Cell Cardiol 32: 273–284, 2000
23. Bürger A, Benicke M, Deten A, Zimmer H-G: Catecholamines stimulate interleukin-6 synthesis in rat cardiac fibroblasts. Am J Physiol 281: H14–H21, 2001
24. Yue P, Massie BM, Simpson PC, Long CS: Cytokine expression increases in nonmyocytes from rats with postinfarction heart failure. Am J Physiol 275: H250–H258, 1998
25. Johnatty RN, Taub DD, Reeder SP, Turcovski-Corrales SM, Cottam DW, Stephenson TJ, Rees RC: Cytokine and chemokine regulation of proMMP-9 and TIMP-1 production by human peripheral blood lymphocytes. J Immunol 158: 2327–33, 1997
26. Siwik DA, Chang DL, Colucci WS: Interleukin-1beta and tumor necrosis factor-alpha decrease collagen synthesis and increase matrix metalloproteinase activity in cardiac fibroblasts *in vitro*. Circ Res 86: 1259–1265, 2000
27. Ali MH, Schlidt SA, Chandel NS, Hynes KL, Schumacker PT, Gewertz BL: Endothelial permeability and IL-6 production during hypoxia: Role of ROS in signal transduction. Am J Physiol 277: L1057–L1065, 1999

28. Rui T, Cepinskas G, Feng Q, Ho Y-S, Kvietys PR: Cardiac myocytes exposed to anoxia-reoxygenation promote neutrophil transendothelial migration. Am J Physiol 281: H440–H447, 2001
29. Choi I, Kang HS, Yang Y, Pyun KH: IL-6 induces hepatic inflammation and collagen synthesis *in vivo*. Ciln Exp Immunol 95: 530–535, 1994
30. Torre-Amione G, Kapadia S, Benedict C, Oral H, Young JB, Mann DL: Proinflammatory cytokine levels in patients with depressed left ventricular ejection fraction: A report from the Studies of Left Ventricular Dysfunction (SOLVD). J Am Coll Cardiol 27: 1201–1206, 1996

Molecular and Cellular Biochemistry **242:** 57–63, 2003.

Increased salt sensitivity secondary to leptin resistance in SHHF rats is mediated by endothelin

M. Judith Radin,[1] Bethany J. Holycross,[2] Toni M. Hoepf[1] and Sylvia A. McCune[3,4]

[1]*Department of Veterinary Biosciences;* [2]*The Dorothy M. Davis Heart and Lung Research Institute;* [3]*Department of Food Science and Technology, The Ohio State University, Columbus, OH, USA*

Abstract

A link between leptin resistance, obesity, and salt sensitivity has been suggested. SHHF/Mcc-fa^{cp} rats (SHHF) were used to study the effect of gene dosage of a null mutation of the leptin receptor (cp) on salt sensitivity and response to a combined endothelin A and B receptor antagonist (bosentan). Obese (cp/cp), heterozygous (+/cp), and homozygous lean (+/+) male SHHF were fed a low salt diet (0.3% NaCl) for 7 days, followed by a high salt diet (8.0% NaCl) for 7 days. There were no significant differences in systolic blood pressure between genotypes on low salt. In response to high salt, cp/cp had significantly greater systolic pressure than +/cp and +/+. On high salt diet, cp/cp showed a significant increase in 24 h urinary endothelin excretion and increased renal expression of preproendothelin mRNA. There was no effect of high salt diet on renal excretion of nitric oxide (NOx) or on gene expression of endothelial, neuronal, or cytokine-induced nitric oxide synthase isoforms (eNOS, nNOS, iNOS, respectively). Treatment with bosentan prevented the high salt-induced increment in systolic blood pressure in cp/cp. This was associated with a doubling of renal NOx excretion, but without changes in eNOS, nNOS, or iNOS expression. Endothelin receptor antagonism did not normalize systolic pressure in any of the genotypes. Our studies indicate that obesity secondary to leptin resistance (cp/cp) results in increased salt sensitivity that is mediated by endothelin in the SHHF rat. (Mol Cell Biochem **242**: 57–63, 2003)

Key words: hypertension, leptin, endothelin, salt sensitivity, SHHF rat, obesity, nitric oxide

Introduction

Obese patients with type-2 diabetes frequently have increased serum levels of leptin and are resistant to the normal actions of leptin. Studies have associated chronic leptin resistance and hyperleptinemia with the development of hypertension [1, 2], a leading cause of morbidity and mortality in obese patients. The mechanism by which leptin resistance promotes the development of hypertension remains unclear. An exaggerated blood pressure response to salt intake (increased salt sensitivity) is observed in obese humans with type 2 diabetes and likely plays an important role in the development of hypertension, as well as renal and cardiovascular disease in these individuals [3–5]. Obese rodent models also have impaired pressure-natriuresis and increased salt sensitivity [6–8]. Leptin has been shown to have a direct effect on the kidney to increase salt excretion in lean rats [9], and this natriuretic response is absent in obese, hyperleptinemic rat models [3, 10].

The SHHF-Mcc-fa^{cp} rat (SHHF) is a genetic model that has been selectively bred for spontaneous hypertension and heart failure [11]. The fa^{cp} designation refers to the fact some of the colony carry the corpulent gene (cp), an allele of the Zucker fa gene. The SHHF originated from breeding between the Koletsky rat strain, the source of the cp gene, and an inbred spontaneously hypertensive rat (SHR), derived from the Okamoto SHR strain. The cp gene is the result of a nonsense point mutation in the leptin receptor at amino acid +763,

Present address: S.A. McCune, Myogen, Inc., Westminster, CO 80021, USA
Address for offprints: M.J. Radin, Department of Veterinary Biosciences, The Ohio State University, 1925 Coffey Road, Columbus, OH 43210, USA (E-mail: radin.1@osu.edu)

which results in a stop codon in the extracellular domain of the leptin receptor and failure to produce a functional leptin receptor [12]. Homozygosity for the cp gene results in marked hyperleptinemia, type 2 diabetes, obesity, and increased renal endothelin excretion, as well as earlier onset of congestive heart failure compared to homozygous lean (+/+) SHHF [13, 14]. SHHF that are heterozygous for the cp gene have mild hyperleptinemia and insulin resistance, resulting in a lifespan that is intermediate between cp/cp and +/+ [13]. Studies determining whether there is a salt-sensitive component to the hypertension observed in SHHF rats and the possible role of leptin resistance in mediating salt sensitivity have not been previously done. In this study, we examined the effect of dosage of the cp gene on salt sensitivity and the role of endothelin and nitric oxide (NO) in mediating salt sensitive hypertension in the SHHF rat model.

Materials and methods

Animal model

Male SHHF rats (5 months of age) were obtained from the colony at The Ohio State University. Zygosity for the cp gene in phenotypically lean rats was determined, as previously described [12, 15]. In brief, genomic DNA was isolated from skin samples taken prior to the start of the study. Using primers (5′-ATGAATGCTGTGCAGTC-3′ sense; 5′-AAGGTTCTTCCATTCAAT-3′ antisense; Integrated DNA Technologies, Inc.), the region of the leptin receptor spanning the mutation site was amplified. This was followed by restriction enzyme digestion using Tru91 and electrophoresis on 5% agarose gel. The mutant cp gene is cut, resulting in 2 bands at 82 and 39 bp. Therefore, +/+ have one band at 121 bp, which is distinguished from +/cp that have 3 bands at 121, 82, and 39 bp.

Study design

Obese (cp/cp), heterozygous (+/cp), and homozygous lean (+/+) SHHF (n = 10/genotype) were placed in metabolic cages and fed a low (0.3%) salt diet for 7 days (Harlan Tecklad diet LM-485, Madison, WI, USA). Food intake, water intake and urine output were measured. After 7 days, systolic blood pressure was measured by the tail cuff method (Visitech Systems, Apex, NC, USA). Tail vein serum was collected for measurement of leptin and aldosterone concentration. Urine samples on day 7 were frozen at –70°C until assayed. Rats were then placed on a high (8.0%) salt diet (Harlan Tecklad diet TD92012, Madison, WI, USA) for an additional 7 days, and blood pressure and metabolic cage studies repeated. A random subset of rats was euthanized on each salt diet, and the kidneys were excised and snap frozen for later determination of gene expression of preproendothelin, endothelial nitric oxide synthase (eNOS), neuronal NOS (nNOS), and cytokine-induced NOS (iNOS).

An additional group of rats (+/+ n = 9, +/cp n = 9, cp/cp n = 10) were given the high salt diet for 7 days and were treated with a combined ET_A/ET_B receptor blocker in food (bosentan, 100 mg/kg/day; kindly supplied by Dr. Martine Clozel, Actelion, Ltd., Switzerland). Rats were kept in metabolic cages and studies performed as described above.

Biochemical analyses

Commercially available radioimmunoassays were used to determine serum aldosterone (DPC Coat-a-Count, Los Angeles, CA, USA) and serum leptin (Rat leptin RIA, Linco, St. Charles, MO, USA) concentrations. Urinary sodium concentration was determined by flame photometer. Urinary endothelin concentrations were determined using a commercially available radioimmunoassay kit (Amersham, Arlington Heights, IL, USA). Urine samples were extracted according to manufacturer's protocol prior to assay and then corrected for extraction efficiency. Twenty-four hour urinary excretions of sodium and endothelin were then calculated.

Urinary nitric oxide was determined as total nitrite (NOx) following conversion of nitrate to nitrite. Proteins were precipitated by adding 50 µl of 30% $ZnSO_4$/ml of urine and pelleted by centrifuging for 10 min at 10,000 rpm. The supernatant was filtered through a Whatman 4 mm syringe tip filter (1.0 µm pore size) and then diluted with sterile water. In triplicate, 50 µl of a 1:30 dilution of sample was incubated with 10 µl each of 1.0 U/ml nitrate reductase (Sigma, St. Louis, MO, USA), 0.86 mM NADPH, and 0.11 mM FAD, as well as 20 µl of 0.31 M potassium phosphate buffer (pH 7.5) for 1 hour at room temperature in a 96 well plate. Griess reagent (200 µl of a 1:1 mixture of 1% sulfanilamide in 5% H_3PO_4 and 0.1% N-(1-naphthyl)ethylenediamine (NEDA) in water) was added to each well and incubated at room temperature for 10 min. Absorbance at 570 nm was recorded using a Dynatech MR 700 plate reader. Concentration of NOx in the media was calculated from a standard curve constructed of 0–50 µM sodium nitrate that was also subjected to nitrate reductase for 1 h. Urinary excretion of NOx was calculated as µmoles/24 h.

Gene expression

Messenger RNA expression of preproendothelin, eNOS, nNOS, and iNOS was quantified in the kidney in rats given low and high salt diets and in rats subjected to high salt diet plus bosentan treatment. RNA was isolated using the TotallyRNA Kit (Ambion, Austin, TX, USA).

For preproendothelin, 250 ng of total RNA was used in the TITANIUMOne-Step RT-PCR kit (Clontech, Palo Alto, CA, USA). A 368 bp fragment of rat preproendothelin was amplified under the following conditions: 54°C 1 h, 94°C for 5 min then 30 cycles of 94°C for 45 sec, 58°C for 45 sec, and 72°C for 2 min using sense (5′-TCTCTGCTGTTTGTGGCT-TTC-3′) and antisense (5′-TCGGAGTTCTTTGTCTGT-TTG-3′) primers [16]. Each sample also contained 4 µl of a 7:3 competimer: primer mix from the ClassicQuantumRNA 18s Internal Standard kit (Ambion, Austin, TX, USA) which amplifies a 489 bp fragment of rat 18s RNA. Band intensity of the ET-1 was divided by band intensity of18s and reported as ET-1/18s ratios.

A similar protocol was used to quantify eNOS, nNOS, and iNOS expression. Total RNA (1 µg) was reverse transcribed for 1 h at 42°C using the Advantage RT for PCR kit (Clontech, Palo Alto, CA, USA). The reaction was terminated by heating for 5 min at 95°C. The 20 µl of RT product was diluted to 100 µl, then 15 µl was used in the Advantage PCR kit (Clontech, Palo Alto, CA, USA), utilizing 1 µl each of 45 µM eNOS sense (5′-CTGCTGCCCGAGATATCTTC-3′) and antisense (5′-CAGGTACTGCAGTCCCTCCT-3′) primers [17]. eNOS product (228 bp) was generated by heating to 94°C for 45 sec, 58°C for 45 sec then 72°C for 2 min for 32 cycles. Each tube also contained 4 µl of a 9:1 18s universal competimer:primer mix (Ambion, Austin, TX, USA), which generated a 315 bp product. Densitometric analysis was performed after subjecting the PCR product to electrophoresis on a 2.2% agarose gel and eNOS:18s ratio was calculated for each sample. For nNOS, primers were 5′-GGCACTGGC-ATCGCACCCTT-3′ (sense) and 5′-CTTTGGCCTGTCC-GGTTCCC-3′ (antisense) which generated a 213 bp PCR product [17]. For iNOS, 35 cycles were performed using primers 5′-TACATGGGCACCGAGATTGG-3′ (sense) and 5′-TAGAGGCGTAGCTGAACAAGG-3′ (antisense) which generated a 580 bp product [18].

Statistical analysis

Analysis of variance (ANOVA) with Tukey's *post hoc* test when $p < 0.05$ was used to compare genotypes on each diet and with bosentan treatment. Paired *t*-tests were used to compare within a genotype between low and high salt diets.

Results

Obese (cp/cp) SHHF rats weighed more and had significantly greater serum leptin concentrations than wildtype (+/+) on both low and high salt diets (Table 1). +/cp rats had similar body weights to +/+. Serum leptin concentrations were significantly greater in +/cp than +/+ on high salt diet and tended to be higher on the low salt diet, compatible with mild leptin resistance. There was no difference in food intake between genotypes or diets (data not shown).

There was no difference in systolic blood pressure between genotypes on the low salt diet (Fig. 1). Systolic blood pressure significantly increased by approximately 20 mmHg in the cp/cp in response to the high salt diet (Fig. 1), compatible with salt sensitivity in this genotype. Systolic blood pressure remained unchanged in response to high salt in the +/+ and +/cp rats, compatible with salt insensitivity (Fig. 1).

All 3 genotypes showed suppression of aldosterone in response to the high salt diet; however, serum aldosterone concentrations in cp/cp were about twice that observed in +/+ or +/cp on both diets (Table 1). Urinary endothelin excretion was similar between genotypes on the low salt diet (Table 1). The cp/cp showed a significantly greater increase in urinary endothelin excretion in response to high salt compared to either +/+ or +/cp. This was associated with a significant increase in renal preproendothelin gene expression in the cp/cp, which remained unchanged in the +/+ and +/cp groups (Fig. 2). These data suggest that increased salt intake may drive endothelin production in the leptin resistant cp/cp and may contribute to the increase in systolic pressure observed in this group. There was no difference in 24 h urinary excretion of NOx between genotypes on either diet (Table 1). No differences were observed between genotypes for expression of eNOS , nNOS, or iNOS mRNA (Fig. 2).

Treatment with bosentan eliminated the rise in systolic pressure observed in cp/cp in response to the high salt diet (Fig. 3). Blood pressure was not normalized by bosentan

Table 1. Effect of low and high salt diet on body weight and hormonal status in homozygous lean (+/+), heterozygous (+/cp) and obese (cp/cp) SHHF rats

	Low salt diet			High salt diet		
	+/+ (n = 10)	+/cp (n = 10)	cp/cp (n = 10)	+/+ (n = 9)	+/cp (n = 9)	cp/cp (n = 10)
Body weight (g)	379 ± 6	371 ± 4	534 ± 8*	371 ± 6	371 ± 5	542 ± 8*
Fed serum leptin (ng/ml)	3.9 ± 0.6	4.9 ± 0.8	47.9 ± 2.3*	2.1 ± 0.3†	4.2 ± 0.8*	38.9 ± 1.5*
Serum aldosterone (pg/ml)	107.0 ± 17.0	109.8 ± 13.9	209.4 ± 14.3*	42.1 ± 11.1†	31.9 ± 6.5†	81.1 ± 8.1*†
Urinary endothelin excretion (fmol/24 h)	249 ± 34	196 ± 19	311 ± 78	728 ± 45†	703 ± 39†	1044 ± 49*†
Urinary NOx excretion (ìmoles/24 h)	7.34 ± 1.01	4.57 ± 0.43	6.59 ± 1.06	7.82 ± 0.99	7.48 ± 1.04†	7.86 ± 0.98

*Significantly different from +/+ within a diet. †Significantly different between low and high salt diet for a genotype.

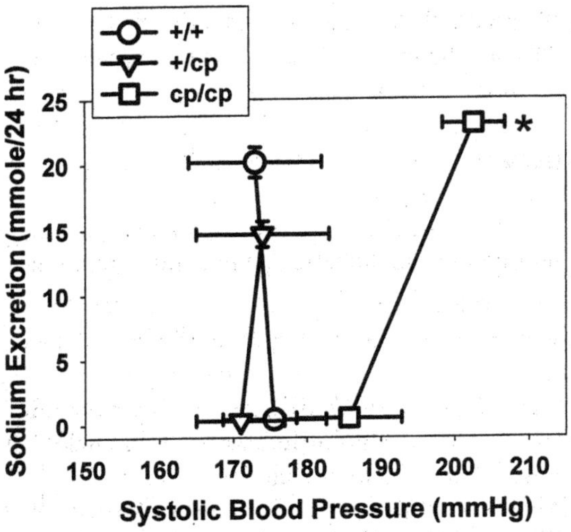

Fig. 1. Systolic blood pressure significantly increased in obese (cp/cp) SHHF rats in response to the high salt diet. No change in systolic blood pressure occurred with the high salt diet in either homozygous (+/+) or heterozygous (+/cp) SHHF. *p < 0.05 compared to +/+ and +/cp rats, n = 10/genotype.

treatment in any of the genotypes and the systolic pressure remained hypertensive, similar to that observed in +/+ on either low or high salt diet. This suggests that combined ET_A/ET_B inhibition is able to eliminate the salt sensitive component of the hypertension in the cp/cp SHHF rats, but does not affect the basal hypertension in this rat strain. Expression of preproendothelin mRNA on the high salt diet was unchanged by treatment with bosentan in the 3 genotypes and remained significantly greater in the cp/cp compared to the other groups (Fig. 2).

Urinary NOx excretion was doubled in cp/cp in response to bosentan treatment, while remaining unchanged in the +/+ and +/cp groups (Fig. 4). There was no change in eNOS, nNOS, or iNOS expression in any of the genotypes in response to high salt or high salt plus bosentan (Fig. 2).

Discussion

Obesity has been linked to the development of salt-sensitive hypertension in both humans and rodent models [3–7]. The

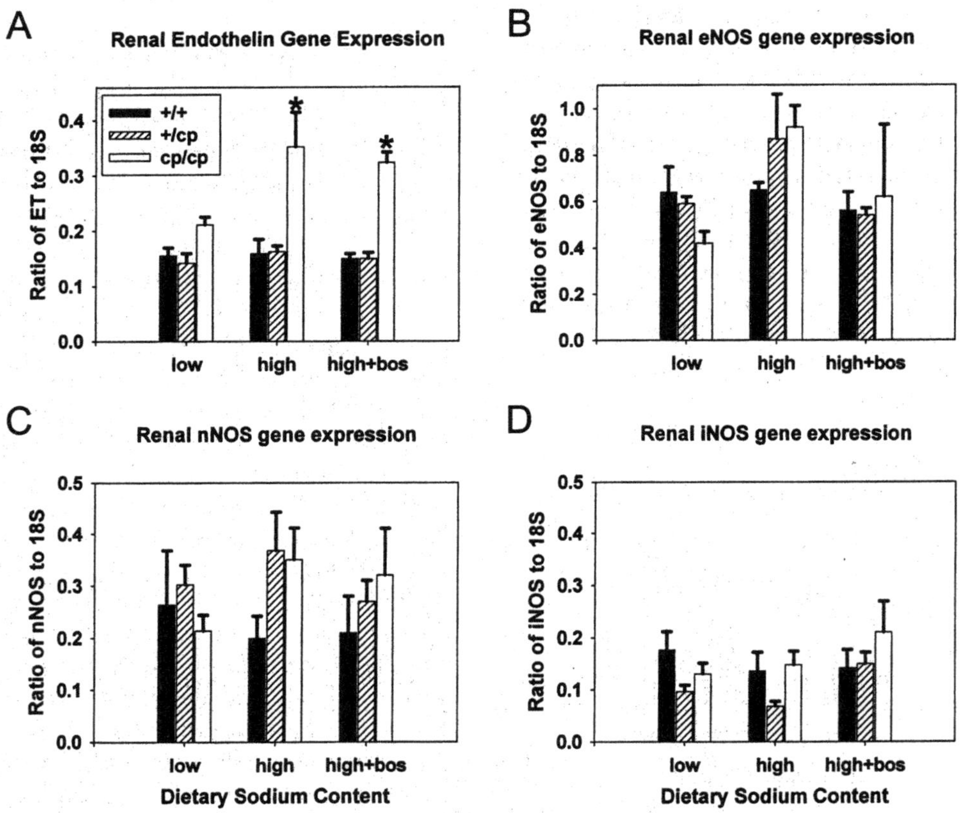

Fig. 2. (A) The high salt diet resulted in a significant increase in renal expression of preproendothelin gene expression in obese (cp/cp), but did not affect gene expression in either homozygous (+/+) or heterozygous (+/cp) SHHF (n = 5/genotype). Treatment with bosentan did not change the effect of a high salt diet on preproendothelin gene expression in any of the genotypes (high + bos, n = 5/genotype). Expression of preproendothelin mRNA remained significantly increased in the cp/cp rats receiving bosentan. *p < 0.05 compared to +/+ and +/cp rats. Neither diet nor bosentan treatment with high salt diet (high + bos) affected renal gene expression of (B) eNOS, (C) nNOS, or (D) iNOS in any of the genotypes (n = 4–5/genotype).

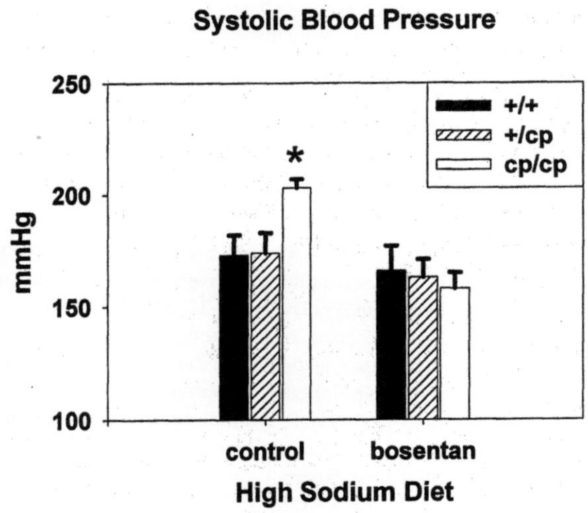

Fig. 3. Treatment with bosentan eliminated the increase in systolic blood pressure observed in obese (cp/cp) SHHF rats (n = 10) in response to the high salt diet, but did not affect systolic pressure in the homozygous (+/+, n = 9) or heterozygous (+/cp, n = 9) SHHF. All 3 genotypes remained hypertensive, despite bosentan treatment.

mechanism underlying salt sensitive hypertension may be multifactorial and roles have been suggested for endothelin and NO [6, 19–25]. Recently, studies have linked elevated levels of leptin to the development of hypertension [1–3, 26]. Leptin is a hormone produced by adipocytes and has a primary role in regulation of food intake and energy metabolism [27, 28]. However, as individuals gain weight, circulating leptin levels often increase, accompanied by a decrease in responsiveness to the normal actions of leptin [3]. Leptin has been shown to have a direct effect on the kidney to increase salt excretion in lean rats [9], and this natriuretic response is absent in obese, hyperleptinemic rat models [3, 10]. Leptin resistant, obese (cp/cp) SHHF show pronounced salt sensitivity when compared to leptin sensitive, +/+ and partially leptin resistant +/cp SHHF. Obese Zucker rats, which are also hyperleptinemic and diabetic due to a mutation in the leptin receptor gene, show similar impaired pressure natriuresis, as well as modest increases in blood pressure in response to high salt diets [6, 7].

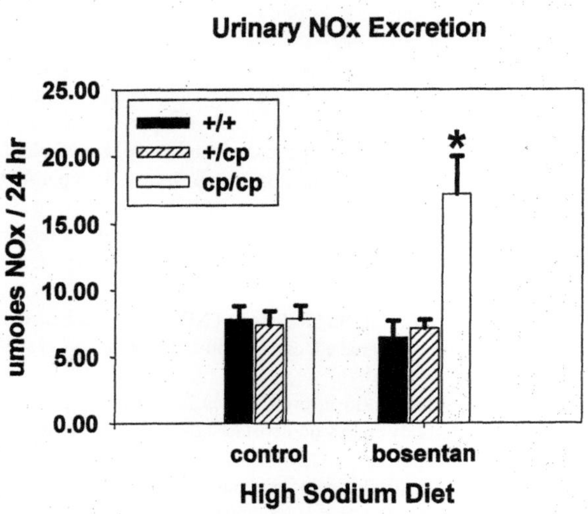

Fig. 4. Twenty-four hour urinary NOx excretion was significantly increased in obese (cp/cp, n = 10) SHHF in response to bosentan treatment and high salt diet. Bosentan treatment did not affect urinary NOx excretion in response to high salt diet in either homozygous (+/+, n = 9) or heterozygous (+/cp, n = 9) SHHF. *$p < 0.05$ compared to +/+ and +/cp rats and to cp/cp on high salt diet alone (control).

The elevation in blood pressure observed in cp/cp in response to the high salt diet is associated with a significant rise in urinary excretion of endothelin as well as increased renal expression of preproendothelin mRNA. Ability of bosentan to eliminate the salt-sensitive component of the hypertension in the cp/cp SHHF suggests that endothelin mediates salt-sensitivity in the SHHF rat model. Other studies have provided evidence that endothelin may mediate salt sensitive hypertension in other rodent models, including the Dahl, stroke-prone SHR (SHR-SP) and deoxycorticosterone acetate-salt (DOCA-salt) [19–21, 29]. Salt sensitive hypertension in these models is also responsive to endothelin receptor antagonism.

High concentrations of endothelin promote renal sodium retention, which is normally opposed by counter-regulatory generation of NO [30, 31], a function that appears to be lacking in the leptin resistant cp/cp. While obese SHHF increased renal endothelin production, they failed to show a compensatory increase in either NOx production or expression of eNOS, nNOS, or iNOS mRNA in response to increased salt intake. Impaired NO metabolism has also been observed in other rodent models of salt sensitive hypertension [6, 22–24]. Treatment with bosentan and elimination of the salt sensitive hypertensive component in the cp/cp SHHF was associated with a marked increase in renal NOx excretion. The increase in urinary NOx excretion in response to bosentan was not explained by enhanced gene expression of any of the NOS isoforms. Alternative mechanisms that may result in increased NOx generation may include increases in NOS activity or altered post-translational stability.

While it was anticipated that the mildly leptin resistant +/cp rats might display some components of salt sensitivity, this was not observed in this study. The +/cp responded to increased salt intake in a similar manner to the +/+. This may be because the degree of leptin resistance that naturally occurs in the +/cp fails to have a significant effect on renal handling of sodium or does not result in sufficient stimulation of the endothelin system. While heterozygous (+/cp) SHHF have mild hyperleptinemia and insulin resistance, they do not show the pronounced hyperinsulinemia and metabolic alterations observed in the homozygous obese (cp/cp) SHHF. It is possible the salt-induced endothelin and pressor response observed in the cp/cp was due to the secondary, pronounced insulin resistance that occurs in the cp/cp rats [32].

Interestingly, bosentan treatment did not normalize blood pressure in any of the SHHF genotypes. It only eliminated the salt sensitive component in the cp/cp group. Like SHR, lean male SHHF do not show lowering of blood pressure in response to endothelin receptor antagonism [29, 33]. This is not unexpected as SHHF are derived, in part, from the SHR strain. The non-salt sensitive component of the hypertension in the SHHF strain appears mediated by other mechanisms [14, 34].

In conclusion, obesity and leptin resistance in the SHHF rat (cp/cp) is associated with increased salt sensitivity that appears mediated by increased renal production of endothelin. Blockade of the renal endothelin system allowed increased generation of NO, which may have played a role in the antihypertensive effects of bosentan.

Acknowledgements

Research funding granted by the Central Ohio Diabetes Association (CODA). A United Way Member Agency, and independent diabetes association, CODA is NOT affiliated with the American Diabetes Association.

References

1. Hall JE, Hildebrandt DA, Kuo J: Obesity hypertension: Role of leptin and sympathetic nervous system. Am J Hypertens 14: 103S–115S, 2001
2. Shek EW, Brands MW, Hall JE: Chronic leptin infusion increases arterial pressure. Hypertension 31: 409–414, 1998
3. Haynes WG, Morgan DA, Walsh SA, Sivitz WI, Mark AL: Cardiovascular consequences of obesity: Role of leptin. Clin Exp Pharmacol Physiol 25: 65–69, 1998
4. Hall JE: Pathophysiology of obesity hypertension. Curr Hypertens Rep 2: 139–147, 2000
5. Thakur V, Richards R, Reisin E: Obesity, hypertension, and the heart. Am J Med Sci 321: 242–248, 2001
6. Fujiwara K, Hayashi K, Matsuda H, Kubota E, Honda M, Ozawa Y, Saruta T: Altered pressure natriuresis in obese Zucker rats. Hypertension 33: 1470–1475, 1999
7. Carlson SH, Shelton J, White CR, Wyss JM: Elevated sympathetic activity contributes to hypertension and salt sensitivity in diabetic obese Zucker rats. Hypertension 35: 403–408, 2000
8. Suzuki H, Ikenaga H, Hayashida T, Otsuka K, Kanno Y, Ohno Y, Ikeda H, Saruta T: Sodium balance and hypertension in obese and fatty rats. Kid Int 49(suppl 55): S150–S153, 1996
9. Jackson EK, Li P: Human leptin has natriuretic activity in the rat. Am J Physiol 272: F333–F338, 1997
10. Villarreal D, Reams G, Freeman RH, Taraben A: Renal effects of leptin in normotensive, hypertensive, and obese rats. Am J Physiol 275: R2056–R2060, 1998
11. McCune SA, Jenkins JE, Stills HF, Park S, Radin MJ, Jurin RR, Hamlin RE: Renal and heart function in the SHHF/Mcc-cp rat. In: E. Shafrir (ed). Frontiers in Diabetes Research. Lessons from Animal Diabetes III. Smith Gordon, Inc., London, 1990, pp 397–401
12. Ishizuka T, Ernsberger P, Liu S, Bedol D, Lehman TM, Koletsky RJ, Friedman JE: Phenotypic consequences of a nonsense mutation in the leptin receptor gene (fa^k) in obese spontaneously hypertensive Koletsky rats (SHROB). J Nutr 128: 2299–2306, 1998
13. McCune SA, Radin MJ, Jenkins JE, Chu Y, Park S, Peterson RG: SHHF/Mcc-facp rat model: Effects of gender and genotype on age of expression of metabolic complications and congestive heart failure and on response to drug therapy. In: E. Shafrir (ed.). Lessons from Animal Diabetes V. Smith-Gorden, London, 1995, pp 255–270
14. Sharkey LC, Holycross BJ, McCune SA, Radin MJ: Obese female SHHF/Mcc-*fa^cp^* rats resist antihypertensive effects of renin-angiotensin system inhibition. Clin Exp Hypertens 23: 227–239, 2001
15. Takaya K, Ogawa Y, Hiraoka J, Hosoda K, Yamori Y, Nakao K, Koletsky RJ: Nonsense mutation of leptin receptor in the obese spontaneously hypertensive Koletsky rat. Nature Gen 14: 130–131, 1996
16. Lepailleur-Enouf D, Egidy F, Philippe M, Louedec L, Henry J-P, Mulder P, Michea J-B: Pulmonary endothelinergic system in experimental congestive heart failure. Cardiovasc Res 49: 330–339, 2001
17. Mattson DL, Wu F: Nitric oxide synthase activity and isoforms in rat renal vasculature. Hypertension 35: 337–341, 2000
18. Goettsch W, Lattmann T, Amann K, Szibor M, Morawietz H, Münter K, Müller S, Shaw S, Barton M: Increased expression of endothelin-1 and inducible nitric oxide synthase isoform II in aging arteries *in vivo:* Implications for atherosclerosis. Biochem Biophys Res Commun 280: 980–913, 2001
19. Kassab S, Miller MT, Novak, J, Reckelhoff J, Clower B, Granger JP: Endothelin-A receptor antagonism attenuates the hypertension and renal injury in Dahl salt-sensitive rats. Hypertension 31: 397–402, 1998
20. Rothermund L, Luckert S, Kobmehl P, Paul M, Kreutz R: Renal endothelin ET_A/ET_B receptor imbalance differentiates salt-sensitive from salt-resistant spontaneous hypertension. Hypertension 37: 275–280, 2001
21. Lariviere R, Day R, Schiffrin EL: Increased expression of endothelin-1 gene in blood vessels of deoxycorticosterone acetate-salt hypertensive rats. Hypertension 21: 916–920, 1993
22. Tan DY, Meng S, Manning RD Jr: Role of neuronal nitric oxide synthase in Dahl salt-sensitive hypertension. Hypertension 33: 456–461, 1999
23. Tan DY, Cason GW, Manning RD Jr: Mechanisms of salt-sensitive hypertension: Role of inducible nitric oxide synthase. Am J Physiol 279: R2297–R2303, 2000
24. Barton M, Vos I, Shaw S, Boer P, D'Uscio LV, Grone HJ, Rabelink TJ, Lattmann T, Moreau P, Luscher TF: Dysfunctional renal nitric oxide synthase as a determinant of salt-sensitive hypertension: Mechanisms of renal artery endothelial dysfunction and role of endothelin for vascular hypertrophy and glomerulosclerosis. J Am Soc Nephrol 11: 835–845, 2000
25. Ni Z, Oveisi F, Vaziri ND: Nitric oxide synthase isotype expression in salt-sensitive and salt-resistant Dahl rats. Hypertension 34: 552–557, 1999
26. Kuo JJ, Jones OB, Hall JE: Inhibition of NO synthesis enhances chronic cardiovascular and renal actions of leptin. Hypertension 37: 670–676, 2001
27. Houseknecht LK, Portocarrero CP: Leptin and its receptors: Regulators of whole-body energy homeostasis. Dom An Endocrinol 15: 457–475, 1998
28. Heskha JT, Jones PJH: A role for dietary fat in leptin receptor, OB-Rb, function. Life Sci 69: 987–1003, 2001
29. Schiffrin EL: Role of endothelin-1 in hypertension and vascular disease. Am J Hypertens 14: 83S–89S, 2001
30. Fortepiani LA, Janvier JJ, Ortiz MC, Atucha MN, Garcia-Estan J: Effect of endothelin blockade on pressure natriuresis in nitric oxide-deficient hypertensive rats. J Hypertens 17: 287–291, 1999

31. Kotelevtsev Y, Webb DJ: Endothelin as a natriuretic hormone: The case for a paracrine action mediated by nitric oxide. Cardiovasc Res 51: 481–488, 2001
32. Walker AB, Dores J, Buckingham RE, Savage MW, Williams G: Impaired insulin-induced attenuation of noradrenaline-mediated vasoconstriction in insulin-resistant obese Zucker rats. Clin Sci 93: 235–241, 1997
33. McCune SA, Shiry LJ, Radin MJ, Holycross BJ, Loeffler B-M, Lindpainter K: Bosentan slows progression of heart failure in male SHHF rats. J Mol Cell Cardiol 32: A68, 2000
34. Holycross BJ, Summers BL, Dunn RB, McCune SA: Plasma renin activity in heart failure prone rats (SHHF/Mcc-facp). Am J Physiol 273: H228–H233, 1997

Molecular and Cellular Biochemistry **242**: 65–70, 2003.

Cloning and bacterial expression of postnatal mouse heart FGF-16

David P. Sontag and Peter A. Cattini

Division of Molecular Endocrinology, Department of Physiology, University of Manitoba, Winnipeg, Manitoba, Canada

Abstract

Fibroblast growth factor-16 (FGF-16) has been reported as the sixteenth member of the heparin sulphate proteoglycan binding growth factor family, which includes acidic and basic FGFs (FGF-1 and FGF-2), based on sequence similarity. The sequences of human (h) and rat (r) FGF-16 complimentary DNA (cDNA) sequences are known. Rat FGF-16 is expressed in brown adipose tissue during embryonic development but also shows some specificity for the postnatal heart. In spite of the importance of other FGF family members in cardiac physiology, there is scant information about FGF-16 function. As a first step towards exploiting mouse genetics in this regard, we have used reverse transcriptase-polymerase chain reaction and primers based on the rFGF-16 sequence to clone the adult mouse (m) FGF-16 cDNA. An mFGF-16 cDNA of 624 base pairs was generated. Based on sequence analysis, mFGF-16 and hFGF-16 share at least 95.2 and 99% nucleotide and amino acid similarity, respectively. In terms of other family members, FGF-16 is most closely related to FGF-9. When used as a radiolabeled probe, the mFGF-16 cDNA detected a single 1.8 kilobase transcript in adult mouse heart RNA. The mFGF-16 cDNA was also used to generate an amino-terminal poly-histidine tagged FGF-16 protein in bacteria. Using sodium dodecyl sulphate-polyacrylamide gel electrophoresis and taking into account the poly-histidine tag, an FGF-16 protein of 26.3 kDa was detected. The generation of cardiac mFGF-16 cDNA and a purified FGF-16 protein preparation are seen as important tools in the further characterization of FGF-16 expression and function in the mammalian heart. (Mol Cell Biochem **242**: 65–70, 2003)

Key words: FGF-16, mouse heart, cDNA, transcript, fusion protein

Introduction

There are 23 reported members of the fibroblast growth factor (FGF) family, which are all characterized by the presence of a heparan sulphate proteoglycan-binding domain in their sequence [1, 2]. FGFs can regulate many of the cellular processes that occur during development, including chemotaxis, differentiation, migration and proliferation [3]. Although not limited to the cardiovascular system, this is clearly an important site of activity for this family of growth factors. FGF-1, FGF-2 and FGF-4 can promote cardiogenesis in explanted precardiac mesoderm [4] suggesting some overlap in their biological function. However, they are not necessarily interchangeable as FGF-7 is unable to induce the same effects [5]. Also, distinct phenotypes in mice were generated with complete and conditional 'knockout' of FGF-2 function, including effects on vascular tone [6–8], heart development [9] and response to hypertension [10]. Thus, in spite of the presence of other FGF family members and the likelihood of some redundancy, such as FGF-1 which shares FGF-2's angiogenic properties [11], distinct roles are still evident. In terms of their possible use as therapeutic agents, FGFs have long been studied for possible benefit in response to cardiac ischemia through their cardioprotective and angiogenic properties. A significant increase in cardiac myocyte viability as well as functional recovery of the heart is observed when FGF-2 is administered prior to ischemia-reperfusion injury [12–15], or even during reperfusion injury [16]. With regard to more chronic FGF-2 administration, infarct size can be reduced after damage has occurred through angiogenesis and the remodeling process [17–21]. Beneficial effects have also been proven in clinical trials where patients receiving FGF at the time of coronary artery bypass surgery demonstrate increased revascularization [20, 22, 23]. To date, most studies

Address for offprints: P.A. Cattini, Department of Physiology, University of Manitoba, 730 William Avenue, Winnipeg, Manitoba, Canada R3E 3J7 (E-mail: Peter_Cattini@umanitoba.ca)

have focused on the prototypic members of the FGF family, FGF-1 and FGF-2, which can be found in numerous tissues and throughout all stages of development. Less attention has been paid to FGF-12 and FGF-16, which have both been reported to show some specificity through their expression in the postnatal heart [24, 25]. FGF-12, however, is expressed broadly in embryonic tissues including the central and peripheral nervous system, as well as connective tissue of the developing skeleton [26]. In contrast, FGF-16 expression is limited to brown adipose tissue in the embryonic rat, where levels decrease at birth with a corresponding induction in the heart [24, 27]. This expression pattern raises the possibility of a role in defining the transition from embryonic to postnatal myocardium.

Transgenic animals, and particularly mice, have now become a routine part of deducing gene function. With this in mind, we have cloned the FGF-16 complimentary DNA (cDNA) from adult mouse heart. In addition to reporting the sequence and making comparisons with other FGF family members, we have used this cDNA to characterize the FGF-16 RNA transcript from adult mouse heart as well as express and purify an FGF-16 protein using a bacterial expression system.

Materials and methods

Cloning of FGF-16

RNA was isolated from the heart of an 8 week-old mouse using TRI-zol (Gibco BRL, Burlington, ON, Canada), quantified and subjected to RT-PCR reaction according to the manufacturer's directions (Gibco BRL). Primers utilized were designed based on sequence reported for rat brain FGF-16 cDNA sequence (GenBank accession number AB002561). Additional terminal sequences coding for *Hin*dIII, *Xho*I, and *Eco*RV restriction endonuclease sites were also added for future subcloning purposes. Primer sequences were as follows: Forward 5′-TTTAAGCTTCTCGAGATGGCGGAGGTCGG-GGGCGTCTTTGCC-3′ and Reverse 5′-TTTGATATCTTA-CCTATAGCGGAAGAGGTCTCTGG-3′. The product was then directly inserted into the TA cloning vector pCR2.1 (Invitrogen, Burlington, ON, Canada). The insert was sequenced using an ABI Prism 310 DNA Genetic Analyzer Sequencing and primers corresponding to flanking T7 promoter and M13 reverse primer sequences, which are present in the pCR2.1 vector.

RNA blotting

Total RNA was isolated from hearts taken from 8 week-old CD-1 mice using guanidine isothiocyanate method [28]. Total RNA (30–35 μg) was denatured with formaldehyde and resolved by electrophoresis through a 1.5% agarose gel. RNA was then blotted to nitrocellulose, cross-linked by baking for 1 h at 80°C. Blots were prehybridized for 24 h at 42°C with a solution containing 50% deionized formamide, 0.75 M sodium chloride, 75 mM sodium citrate, 10% Denhardt's solution, 50 mM sodium phosphate pH 6.5, 0.1 g sodium dodecyl sulfate (SDS), 10 mg/100 ml tRNA, 50 mg/100 ml salmon sperm DNA. Blots were then hybridized with a similar solution containing dextran sulphate (10 g/100 ml) and the FGF-16 cDNA probe, but without SDS, tRNA and salmon sperm DNA. The mouse FGF-16 cDNA was radiolabelled to a specific activity of about 2×10^6 cpm/ng with ^{32}P using the random prime protocol supplied with the Prime-a-Gene kit (Promega, Madison, WI, USA). After 24 h hybridization at 42°C, blots were washed two times with 0.1 × SSC (15 mM sodium chloride and 16.6 mM sodium citrate) and 0.1% SDS for 10 min at 65°C.

Bacterial expression of FGF-16 protein

The FGF-16 cDNA was excised from the pCR2.1 vector and inserted into the pET19b expression vector at the *Xho*I restriction site, which allowed production of a fusion protein containing an amino-terminal poly histidine fusion tag (Novagen, Madison, WI, USA). The vector and fragment were then partially sequenced to verify that the FGF-16 cDNA was in the correct reading frame. BL21(DE3)pLyS bacteria (Novagen) were then transformed with the construct, a positive colony identified, a culture grown and then induced with IPTG as per the manufactures directions. The protein was then purified from a bacterial pellet using a nickel affinity column to retain the his-mFGF16 fusion protein, which was then eluted as per the manufacturers directions.

Results

A reverse transcriptase-PCR strategy using rFGF-16 sequences as primers was used to generate an FGF-16 cDNA clone from total RNA isolated from 8 week CD-1 mouse hearts. The reaction amplified a product of 648 base pairs (bp) that was consistent with the size reported for hFGF-16 and rFGF-16 (Fig. 1A). No PCR product was observed from a reaction done in parallel lacking reverse transcriptase, indicating that the PCR product was not due to DNA contamination.

For sequencing and identification of the amplified product, the 648 bp fragment was subcloned directly into the plasmid pCR2.1, which contains flanking M13 and T7 primer sites to facilitate sequencing of the insert. The sequence con-

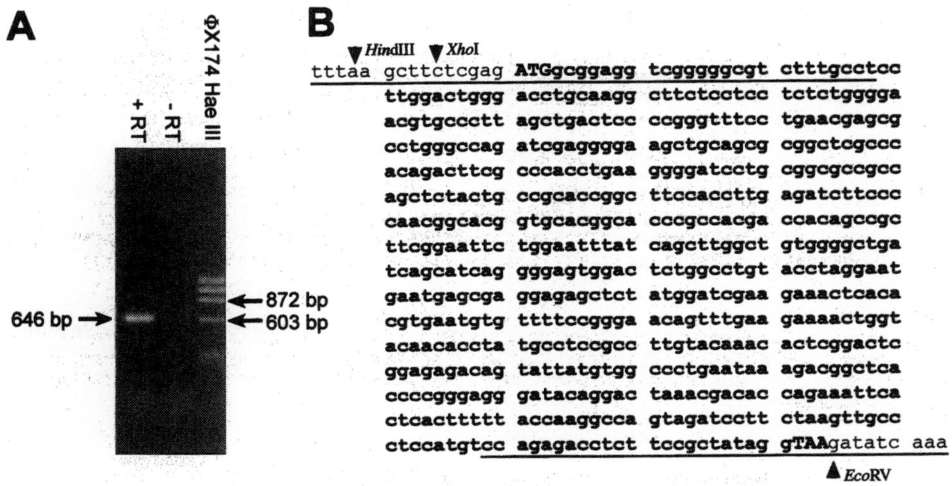

Fig. 1. Reverse transcriptase (RT) PCR amplification of FGF-16 RNA from adult (8 week old) mouse heart and the nucleotide sequence of the FGF-16 cDNA product generated. (A) RNA was treated with (+) or without (–) RT before amplification with primers based on the rat brain FGF-16 cDNA sequence. The reaction was subjected to electrophoresis in a 1% agarose gel and the product (arrowhead) visualized by ethidium bromide staining. A ϕX174/*Hae*III marker was used to estimate the size of the product. Units for the bands indicated are base pairs. (B) The mFGF-16 cDNA sequence (GenBank accession number AF292104) includes the primers used for PCR (underlined), which are based on the rat brain FGF-16 cDNA sequence. The primers also included restriction endonuclease sites, as indicated, to facilitate cloning upstream and downstream of the initiator methionine residue (ATG) and stop codon (TAA), respectively.

firmed the presence of a 624 bp mouse (m) FGF-16 cDNA (GenBank accession number AF292104) in the fragment (Fig. 1B). A comparison with human heart FGF-16 (GenBank accession number AB009391) and rat brain FGF-16 cDNAs, revealed 95.2 and 97.9% sequence similarity, respectively.

The mFGF-16 cDNA was then used as a radiolabeled probe to characterize adult (8 weeks) mouse heart FGF-16 RNA by RNA (Northern) blotting. The blot revealed a single transcript of about 1.79 kb, with mobility slightly higher than 18S ribosomal RNA (Fig. 2).

Further conservation of FGF-16 between species was observed when the mFGF-16 nucleotide sequence was translated (Fig. 3). The similarity between mFGF-16 and hFGF-16 or rFGF-16 increased to 99% at the amino acid level. The only differences observed were a Q^{16} vs. H^{16} and R^{205} vs. H^{205} between the mouse and human, and F^{124} vs. Y^{124} and T^{150} vs. A^{150} for the rat, as denoted by a hyphen in the consensus sequence.

The mouse FGF-16 cDNA was then inserted into a bacterial expression vector, which produced mFGF-16 with a polyhistidine fusion (His) tag at the amino-terminal end. The His-tag was then used to purify the protein using a nickel affinity column from which it was eluted using imidazol. The eluted protein was then subjected to SDS-polyacrylamide gel electrophoresis (SDS-PAGE) in a 15% gel and detected with Coomassie brilliant blue stain in order to examine the size and purity of the protein (Fig. 4). This revealed a prominent band that measured 26.3 kDa when compared to the adjacent broad range markers (New England BioLabs, Mississauga, ON, Canada).

Discussion

The 624 bp adult mouse heart FGF-16 cDNA generated by reverse transcriptase-PCR was consistent with the size of corresponding human and rat cDNAs, and only minor differences in nucleotide and, particularly, amino acid sequences

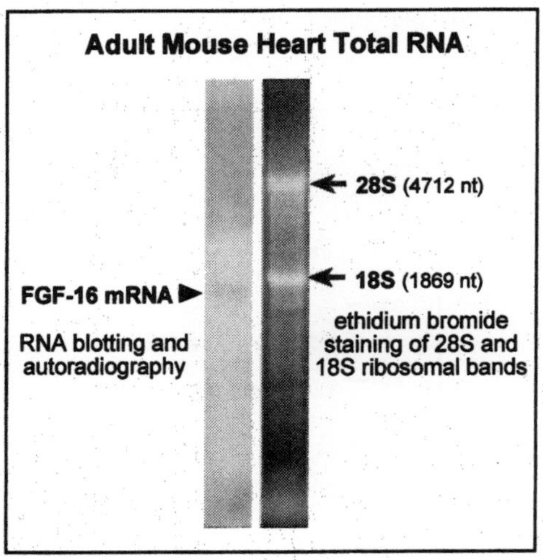

Fig. 2. A FGF-16 transcript of about 1.8 kb is detected in total RNA isolated from an adult mouse heart. RNA was denatured, electrophoresed (1.5% formaldehyde-agarose gel), blotted to nitrocellulose, probed with radiolabeled FGF-16 cDNA and visualized by autoradiography. The size of the FGF-16 transcript detected was estimated based on the mobility of 28S and 18S RNA band, which were visualized in the gel by ethidium bromide staining before transfer to nitrocellulose.

```
                    10        20        30        40        50
                     |         |         |         |         |
RAT        MAEVGGVFASLDWDLQGFSSSLGNVPLADSPGFLNERLGQIEGKLQRGSPTDFAHLKGIL
MOUSE      MAEVGGVFASLDWDLQGFSSSLGNVPLADSPGFLNERLGQIEGKLQRGSPTDFAHLKGIL
HUMAN      MAEVGGVFASLDWDLHGFSSSLGNVPLADSPGFLNERLGQIEGKLQRGSPTDFAHLKGIL
Consensus  :::::::::::::::-::::::::::::::::::::::::::::::::::::::::::::

                    70        80        90       100       110
                  **|         |         |         |         |
RAT        RRRQLYCRTGFHLEIFPNGTVHGTRHDHSRFGILEFISLAVGLISIRGVDSGLYLGMNER
MOUSE      RRRQLYCRTGFHLEIFPNGTVHGTRHDHSRFGILEFISLAVGLISIRGVDSGLYLGMNER
HUMAN      RRRQLYCRTGFHLEIFPNGTVHGTRHDHSRFGILEFISLAVGLISIRGVDSGLYLGMNER
Consensus  ::::::::::::::::::::::::::::::::::::::::::::::::::::::::::::

                   130       140       150       160       170
                     |         | *       |         |     ** | *
RAT        GELFGSKKLTRECVFREQFEENWYNTYASTLYKHSDSERQYYVALNKDGSPREGYRTKRH
MOUSE      GELYGSKKLTRECVFREQFEENWYNTYASALYKHSDSERQYYVALNKDGSPREGYRTKRH
HUMAN      GELYGSKKLTRECVFREQFEENWYNTYASTLYKHSDSERQYYVALNKDGSPREGYRTKRH
Consensus  :::-:::::::::::::::::::::::::-::::::::::::::::::::::::::::::

                   190       200
           **        |         |
RAT        QKFTHFLPRPVDPSKLPSMSRDLFRYR
MOUSE      QKFTHFLPRPVDPSKLPSMSRDLFRYR
HUMAN      QKFTHFLPRPVDPSKLPSMSRDLFHYR
Consensus  ::::::::::::::::::::::::-::
```

Fig. 3. Translation of the adult mouse FGF-16 cDNA sequence and comparison with reported human and rat FGF-16 sequences. A consensus sequence is also shown with identical residues marked with a colon (:) and differences with a hyphen (–). Residues that are likely involved in heparin-FGF-16 interaction are indicated (*) based on those reported for the highly related FGF-9 protein [2].

were observed (Figs 1 and 3). The sequences at either end of the cDNA cannot be considered rigorously established (upstream flanking 27 bp and downstream flanking 26 bp), as they were generated by introduction of the 'primers', and are rat FGF-16 cDNA in origin. There is, however, no difference between human and rat FGF-16 amino acid sequences at the amino-end and only one amino acid change (amino acid position 205) at the carboxyl-end of the protein (Fig. 3), suggesting that, with the possible exception of this one amino acid, even this sequence is likely conserved between the rat and the mouse.

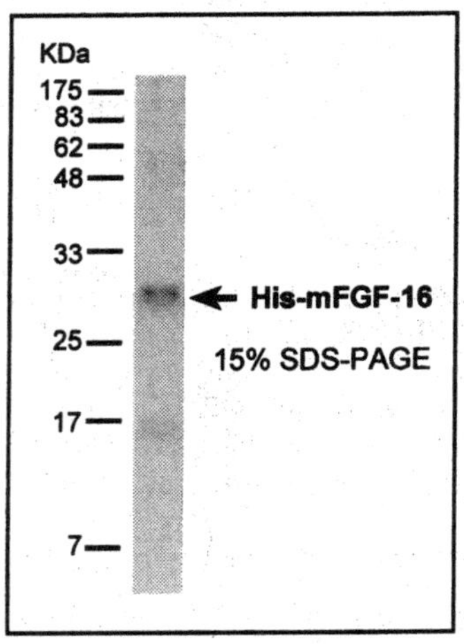

Fig. 4. Detection of His-tagged mFGF-16 protein (His-mFGF-16) protein by SDS-PAGE. Following nickel affinity column purification, the eluted protein was resolved by SDS-PAGE in a 15% gel and detected with Coomassie brilliant blue stain. The sizes of broad range markers (kDa) used to estimate the mass of His-mFGF-16 are indicated.

In terms of the mouse FGF family, FGF-16 protein is most closely related to FGF-9 (72%, GenBank accession number U33535) and shares only 32 and 15% amino acid similarity with the prototypic members FGF-1 (GenBank accession number U67610) and FGF-2 (GenBank accession number AF311741), respectively. All members of the FGF family are described as having a high affinity for heparin, and the amino acids involved in FGF-16/heparin interaction are best postulated based on what has been reported for FGF-9. This is the most closely related to FGF-16 of all members of the FGF family, and its interaction with heparin has been modeled [2]. The amino acids of FGF-9 which interact with heparin are identical to and align perfectly with those of FGF-16 at residues R^{68}, T^{69}, N^{142}, N^{166}, K^{167}, R^{172}, Q^{180} and K^{181} (asterisks, Fig. 3), suggesting similar if not identical heparin binding properties of these two family members.

RNA blot analysis using the mFGF-16 cDNA as a probe revealed a single transcript of about 1.8 kb (Fig. 2). The size is consistent with that previously reported with poly-A enriched rat RNA [23, 26]. Thus, our data suggest that poly-A enriched RNA is not necessary for detection of cardiac FGF-16 mRNA, at least in the mouse. This relative ease of detection should facilitate characterization of FGF-16 RNA expression in mouse heart development and models of disease.

Based on SDS-PAGE, our His-tagged FGF-16 measured about 26.3 kDa. The His-tag is expected to add about 3.1 kDa to the size of mFGF-16 protein, which based on amino acid sequence is predicted to be 23.7 kDa. The availability of a tagged mFGF-16 protein preparation provides an additional tool to assist in the elucidation of FGF-16 function. Specifically, this might involve the characterization of membrane receptor binding or FGF-16 antibodies. A previous report of the purification of recombinant rat FGF-16 revealed the presence of a cleavage sequence close to the amino terminus at residue 34 [29]. We did not observe this truncation, and this most likely relates to the method of purification. When the cleavage site was observed, heparin affinity of FGF-16 was used to purify the protein, and this resulted in both a "full length" FGF-16 as well as a truncated form. As we only selected for the amino terminal His-tag, our extract would not retain any cleaved form of FGF-16. Also, as a bacterially expressed protein, it is recognized that our mFGF-16 preparation may not be processed as it would normally be in the adult heart. Secondary modifications that may confer biological activity such as glycosylation, phosphorylation and truncation are not yet known and may be necessary for this growth factor to signal a particular biological effect. However, when a recombinant rFGF-16 fusion protein containing an epitope tag was expressed by insect Sf9 cells, there was good agreement between the predicted size (26.3 kDa) and that observed by protein blotting (26 kDa) [24]. Thus, given the similarity between the predicted sizes of the proteins in both prokaryotic and eukaryotic systems, the involvement of extensive post-translational modification is less compelling. It is also worth noting that, for similar reasons to those of our purified mFGF-16, no truncation of rFGF-16 was detected after expression in the insect cells, as the epitope for the antibody was located at the amino terminus of the fusion protein [24].

In summary, FGFs as well as other growth factors are seen as potentially important therapeutic agents in the treatment of heart disease as well as other medical conditions [30–32]. The pattern of expression and specifically its induction in the postnatal heart dictate further characterization of FGF-16. Transgenic mice will no doubt play an important role in this study. To aid in this process, we have cloned an mFGF-16 cDNA and provide evidence that its structure, nucleotide and particularly amino acid sequence, is highly conserved among human, rat and mouse species. Furthermore, we have used this cDNA to detect a single 1.8 kb transcript in the adult mouse heart and expressed a 26.3 kDa His-tagged FGF-16 fusion protein in bacteria.

Acknowledgement

This work was supported by a grant from the Heart and Stroke Foundation of Manitoba.

References

1. Yamashita T, Yoshioka M, Itoh N: Identification of a novel fibroblast growth factor, FGF-23, preferentially expressed in the ventrolateral thalamic nucleus of the brain. Biochem Biophys Res Commun 277: 494–498, 2000
2. Faham S, Hileman RE, Fromm JR, Linhardt RJ, Rees DC: Heparin structure and interactions with basic fibroblast growth factor. Science 271: 1116–1120, 1996
3. Szebenyi G, Fallon JF: Fibroblast growth factors as multifunctional signaling factors. Int Rev Cytol 185: 45–106, 1999
4. Sugi Y, Lough J: Activin-A and FGF-2 mimic the inductive effects of anterior endoderm on terminal cardiac myogenesis *in vitro*. Dev Biol 168: 567–574, 1995
5. Barron M, Gao M, Lough J: Requirement for BMP and FGF signaling during cardiogenic induction in non-precardiac mesoderm is specific, transient, and cooperative. Dev Dyn 218: 383–393, 2000
6. Zhou M, Sutliff RL, Paul RJ, Lorenz JN, Hoying JB, Haudenschild CC, Yin M, Coffin JD, Kong L, Kranias EG, Luo W, Boivin GP, Duffy JJ, Pawlowski SA, Doetschman T: Fibroblast growth factor 2 control of vascular tone. Nat Med 4: 201–207, 1998
7. Ortega S, Ittmann M, Tsang SH, Ehrlich M, Basilico C: Neuronal defects and delayed wound healing in mice lacking fibroblast growth factor 2. Proc Natl Acad Sci USA 12: 5672–5677, 1998
8. Dono R, Texido G, Dussel R, Ehmke H, Zeller R: Impaired cerebral cortex development and blood pressure regulation in FGF-2-deficient mice. EMBO J 17: 4213–4225, 1998
9. Leconte I, Fox JC, Baldwin HS, Buck CA, Swain JL: Adenoviral-mediated expression of antisense RNA to fibroblast growth factors disrupts murine vascular development. Dev Dyn 213: 421–430, 1998
10. Pellieux C, Foletti A, Peduto G, Aubert J, Nussberger J, Beerman F, Brunner H, Pedrazzini T: Dilated cardiomyopathy and impaired cardiac hypertrophic response to angiotensin II and mice lacking FGF-2. J Clin Invest 108: 1843–1851, 2001
11. Fernandez B, Buehler A, Wolfram S, Kostin S, Espanion G, Franz WM, Niemann H, Doevendans PA, Schaper W, Zimmermann R: Transgenic myocardial overexpression of fibroblast growth factor-1 increases coronary artery density and branching. Circ Res 87: 207–213, 2000
12. Padua RR, Sethi R, Dhalla NS, Kardami E: Basic fibroblast growth factor is cardioprotective in ischemia-reperfusion injury: Mol Cell Biochem 143: 129–135, 1995
13. Cuevas P, Carceller F, Lozano RM, Crespo A, Zazo M, Gimenez-Gallego G: Protection of rat myocardium by mitogenic and non-mitogenic fibroblast growth factor during post-ischemic reperfusion. Growth Factors 15: 29–40, 1997
14. Padua RR, Merle PL, Doble BW, Yu CH, Zahradka P, Pierce GN, Panagia V, Kardami E: FGF-2-induced negative inotropism and cardioprotection are inhibited by chelerythrine: Involvement of sarcolemmal calcium-independent protein kinase C. J Mol Cell Cardiol 12: 2695–2709, 1998
15. Jiang Z, Padua RR, Ju H, Doble BW, Jin Y, Hao J, Cattini PA, Dixon IM, Kardami E: Acute protection of the ischemic heart by FGF-2; involvement of FGF-2 receptors and protein kinase C. Am J Physiol Heart Circ Physiol 282: H1071–H1080, 2002
16. Sheikh F, Sontag DP, Fandrich RR, Kardami E, Cattini PA: Overexpression of FGF-2 increases cardiac myocyte viability after injury in isolated mouse hearts. Am J Physiol Heart Circ Physiol 280: H1039–H1050, 2001
17. Yanagisawa-Miwa A, Uchida Y, Nakamura F, Tomaru T, Kido H, Kamijo T, Sugimoto T, Kaji K, Utsuyama M, Kurashima C, Ito H: Salvage of infarcted myocardium by angiogenic action of basic fibroblast growth factor. Science 257: 1401–1403, 1992
18. Unger EF, Banai S, Shou M, Lazarous DF, Jaklitsch MT, Scheinowitz M, Correa R, Klingbeil C, Epstein SE: Basic fibroblast growth factor enhances myocardial collateral flow in a canine model. Am J Physiol 266: H1588–H1595, 1994

19. Harada K, Grossman W, Friedman M, Edelman ER, Prasad PV, Keighley CS, Manning WJ, Sellke FW, Simons M: Basic fibroblast growth factor improves myocardial function in chronically ischemic porcine hearts. J Clin Invest 94: 623–630, 1994
20. Laham RJ, Hung D, Simons M: Therapeutic myocardial angiogenesis using percutaneous intrapericardial drug delivery. Clin Cardiol 22(suppl 1): I6–I9, 1999
21. Horrigan MC, Malycky JL, Ellis SG, Topol EJ, Nicolini FA: Reduction in myocardial infarct size by basic fibroblast growth factor following coronary occlusion in a canine model. Int J Cardiol 68(suppl 1): S85–S91, 1999
22. Sellke FW, Laham RJ, Edelman ER, Pearlman JD, Simons M: Therapeutic angiogenesis with basic fibroblast growth factor: Technique and early results. Ann Thorac Surg 65: 1540–1548, 1998
23. Udelson JE, Dilsizian V, Laham RJ, Chronos N, Vansant J, Blais M, Galt JR, Pike M, Yoshizawa C, Simons M: Therapeutic angiogenesis with recombinant fibroblast growth factor-2 improves stress and rest myocardial perfusion abnormalities in patients with severe symptomatic chronic coronary artery disease. Circulation 102: 1605–1610, 2000
24. Miyake A, Konishi M, Martin FH, Hernday NA, Ozaki K, Yamamoto S, Mikami T, Arakawa T, Itoh N: Structure and expression of a novel member, FGF-16, of the fibroblast growth factor family. Biochem Biophys Res Commun. 243: 148–152, 1998
25. Kok LDS, Tsui SKW, Waye M, Liew CC, Lee CY, Fung KP: Cloning and characterization of a cDNA encoding a novel fibroblast growth factor preferentially expressed in human heart. Biochem Biophys Res Commun 255: 717–721, 1999
26. Hartung H, Feldman B, Lovec H, Coulier F, Birnbaum D, Goldfarb M: Murine FGF-12 and FGF-13: Expression in embryonic nervous system, connective tissue and heart. Mech Dev 64: 31–39, 1997
27. Konishi M, Mikami T, Yamasaki M, Miyake A, Itoh N: Fibroblast growth factor-16 is a growth factor for embryonic brown adipocytes. J Biol Chem 275: 12119–12122, 2000
28. Chomczynski P, Sacchi N: Single-step method of RNA isolation by acid guanidinium thiocyanate-phenol-chloroform extraction. Anal Biochem 162: 156–159, 1987
29. Danilenko DM, Montestruque S, Philo JS, Li T, Hill D, Speakman J, Bahru M, Zhang M, Konishi M, Itoh N, Chirica M, Delaney J, Hernday N, Martin F, Hara S, Talvenheimo J, Narhi LO, Arakawa T: Recombinant rat fibroblast growth factor-16: Structure and biological activity. Arch Biochem Biophys 361: 34–46, 1999
30. Fu X, Shen Z, Chen Y, Xie J, Guo Z, Zhang M, Sheng Z: Randomised placebo-controlled trial of use of topical recombinant bovine basic fibroblast growth factor for second-degree burns. Lancet 352: 1661–1664, 1998
31. Payne WG, Ochs DE, Meltzer DD, Hill DP, Mannari RJ, Robson LE, Robson MC: Long-term outcome study of growth factor-treated pressure ulcers. Am J Surg 181: 81–86, 2001
32. Freedman SB, Isner JM: Therapeutic angiogenesis for ischemic cardiovascular disease. J Mol Cell Cardiol 33: 379–93, 2001

Molecular and Cellular Biochemistry **242:** 71–79, 2003.

An ischemic β-dystroglycan (βDG) degradation product: Correlation with irreversible injury in adult rabbit cardiomyocytes

Stephen C. Armstrong, Carole A. Latham and Charles E. Ganote
Departments of Pathology, Veterans Administration Hospital and East Tennessee State University, Johnson City, TN, USA

Abstract

A loss of sarcolemmal dystrophin was observed by immuno-fluorescence studies in rabbit hearts subjected to *in situ* myocardial ischemia and by immuno-blotting of the Triton soluble membrane fraction of isolated rabbit cardiomyocytes subjected to *in vitro* ischemia. This ischemic loss of dystrophin was a specific event in that no ischemic loss of sarcolemmal α-sarcoglycan, γ-sarcoglycan, αDG, or βDG was observed. The maintenance of sarcolemmal βDG (43 Kd) during ischemia was interesting in that dystrophin binds to the C-terminus of βDG. However, during late *in vitro* ischemia, a 30 Kd band was observed that was immuno-reactive for βDG. Additionally, this 30 Kd-βDG band was observed in rabbit myocardium subjected to autolysis. Finally, the 30 Kd-βDG was observed in the purified sarcolemmal fraction of rabbit cardiomyocytes subjected to a prolonged period of *in vitro* ischemia, confirming the sarcolemmal localization of this band. The potential patho-physiologic significance of this band was indicated by the appearance of this band at 120–180 min of *in vitro* ischemia, directly correlating with the onset of irreversible injury, as manifested by osmotic fragility. Additionally the appearance of this band was significantly reduced by the endogenous cardioprotective mechanism, *in vitro* ischemic preconditioning, which delays the onset of osmotic fragility. In addition to dystrophin, βDG binds caveolin-3 and Grb-2 at its C-terminus. The presence of Grb-2 and caveolin-3 in the membrane fractions of oxygenated and ischemic cardiomyocytes was determined by Western blotting. An increase in the level of membrane Grb-2 and caveolin-3 was observed following ischemic preconditioning as compared to control cells. The formation of this 30 Kd-βDG degradation product is potentially related to the transition from the reversible to the irreversible phase of myocardial ischemic cell injury and a decrease in 30 Kd-βDG might mediate the cardioprotection provided by ischemic preconditioning. (Mol Cell Biochem **242**: 71–79, 2003)

Key words: myocardial ischemia, ischemic preconditioning, sarcolemmal proteins, volume regulation, isolated cardiomyocytes

Introduction

Myocardial ischemic cell death is defined by sarcolemmal membrane rupture. The events that lead to this terminal state can be defined as the reversible and irreversible phases, the latter in which ischemic cells will die despite a return to an oxygenated state. The transition to this irreversible phase is accompanied by a breakdown in the sarcolemmal membrane [1–3]. The specific molecular events that lead to this sarcolemmal membrane breakdown have not been clearly defined. Additionally there is a cell swelling component of irreversible myocardial ischemic cell injury. The correlation between ischemic myocardial cell swelling and the loss of sarcolemmal membrane integrity was demonstrated in anoxic myocardial slices subjected to a hypo-osmotic incubation and examined by electron microscopy [4]. In a porcine *in situ* myocardial model, cell swelling is associated with a 50 mOsm increase in the osmolar load of the cardiomyocytes [5]. These findings have been validated in isolated rat hearts, in which a spike of myoglobin release is observed during an anoxic incubation, following a hypotonic challenge [3]. This precipitous loss of cell membrane integrity is termed osmotic fragility. Osmotic fragility has been observed in ischemic [6] or anoxic cardiomyocytes [7, 8]. In intact hearts, swelling or

Address for offprints: S.C. Armstrong, Department of Pathology, P.O. Box 70568, James H. Quillen College of Medicine, East Tennessee State University, Johnson City, TN 37614, USA (E-mail: armstron@mail.etsu.edu)

induction of reoxygenation contracture ruptures cell membranes. A reduction of swelling with hypertonic buffers prevents membrane rupture [5, 9].

The dystrophin associated protein complex, in skeletal and cardiac muscle, is composed of αDG, a heavily glycosylated extrinsic protein which binds (via its N-terminus) to laminin extracellularly [10] and βDG, a transmembrane protein that binds to αDG at its N-terminal extracellular region and intracellularly to dystrophin at its C-terminus [11]. The βDG-dystrophin interaction has been demonstrated ultrastructurally in cardiac muscle [12]. Dystrophin binds to F-actin at its C-terminus, completing a linkage between the actino-myosin complex and the extracellular matrix. Associated with this protein complex, are the α-, β-, γ-, and δ-sarcoglycans, whose functions are not clearly defined [13]. The interaction between dystrophin and βDG is mediated by a Pro-Pro-Pro-Tyr motif on βDG and a WW domain on dystrophin [14]. The interaction between utrophin or dystrophin and the Pro-Pro-Pro-Tyr motif of βDG is negatively regulated by tyrosine phosphorylation of this motif [15, 16]. In addition to dystrophin, βDG can also bind to either caveolin-3 or Grb-2 at the Pro-Pro-Pro-Tyr motif [17, 18]. Caveolin-3 is a structural component of caveolae. Additionally, it has several signal transduction functions, specifically inhibition of Src kinase [19, 20]. Grb-2 mediates multiple signal transduction, including binding to focal adhesion kinase (FAK) [21, 22]. Thus it appears that, in addition to the mechanical linkage provided by the α-/βDG-dystrophin binding, there is a potential for a signal transduction mechanism.

We have observed that there is a loss of dystrophin from the membrane fraction of adult rabbit cardiomyocytes during early *in vitro* ischemia [23]. The ischemic loss of dystrophin is a specific event in that no ischemic loss of the dystrophin associated proteins α-sarcoglycan, βDG or the membrane protein Na^+/Ca^{2+} exchanger was observed. However, during late *in vitro* ischemia, a 30 Kd band appeared in the membrane fraction and in microsomes, that was immuno-reactive for βDG, similar to the 30 Kd-βDG degradation product observed in carcinoma cells [24]. In this study, the levels of this putative βDG degradation product were quantified in the membrane fraction and microsomes of adult rabbit cardiomyocytes and correlated to the onset of irreversible injury and its delay by ischemic preconditioning.

Materials and methods

Cell isolation

Adult rabbit cardiomyocytes were prepared by collagenase perfusion as previously described [25] according to the methods of Hohl *et al.* [26]. The investigation conforms with the 'Guide for the Care and Use of Laboratory Animals' published by the US National Institutes of Health (NIH publication No 85-23, revised 1985). Hearts were excised, cannulated, attached to a recirculating Langendorff apparatus and perfused at 37°C with a nominally calcium-free buffer containing in mM, NaCl (125), KCl (4.75), KH_2PO_4 (1.2), $MgCl_2$ (1.2), HEPES (30) bovine serum albumin (BSA, Pentax fraction V) (0.1%), glucose (11), taurine (58.5) creatine (24.9), EGTA (0.02), L-glutamine (0.682) including basal medium, Eagle (BME) and modified Eagle's medium non-essential amino acid solutions and BME vitamins, diluted 1:50, 1:100, and 1:100, respectively. Collagenase (Worthington, Type II) was added to 200 units/ml and perfusion continued until hearts softened (15–20 min). Ventricles were then minced and post-incubated for 10 min at 37°C. Cells were dispersed with a large bore pipette, followed by filtration through nylon mesh. Cells were washed by a brief 20 × g centrifugation and resuspension in Krebs-Henseleit buffer, containing in mM, NaCl (125), KCl (4.75), KH_2PO_4 (1.2), $MgCl_2$ (1.2), HEPES (30), glucose (11), and 2% bovine serum albumin (BSA, Pentex Fraction V) supplemented with glutamine, vitamins, creatine, taurine and amino acids. Cells were incubated for 30 min at 37°C, followed by calcium addition to a final concentration of 1.25 mM. Calcium tolerant cells were harvested by two brief centrifugations, discarding the supernatants. Cells were resuspended in wash solution, buffered to pH 7.4. Isolates averaged 85–90% viability. A separate isolate was used for each experiment.

Ischemia model and assessment of cell injury

The initial cell suspension was equally divided into the experimental groups. Each group of cells were transferred to a 15 ml conical tube, centrifuged and resuspended into fresh media. An aliquot of each cell suspension was then placed in a 1.8 ml microcentrifuge tube and centrifuged into a pellet. In the experiments following a preconditioning protocol a 15 min post-incubation period preceded resuspension for final ischemic pelleting. Excess supernatant was removed to leave a fluid layer above the pelleted cells of about one third the volume of the pellet. After layering with mineral oil, the cell pellets were incubated without agitation at 37°C. For the ischemic preconditioning experiments, the pellets were incubated for 10 min under oil and then resuspended for a 15 min post-incubation period prior to the final ischemic pelleting. A 25 μl sample of the final cell pellet was removed through the oil layer, with care to prevent introduction of air, at the appropriate time points and resuspended for 3–5 min in 200 μl of hypotonic (85 mOsm) buffer (to determine osmotic fragility) containing 3 mM amytal as a mitochondrial inhibitor, to preclude cell rounding due to reoxygenation during cell swelling. A 25 μl sample was then mixed with an equal volume of counting media (0.5% glutaraldehyde in

85 mOsM modified Tyrodes solution, with reduced NaCl, containing 1% trypan blue). Microscopic examination determined the morphology (rod, round or square) and the permeability of the cells to trypan blue as described previously [27].

Autolysis of ventricular tissue

Rabbit hearts were excised and placed in ice-cold Krebs-Henseleit-HEPES buffer in the absence of calcium to induce rapid cardioplegia. The aorta was then cannulated and flushed with 60 ml of this buffer. Left ventricular tissue was divided into eight sections, one of which was immediately washed with cold PBS and snap frozen with powdered dry ice for the 0 min time point. *In vitro* autolysis was initiated by a modification of an established model [1], with incubation of the remaining myocardial sections on saline moistened gauze, tightly overlaid with Parafilm in a capped 50 ml tube at 37°C. At various time points of autolysis, a tissue section was removed, washed and snap frozen. Cell fractionation was performed as described below with the modification that the tissue was initially homogenized with a Tissumizer (2 × 10 sec rapid bursts) on ice in 0.05% digitonin.

Sample preparation and Western blots

Microsomes were prepared by a modification of the method of Ohlendieck [28]. Briefly, oxygenated and ischemic cardiomyocytes were homogenized in buffer (20 mM sodium pyrophosphate, 20 mM sodium phosphate monohydrate 1 mM $MgCl_2$, 0.303 M sucrose, 0.5 mM EDTA, pH 7.0) in the presence of protease inhibitors. The homogenate was subjected to a 30 min, 10,000 × g centrifugation and the supernatant of this first step was centrifuged for 30 min at 30,000 × g. The pellet of the second step was the heavy microsomes and the supernatant of this step was centrifuged for 30 min at 142,000 × g to obtain the light microsomes. The heavy and light microsomes were incubated for 30 min at 4°C with 0.6 M KCl and re-centrifuged to obtain the KCl washed microsomes. A purified sarcolemmal fraction was obtained from light microsomal fractions obtained from oxygenated and 180 min ischemic cardiomyocytes by sucrose density gradient centrifugation [28]. Microsomes were solubilized in 2% SDS and analyzed by Western blotting as described below. Cell fractionation was performed with oxygenated cells or ischemic pellets, washed in PBS and snap frozen in dry ice. Separate aliquots of oxygenated or ischemic pellets were sampled prior to cell harvest to document cell viability and osmotic fragility. Membrane and cytoskeletal fractions were prepared by the method of Pelech [29]. Lysis buffers (pH 7.4) containing (in mM) Tris HCl (20), NaCl (137), NaF (50), EDTA (5), PMSF (1), Na pyrophosphate (10), sodium orthovanadate (1), 2 μg/ml leupeptin and either 1% Triton X-100 or 0.05% digitonin were prepared and chilled on ice. Digitonin lysis buffer was added to frozen cell pellets in microcentrifuge tubes, vortexed and rocked for 5 min at 4°C. Samples were then centrifuged for 2 min at 15,000 × g. The supernatant containing the cytosolic fraction was transferred to another microcentrifuge tube on ice. Cold Triton X-100 lysis buffer was added to the pellet, vortexed for 1 min and centrifuged at 15,000 × g for 5 min. Supernatants containing the particulate (membrane) fraction were transferred to microcentrifuge tubes and placed on ice. Pellets were solubilized with 2% SDS to obtain a cytoskeletal fraction. Aliquots of cytosolic and particulate fractions were diluted with equal volumes of sample buffer containing 2% SDS, 20% glycerol, 40 mM DTT, 0.1% bromophenol blue, and 200 mM Tris HCl (pH 6.8), placed in a boiling water bath for 5 min and stored at –20°C. The protein concentration in samples was determined by the BioRad DC kit. One dimensional gel electrophoresis was run using 4.25% stack gels and 12% separating gels run at a constant current of 50 mA prior to immuno-blotting.

Immunoblotting

Proteins were transferred from 1D-SDS-PAGE to nitrocellulose paper in buffer containing 25 mM Tris HCl and 192 mM glycine and the blots were incubated in blocking buffer (pH 7.5), containing 5% milk protein, 20 mM Tris HCl, 137 mM NaCl, and 0.1% Tween 20 (TBS-Tween) for 2 h at room temperature. Primary monoclonal antibodies for the C-terminus of βDG (Novocastra), Grb2 and caveolin-3 (Transduction Labs) were used at 1:100, 1:500 and 1:500 dilutions, respectively. Blots were incubated with primary antibodies overnight at 4°C. After 3 × 20 min washes in TBS-Tween, blots were transferred to blocking buffer containing goat anti-mouse peroxidase conjugate (1:2500; Sigma), incubated for 1 h at room temperature followed by 3 × 20 min washes in TBS-Tween. Following incubation in LumiGlo substrate solution (Kirkegard and Perry, Gaithersburg, MD, USA) blots were exposed to Kodak X-Omat LS imaging film for 30 min. Films were developed and analyzed by densitometry using a BioImage Whole Band Analyzer (Ann Arbor, MI, USA) gel documentation image analysis system running on a Sun IPC work station.

Experimental protocols

Control cells were pre-incubated in calcium-containing oxygenated buffer for 10 min, washed, post-incubated for 15 min, rewashed and finally pelleted to initiate prolonged *in vitro* ischemia. Separate ischemic cell pellets were established in microcentrifuge tubes, incubated at 37°C and harvested various time points. The '0 min' time point consisted of cells harvested immediately after post-incubation washes.

Cells were preconditioned (IPC) by a 10 min period of *in vitro* ischemic pelleting, followed by resuspension of cells in oxygenated buffer. A single wash was followed by an oxygenated 15 min post-incubation period. Paired controls for this group consisted of cells from the same isolate incubated in oxygenated buffer for 10 min but otherwise subjected to the same wash and post-incubation protocol, prior to final ischemic pelleting.

Statistics

Data was analysed by ANOVA (SupraANOVA II, Abacus Concepts, Inc., Berkeley, CA, USA). A $p < 0.05$ was considered significant.

Results

Ischemic production of a βDG degradation band

The levels of sarcolemmal βDG (43 Kd) were quantitated by immuno-blotting in the Triton soluble membrane fractions of oxygenated adult rabbit cardiomyocytes and in cells subjected to *in vitro* ischemia for 30–180 min (Fig. 1A). In contrast to the loss of dystrophin that we have reported [23], βDG was maintained in the sarcolemma during *in vitro* ischemia in rabbit cardiomyocytes, at levels similar to that observed in oxygenated cells (Fig. 1B). Significant levels of βDG were also observed in the SDS soluble, cytoskeletal fraction of oxygenated rabbit cardiomyocytes and during early (75 min) *in vitro* ischemia (Fig. 1B). However, during late (120–180 min) ischemia, this cytoskeletal βDG was lost with a corresponding appearance of a 30 Kd band, in the membrane fraction, that was immuno-reactive for βDG (30 Kd-βDG) (Fig. 1B). Quantitative analysis of the membrane 30 Kd-βDG degradation product, expressed as a percentage of membrane 43 Kd-βDG, revealed that the appearance of 30 Kd-βDG was coincident with the onset of irreversible ischemic injury, as defined by osmotic fragility (Fig. 1C). The overlapping of the cell injury curves, defined by the percentage of contracted cells (square morphology) that are unable to exclude trypan blue, with the levels of 30 Kd-βDG, was observed in both the control group and in cells subjected to IPC, prior to the onset of the prolonged ischemic period. The delay in the onset of osmotic fragility in the IPC group was associated with a decrease in the levels of the 30 Kd-βDG band observed. A significant decrease in the levels of 30 Kd-βDG by IPC was observed at 120 and 180 min ($p < 0.0207$ and 0.0408, respectively) as compared to controls, corresponding to the significant decrease in osmotic fragility at 120 min ($p < 0.0077$).

Sarcolemmal localization of the 30 Kd-βDG degradation product

The Triton soluble, membrane fraction represents the sum of several cellular compartments including the sarcolemmal membrane. The sarcolemmal localization of the 43 and 30 Kd-βDG bands was confirmed by Western blot analysis of

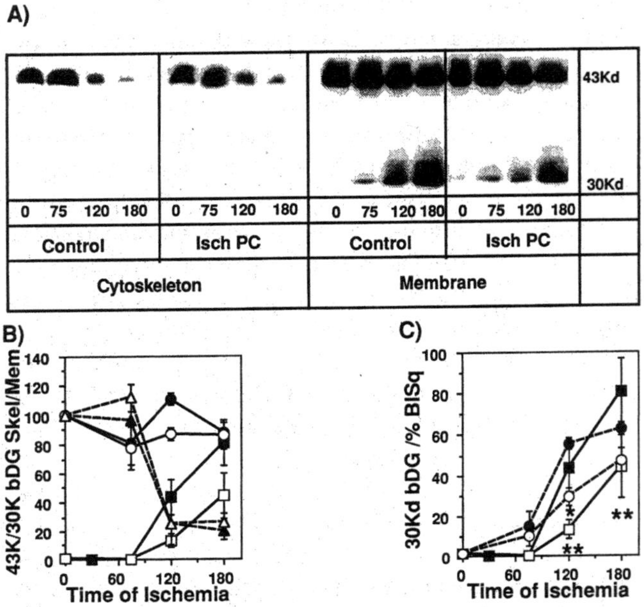

Fig. 1. (A) Western blot analysis of βDG in the membrane and cytoskeletal fraction of cardiomyocytes subjected to oxygenation and *in vitro* ischemia, in the presence or absence of *in vitro* ischemic preconditioning (IPC). (B) Quantitative analysis demonstrates that 43 Kd-βDG was maintained in the membrane of ischemic cardiomyocytes in the control (solid line, closed circle) and IPC groups (solid line, open circle). The cytoskeletal fraction contained 43 Kd-βDG, with an equivalent loss observed during late ischemia in the control (dashed line, closed triangle) and IPC groups (dashed line, open triangle). This loss corresponded to the appearance of a 30 Kd band, in the membrane fraction, that is immuno-reactive for βDG. The level of this 30 Kd-βDG band was decreased in the IPC group (solid line, open square) as compared to the control group (solid line, closed square). (C) The levels of the membrane 30 Kd-βDG band in the control (solid lines, closed square) and IPC groups (solid lines, open square) is compared to the percentage of osmotically fragile cells in the control (dashed line, closed circle) and IPC group (dashed line, open circle) to demonstrate the correspondence between the onset of osmotic fragility and the appearance of the membrane 30 Kd-βDG band. A significant decrease of cell death by IPC was observed at 120 min as compared to the control group ($p < 0.0077$*). The level of the 30 Kd-βDG band was significantly decreased in the IPC group (solid line, open square) at 120 and 180 min ($p < 0.0276$ and 0.0408, respectively**) as compared to the control group (solid line, closed square). This quantitative analysis expresses the levels of membrane and cytoskeletal 43 Kd-βDG as a percentage of the levels of 43 Kd-βDG in the respective cellular fractions of the oxygenated cells in the control and IPC groups. The membrane 30 Kd-βDG degradation product is expressed as a percentage of membrane 43 Kd-βDG in the control and IPC groups, at each time point. Cell death is expressed as the percentage of contracted (square) cells that are unable to exclude trypan blue ($N = 6$).

βDG in microsomal fractions of oxygenated and ischemic rabbit cardiomyocytes (Fig. 2). The levels of βDG (43 Kd band) was maintained during 180 min of *in vitro* ischemia as compared to the oxygenated (0 min) cells with an enrichment of the 43 Kd band in the light and sucrose gradient purified sarcolemmal fractions, as compared to the initial homogenate, supernatant and heavy microsome fractions, confirming the sarcolemmal localization of this band. The 30 Kd-βDG band was observed during 180 min of *in vitro* ischemia, as compared to the relative absence of this band in oxygenated cells. There was an enrichment of this band in the heavy, light and sucrose gradient purified sarcolemmal fractions, as compared to the initial homogenate and supernatant fractions. The 30 Kd-βDG band resolved as a doublet in the microsomal fractions, indicating multiple levels of degradation.

Autolytic production of 30 Kd-βDG in rabbit myocardium

The patho-physiologic relevance of the production of the ischemic 30 Kd-βDG in isolated rabbit cardiomyocytes was assessed by Western blot analysis of βDG content in the cytoskeletal and membrane fractions of rabbit myocardium subjected to autolysis (Fig. 3A). High levels of βDG (43 Kd) were maintained, in the membrane fraction, throughout autolysis, with no significant loss as compared to rabbit myocardium prior to autolysis (Fig. 3B). There was a complete loss of βDG (43 Kd) from the cytoskeletal fraction at 180–240 min autolysis. As observed in the membrane fraction of isolated cardiomyocytes, this loss of βDG (43 Kd) from the cytoskeletal fraction corresponded to the appearance of high levels of the 30 Kd-βDG band during late autolysis, demonstrating the potential patho-physiologic relevance of the 30 Kd-βDG band (Fig. 3B). An intermediate localization of low levels of the 30 Kd-βDG band was detected in the cytoskeletal fraction of rabbit myocardium, during late autolysis.

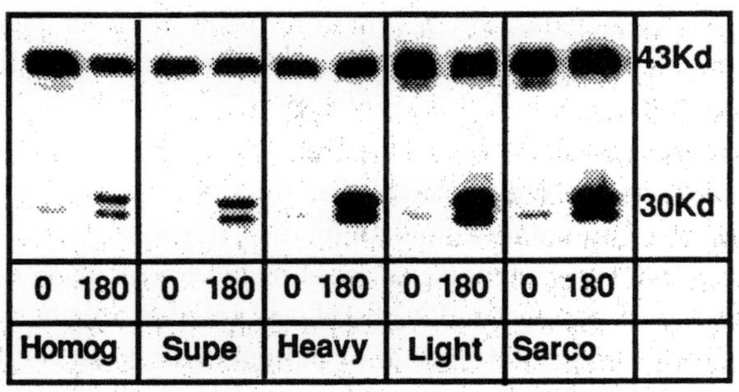

Fig. 2. Western blot analysis of βDG in microsomal fractions of oxygenated and ischemic rabbit cardiomyocytes. The levels of 43 Kd-βDG were maintained during 180 min of *in vitro* ischemia as compared to the oxygenated (0 min) cells with an enrichment of the 43 Kd-βDG in the light and sucrose gradient purified sarcolemmal fractions, as compared to the initial homogenate, supernatant and heavy microsome fractions, confirming the sarcolemmal localization of this band. The 30 Kd-βDG band was observed only in the ischemic cells with an enrichment in the heavy and light microsome and in the sucrose gradient purified sarcolemmal fractions. The 30 Kd-βDG band resolved as a doublet, indicating varying degrees of degradation.

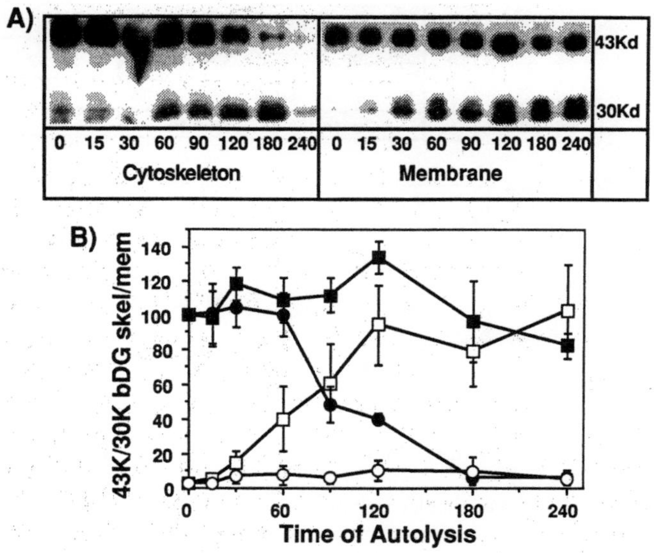

Fig. 3. (A) Western blot analysis of βDG in the cytoskeletal and membrane fractions of rabbit myocardium subjected to *in vitro* autolysis. (B) βDG (43 Kd) was maintained in the membrane (closed squares). There was loss of 43 Kd-βDG from the cytoskeletal fraction during late autolysis (closed circle). This loss corresponded to the appearance of the 30 Kd-βDG band in the membrane fraction (open squares). Low levels of the 30 Kd-βDG band were detected in the cytoskeletal fraction during late autolysis (open circles). This quantitative analysis expresses the levels of membrane and cytoskeletal 43 Kd-βDG as a percentage of the levels of 43 Kd-βDG in the respective cellular fractions of the oxygenated group. Membrane and cytoskeletal 30 Kd-βDG was expressed as a percentage of 43 Kd-βDG in the respective cellular fractions at each time point (N = 3).

Membrane localization of the βDG associated proteins: Caveolin-3 and Grb2

The caveolin-3 and Grb-2 content of the membrane fraction of rabbit cardiomyocytes was quantitated by Western blot analysis (Fig. 4A). Minimal levels of caveolin-3 and Grb-2 were observed in the membrane fraction of oxygenated or ischemic cardiomyocytes, in the absence of ischemic preconditioning. In contrast, high levels of caveolin-3 and Grb-2 were observed in the membrane fraction of cardiomyocytes subjected to ischemic preconditioning. Quantitative analysis of the Western blot studies of caveolin-3 (Fig. 4B) and Grb 2 (Fig. 4C) revealed that the increased membrane association in the ischemic preconditioning group, as compared to the control group, achieved significance at 0 and 120 min ($p < 0.03$ and 0.04 respectively) for caveolin-3 and at 75 and 120

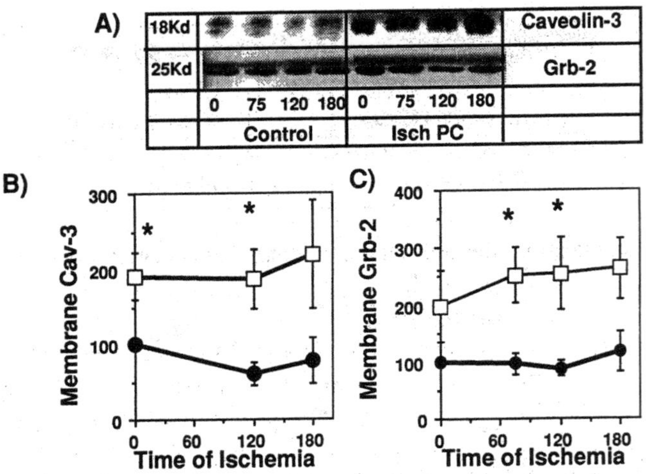

Fig. 4. (A) Western blot analysis of caveolin-3 and Grb-2 content in the membrane fraction of cardiomyocytes subjected to oxygenation or *in vitro* ischemia, in the presence or absence of *in vitro* ischemic preconditioning (IPC). Quantitative analysis of caveolin-3 (B) and Grb-2 (C) content in membrane fractions. IPC (open squares) induced a significant increase in the membrane association of caveolin-3 at 0 and 120 min ($p < 0.03$ and 0.04 respectively*) and Grb2 at 75 and 120 min ($p < 0.0229$ and 0.043 respectively*) as compared to the control groups (closed circles) ($N = 4$). The data is expressed as a percentage of the control oxygenated group.

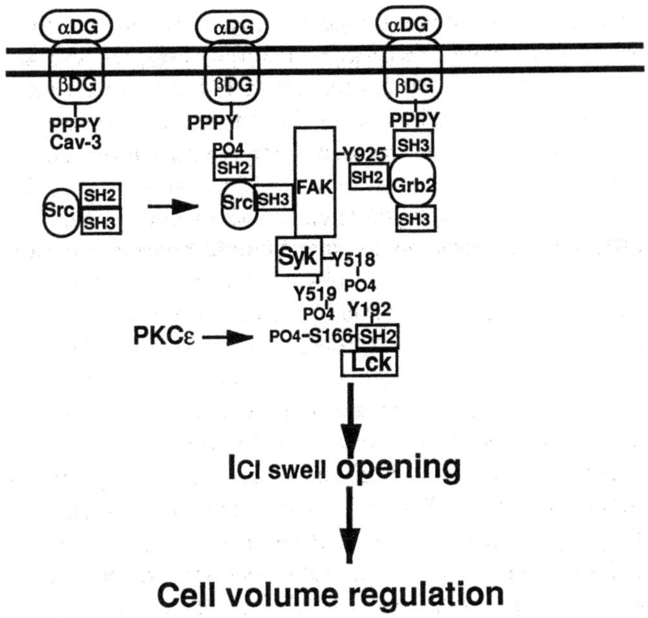

Fig. 5. Schematic diagram of signal transduction molecules associated with the C-terminus of βDG. Grb-2 and Src bind to βDG [38, 43]. The Src mediated tyrosine phosphorylation of the tyrosine kinase, FAK, at Tyr-925, allows it to bind to Grb-2 [22, 40]. Grb-2 is an adaptor protein that links βDG to FAK [21]. The tyrosine kinase Syk associates with FAK [44]. Phosphorylation of Tyr-518 and 519 on Syk allows its association with the SH2 domain of the tyrosine kinase, Lck, at Tyr-192 [45]. Ischemic stress initiates the Syk mediated tyrosine phosphorylation of Tyr-192 and the release and activation of Lck, which facilitates swell-activated chloride channel ($I_{Cl\,swell}$) opening and cell volume regulation [41]. The formation of this multi-complex is enhanced by ischemic preconditioning and the association of these signal transduction molecules with βDG (43 Kd) is competitively inhibited by 30 Kd-βDG. This tyrosine kinase pathway would complement the association between PKCε mediated cardioprotection and Lck activation [42].

min ($p < 0.0229$ and 0.043 respectively) for Grb-2. There was no ischemic increase in the membrane association of caveolin-3 or Grb-2 in the control or ischemic preconditioning group.

Discussion

The primary results presented in this study describe the late ischemic generation of a 30 Kd band that is immuno-reactive for βDG in the membrane and sarcolemmal fractions of adult rabbit cardiomyocytes. The production of the 30 Kd-βDG band directly correlated with the onset of irreversible injury, as determined by osmotic fragility. This correlation was further evidenced by a corresponding decrease in the production of this band and a delay in the onset of osmotic fragility, in cells subjected to ischemic preconditioning. An assessment of the data presented requires consideration of three fundamental questions: (1) Is this βDG immuno-reactive band a valid degradation product of βDG?; (2) does this ischemic βDG degradation product have patho-physiologic relevance?; and (3) What is the mechanism that mediates the contribution of this degradation product to the onset of irreversible ischemic injury? These questions will be addressed sequentially.

The first consideration in the appearance of a immuno-reactive band with a molecular weight distinct from the native protein is the possibility of a cross-reactivity that is unique to the model examined. This 30 Kd-βDG band was previously detected at low levels in rat ventricular and skeletal muscle, with the C-terminus reactive monoclonal antibody utilized in our studies (Dag1/8D5) [12]. Additionally, the 30 Kd-βDG band has been reported in human skeletal muscle homogenates and the membrane fraction of rat glioma cells using several monoclonal antibodies, negating the possibility that this band represents cross-reactive proteins [30]. Furthermore, the 30 Kd-βDG band was noted in Schwann cells and in the skeletal muscle of patients with γ-sarcoglycanopathy (limb-girdle muscular dystrophy type 2C) [31, 32] Finally, the 30 Kd-βDG band was observed in several carcinoma cell lines [24], in the absence of any alternative splicing events, indicating that this band is due to either a proteolytic cleavage or a loss of post-translational processing. Additionally they demonstrated that the transmembrane segment is maintained in 30 Kd-βDG.

Given that the 30 Kd-βDG band is a recognized degradation product of βDG, the patho-physiologic relevance of this band must be considered. The correlation between the onset of irreversible ischemic injury (osmotic fragility) and the appearance of the 30 Kd-βDG band is the primary evidence of the patho-physiologic relevance of this degradation product. This correlation is a unique event in that in other proteins that we have examined there is either no ischemic change, e.g. βDG, α-sarcoglycan [23], γ-sarcoglycan, and αDG (unpublished data) or there is an ischemic loss that significantly precedes the onset of osmotic fragility e.g. dystrophin [23], HSP27 [33] and αβ crystallin [34]. The latter ischemic protein losses occurred concurrently with the onset of cell contracture, at 60 min of *in vitro* ischemia. The production of the 30 Kd-βDG band occurs at 120 min of *in vitro* ischemia, temporally separating this degradation event from cell contracture. Additionally, the late ischemic appearance of the 30 Kd-βDG band was attenuated by ischemic preconditioning, corresponding to the delay in the onset of osmotic fragility by this cardioprotective protocol. This is also a unique event in that this correlation has not been detected in any other protein that we have examined. The cardiomyocytes in the ischemic pellet at 120 min are innately viable in that resuspension of these cells in an isotonic buffer yields a population of cells that predominantly exclude trypan blue. This indicates that the appearance of the 30 Kd-βDG band is not a secondary event to cell degradation, but rather it is a potential indicator of the transition from the reversible to irreversible phase of ischemic cell injury. The loss of cytoskeletal βDG (43 Kd) from cardiomyocytes was equivalent in the control and IPC groups and yet the level of the 30 Kd-βDG band in the membrane fraction was decreased in the IPC group. This suggests that ischemic preconditioning initiates the further degradation of the 30 Kd-βDG band, eliminating this potential mediator of irreversible myocardial ischemic cell injury. It is possible that the tyrosine phosphorylation of 43 Kd-βDG [15, 16] renders it susceptible to proteolysis and degradation. A recent report has indicated that the formation of the 30 Kd-βDG band is mediated by matrix metalloproteinases [35]. This offer the possibility that there is a translocation of 43 Kd-βDG from the cytoskeleton to the membrane with a subsequent proteolysis event.

The late production of the 30 Kd-βDG band in autolytic rabbit myocardium demonstrates that this event is not unique to isolated cardiomyocytes, but instead can be observed in myocardial tissue. This autolytic model has been characterized in canine myocardium [4, 36, 37]. Our studies on HSP27, p38 MAPK, dystrophin and αβ crystallin utilized this rabbit myocardium model , with the results corresponding to those obtained with isolated rabbit cardiomyocytes. In contrast to the consistent absence of the 30 Kd-βDG band in the cytoskeletal fraction of cardiomyocytes, there were low levels of the 30 Kd-βDG band in the cytoskeletal fraction of autolytic myocardium, notable in the blot presented in Fig. 3. The reason for this discrepancy is unclear and the membrane is still a likely site of the proteolytic event.

The correlations between the formation of the 30 Kd-βDG band and the onset of osmotic fragility, and its delay by ischemic preconditioning, might be co-incidental. However, there is sufficient experimental evidence to suggest that the 30 Kd-βDG degradation product does have patho-physiologic relevance to the onset of irreversible myocardial ischemic cell injury. This permits a consideration of the final question: the cellular mechanism that is negatively regulated by the 30 Kd-βDG degradation product. The ability of βDG to bind caveolin-3 and Grb-2 [17, 18, 38] suggests that βDG (43 Kd) is able to initiate signal transduction mechanisms that could be impacted by the 30 Kd-βDG degradation product. The ability of *in vitro* ischemic preconditioning to increase the membrane association of Grb2 and caveolin-3 is supportive of this hypothesis, although no ischemic increase in the membrane association of these proteins was observed in these preliminary studies. The cardioprotection provided by ischemic preconditioning is possibly mediated by a limitation of ischemic cell volume increases by the opening of the swelling activated chloride channel (ICl Swell channel) [39]. It is proposed that, in oxygenated cardiomyocytes, α-/-βDG and dystrophin form a dynamic complex that is a signal transduction module, detecting alterations in cell volume as mechanical perturbations in the sarcolemmal membrane. This stress signal is transmitted intracellularly, to initiate the opening of ion channels and volume regulation (Fig. 5). The ischemic absence of membrane associated dystrophin initiates a compensatory mechanism in which there is an association of caveolin-3 and Grb-2 with βDG. The association of caveolin-3 with βDG would release caveolin-3 from its inhibition of Src kinase [19, 20]. This would allow the Src mediated phosphorylation of FAK at Tyr 925 [22]. The binding of Grb2 to βDG and the binding of Grb-2 to focal adhesion kinase (FAK), at the phospho-Tyr 925 site [22] would permit the formation of the Grb2-FAK complex that has been reported in synaptosomes [21] and in hypoxic cardiomyocytes [40]. FAK would initiate the Syk mediated tyrosine phosphorylation and activation of the tyrosine kinase Lck [44, 45], which would execute its recognized function in facilitating the opening of the swell-activated chloride channel ($I_{Cl\ swell}$), and cell volume regulation subsequent to hypotonic stress [41]. This Lck mediated cell volume regulation would correspond to the report that ischemic preconditioning is dependent upon Lck tyrosine kinase [42]. The βDG (43 Kd)-Grb2-FAK complex would be competitively inhibited by the 30 Kd-βDG degradation product, eliminating this ischemic mode of volume regulation and initiating the irreversible phase of myocardial ischemic cell injury. Alternatively, Src can tyrosine phosphorylate the PPxY motif and bind directly to βDG [43], potentially linking this trans-membrane protein to FAK and the putative volume regulatory pathway described above. This Src-βDG (43 Kd) interaction could be similarly competitively inhibited by the 30 Kd-βDG degradation product. Further investigations are required to define this potential volume regulatory pathway and its role in the cardioprotection provided by ischemic preconditioning.

Acknowledgements

This research was supported by a grant from the National Institutes of Health, HL 51859-04. This material is the result of work supported with resources and the use of facilities at James H. Quillen VA Medical Center.

References

1. Sage MD, Jennings RB: Cytoskeletal injury and sub-sarcolemmal bleb formation in dog heart during *in vitro* total ischemia. Am J Pathol 133: 327–337, 1988
2. Steenbergen CJ, Hill ML, Jennings RB: Cytoskeletal damage during myocardial ischemia: Changes in vinculin immunofluorescence staining during total *in vitro* ischemia in canine heart. Circ Res 60: 478–486, 1987
3. Ganote CE, Vander Heide RS: Cytoskeletal lesions in anoxic myocardial injury: A conventional and high voltage electron microscopic and immunofluorescence study. Am J Pathol 129: 327–344, 1987
4. Steenbergen CJ, Hill ML, Jennings RB: Volume regulation and plasma membrane injury in aerobic, anaerobic and ischemic myocardium *in vitro*: Effect of osmotic swelling on plasma membrane integrity. Circ Res 57: 864–875, 1985
5. Tranum-Jensen J, Janse MJ, Fiolet JWT, Krieger JG, D'Alnoncourt CN, Durrer D: Tissue osmolality, cell swelling, and reperfusion in acute regional myocardial ischemia in the isolated porcine heart. Circ Res 49: 364–381, 1981
6. Armstrong SC, Shivell LC, Ganote CE: Sarcolemmal blebs and osmotic fragility as correlates of irreversible ischemic injury in preconditioned isolated rabbit cardiomyocytes. J Mol Cell Cardiol 33: 149–160, 2001
7. Ganote CE, Vander Heide RS: Irreversible injury of isolated adult rat myocytes: Osmotic fragility during metabolic inhibition. Am J Pathol 132: 212–222, 1988
8. Siegmund B, Koop A, Klietz T, Schwartz P, Piper HM: Sarcolemmal integrity and metabolic competence of cardiomyocytes under anoxia-reoxygenation. Am J Physiol 258 (Heart Circ Physiol 27): H285–H291, 1990
9. Buja LM, Willerson JT: Abnormalities of volume regulation and membrane integrity in myocardial tissue slices after early ischemic injury in the dog. Effects of mannitol, polyethylene glycol, and propranolol. Am J Pathol 103: 79–95, 1981
10. Di Stasio E, Sciandra F, Maras B, Di Tommaso F, Petrucci T, Giardina B, Branacaccio A: Structural and functional analysis of the N-terminal extracellular region of β-dystroglycan. Biochem Biophys Res Commun 266: 274–278, 1999
11. Jung D, Yang B, Meyer J, Chamberlain J, Campbell K: Identification and characterization of the dystrophin anchoring site on β-dystroglycan. J Biol Chem 270: 27305–27310, 1995
12. Stevenson S, Rothery S, Cullen MJ, Severs NJ: Spatial relationship of the C-terminal domains of dystrophin and β-dystroglycan in cardiac muscle support a direct molecular interaction at the plasma membrane interface. Circ Res 82: 82–93, 1998
13. Ohlendieck K: Towards an understanding of the dystrophin-glycoprotein complex: Linkage between the extracellular matrix and the membrane cytoskeleton in muscle fibers. Eur J Cell Biol 69: 1–10, 1996
14. Rentschler S, Linn H, Deininger K, Bedford M, Espanel X, Sudol M: The WW domain of dystrophin requires EF-hands region to interact with β-dystroglycan. Biol Chem 380: 431–442, 1999
15. James M, Nuttall A, Ilsley J, Ottersbach K, Tinsley J, Sudol M, Winder S: Adhesion-dependent tyrosine phosphorylation of β-dystroglycan regulates its interaction with utrophin. J Cell Sci 113: 1717–1726, 2000
16. Ilsley J, Sudol M, Winder S: The interaction of dystrophin with beta-dystroglycan is regulated by tyrosine phosphorylation. Cell Sig 13: 625–632, 2001
17. Sotgia F, Lee J, Das K, Bedford M, Petrucci T, Macioce P, Sargiacomo M, Bricarelli F, Minetti C, Sudol M, Lisanti M: Caveolin 3 directly interacts with the C-terminal tail of β-dystroglycan. Identification of a central WW-like domain within caveolin family members. J Biol Chem 275: 38048–38058, 2000
18. Russo K, Di Stasio E, Macchia G, Rosa G, Brancaccio A, Petrucci T: Characterization of the β-dystroglycan-growth factor receptor 2 (Grb2) interaction. Biochem Biophys Res Commun 274: 93–98, 2000
19. Venema V, Ju H, Zou R, Venema R: Interaction of neuronal nitric-oxide synthase with caveolin-3 in skeletal muscle: Identification of a novel caveolin scaffolding/inhibitory domain. J Biol Chem 272: 28187–28190, 1997
20. Li S, Couet J, Lisanti M: Src tyrosine kinases, Ga subunits and H-ras share a common membrane-anchored scaffolding protein, caveolin: Caveolin binding negatively regulates the auto-activation of Src tyrosine kinases. J Biol Chem 271: 29182–29190, 1996
21. Calvadesi M, Macchia G, Barca S, Defilippi P, Tarone G, Petrucci T: Association of the dystroglycan complex isolated from bovine brain synaptosomes with proteins involved in signal transduction. J Neurochem 72: 1648–1655, 1999
22. Schlaepfer D, Hunter T: Evidence for *in vivo* phosphorylation of the Grb2 SH2-domain binding site on focal adhesion kinase by Src family protein-tyrosine kinases. Mol Cell Biol 16: 5623–5633, 1996
23. Armstrong S, Latham C, Shivell L, Ganote C: Ischemic loss of sarcolemmal dystrophin and spectrin: Correlation with myocardial injury. J Mol Cell Cardiol 33: 1165–1179, 2001
24. Losasso C, Di Tommaso F, Sgambato A, Ardito R, Cittadini A, Giardina B, Petrucci T, Brancaccio A: Anomalous dystroglycan in carcinoma cell lines. FEBS Lett 484: 194–198, 2000
25. Vander Heide RS, Angelo JP, Altschuld RA, Ganote CE: Energy dependence of contraction band formation in perfused hearts and isolated adult myocytes. Am J Pathol 125: 55–68, 1986
26. Hohl CM, Altschuld RA, Brierley GP: Effects of calcium on the permeability of isolated adult rat cells to sodium and potassium. Arch Biochem Biophys 221: 197–205,1982
27. Armstrong SC, Ganote CE: Effects of 2,3-butanedione monoxime (BDM) on contracture and injury of isolated rat myocytes following metabolic inhibition and ischemia. J Mol Cell Cardiol 23: 1001–1014, 1991
28. Ohlendieck K, Ervasti JM, Snook JB, Campbell KP: Dystrophin-glycoprotein complex is highly enriched in isolated skeletal muscle sarcolemma. J Cell Biol 112: 135–148, 1991
29. Pelech SL, Meier KE, Krebs EG: Rapid microassay for protein kinase C translocation in Swiss 3T3 cells. Biochemistry 25: 8348–8353, 1986
30. Pereboev A, Ahmed N, thi Man N, Morris G: Epitopes in the interacting regions of β-dystroglycan (PPxY motif) and dystrophin (WW domain). Biochim Biophys Acta 1527: 54–60, 2001
31. Saito F, Masaki T, Kamakura K, Anderson LVB, Fujita S, Fukuta-Ohi H, Sunada Y, Shimizu T, Matsumara K: Characterization of the transmembrane molecular architecture of the dystroglycan complex in Schwann cells. J Biol Chem 274: 8240–8246, 1999

32. Anderson LVB, Davison K: Multiplex Western blotting system for the analysis of muscular dystrophy patients. Am J Pathol 154: 1017–1022, 1999
33. Armstrong SC, Delacey M, Ganote CE: Phosphorylation state of hsp27 and p38 MAPK during preconditioning and protein phosphatase inhibitor protection of rabbit cardiomyocytes. J Mol Cell Cardiol 31: 555–567, 1999
34. Armstrong SC, Shivell LC, Ganote CE: Differential translocation or phosphorylation of alpha B crystallin cannot be detected in ischemically preconditioned rabbit cardiomyocytes. J Mol Cell Cardiol 32: 1301–1314, 2000
35. Yamada H, Fumiaki S, Fukuta-Ohi H, Zhong D, Hase A, Arai K, Okuyama A, Maekawa R, Shimizu T, Matsumara K: Processing of β-dystroglycan by matrix metalloproteinase disrupts the link between the extracellular matrix and cell membrane via the dystroglycan complex. Hum Mol Genet 10: 1563–1569, 2001
36. Herdson P, Kaltenbach J, Jennings R: Fine structural and biochemical changes in dog myocardium during autolysis. Am J Pathol 57: 539–557, 1969
37. Sage MD, Jennings RB: Subplasmalemmal leptomeres in adult canine myocytes: Putative cytoskeletal function and the effect of total *in vitro* ischemia. Lab Invest 61: 171–176, 1989
38. Yang B, Jung D, Motto D, Meyer J, Koretzky G, Campbell K: SH3 domain-mediated interaction of dystroglycan and Grb2. J Biol Chem 270: 11711–11714, 1995
39. Diaz RJ, Losito VA, Mao GD, Ford MK, Backx PH, Wilson GL: Chloride channel inhibition blocks the protection of ischemic preconditioning and hypo-osmotic stress in rabbit ventricular myocardium. Circ Res 84: 763–775, 1999
40. Seko Y, Takahashi N, Sabe H, Tobe K: Hypoxia induces activation and subcellular translocation of focal adhesion kinase (p125FAK) in cultured rat cardiac myocytes. Biochem Biophys Res Commun 262: 290–296, 1999
41. Lepple-Wienhues A, Szabo I, Laun T, Kaba N, Gulbins E, Lang F: The tyrosine kinase p56lck mediates activation of swelling induced chloride channels in lymphocytes. J Cell Biol 141: 281–286, 1998
42. Ping P, Zhang J, Zheng Y, Li R, Dawn B, Tang X, Takano H, Balafanova Z, Bolli R: Demonstration of selective protein kinase C dependent activation of Src and Lck tyrosine kinases during ischemic preconditioning in conscious rabbits. Circ Res 85: 542–550, 1999
43. Sotgia F, Lee H, Bedford MT, Petrucci T, Sudol M, Lisanti MP: Tyrosine phosphorylation of β-dystroglycan at its WW domain binding motif, PPxY, recruits SH2 domain containing proteins. Biochemistry 40: 14585–14592, 2001
44. Sada K, Minami Y, Yamamura H: Relocation of Syk protein tyrosine kinase to the actin filament network and subsequent association with Fak. Eur J Biochem 248: 827–833, 1997
45. Couture C, Baier G, Oetken C, Williams S, Telford D, Marie-Cardine A, Baier-Bitterlich G, Fischer S, Burn P, A A: Activation of p56lck by p72syk through physical association and N-terminal tyrosine phosphorylation. Mol Cell Biol 14: 5249–5258, 1994

Molecular and Cellular Biochemistry **242**: 81–87, 2003.

Effects of glyburide (glibenclamide) on myocardial function in Langendorff perfused rabbit heart and on myocardial contractility and slow calcium current in guinea-pig single myocytes

Said Y. Khatib[1] and Mark R. Boyett[2]

[1]*Department of Physiology, Faculty of Medicine, Jordan University of Science and Technology, Irbid, Jordan;*
[2]*Department of Physiology, University of Leeds, Leeds, UK*

Abstract

Glyburide, also known as glibenclamide, was shown to have positive inotropic effect in human and animal hearts. The objectives of the present study was to investigate the effects of glyburide on developed left ventricular pressure (DLVP), coronary flow (CF), and heart rate (HR), in isolated rabbit heart as well as its effects on myocardial contractility and L-type calcium current, $i_{Ca,}$ in guinea pig myocytes. Rabbit hearts were mounted on Langendorff apparatus and perfused with an oxygenated Krebs for 30 min until reaching steady state to be followed by 20 min of experimental perfusion divided into 5 min of control perfusion and 15 min of perfusion with Glyburide (10 μM). Ventricular myocytes were isolated by enzymatic dispersion technique and superfused in an oxygenated Tyrode solution. Cells were voltage-clamped at holding potential –40 mV to inactivate Na^+ current and a step depolarizations, 200 msec duration, to 0 mV was applied to elicit i_{Ca}. The contractions of the myocytes were measured by optical methods. Glyburide significantly increased DLVP by 30% and CF by 36% but had no effect on HR. Glyburide increased cell contractility by 7 ± 6, 18 ± 7, 28 ± 9 and 54 ± 15% for 0.1, 1, 10 and 100 μM respectively, $p < 0.001$. Meanwhile it depressed i_{Ca} by 9 ± 6 and 19 ± 8% for 1 and 10 μM respectively. In conclusion, glyburide increased contractility of guinea pig single myocytes and of isolated rabbit heart, as indicated by increased developed left ventricular pressure while it depressed i_{Ca}. It is hypothesized that an elevation in intracellular calcium, which caused increased myocardial contractility, could be attributed to an increase in intracellular Na^+ that could increase intracellular calcium via Na^+/Ca^{2+} exchange. (Mol Cell Biochem **242**: 81–87, 2003)

Key words: glyburide, contractility, calcium current, left ventricular pressure, myocytes

Introduction

The anti-diabetic sulfonylurea drugs, like glyburide, have been established as potent blockers of ATP dependent potassium channel (K^+_{ATP}) in the heart [1]. They have been reported to cause positive inotropic effect in animals and human. Linden and Brooker [2] reported a significant and large increase in contractility after an initial decrease in rabbit atrial muscle. Tolbutamide, first generation of sulfonylurea agent, was shown to have positive inotropic effect and to dramatically improve the recoverability of myocardial function of ischemic myocardium, which was suggested to be a cause of increased cardiac mortality [3, 4]. Glyburide, a second generation of sulfonylurea agent, was shown to have a positive inotropic effect in healthy volunteers. It produced a significant increase in myocardial contractility as indicated by increased ejection fraction and increased cardiac work [5]. We reported a significant positive inotropic action for glyburide as shown by increased left ventricular pressure produced by 10 μM, in Langendorff isolated rabbit heart under normoxic

Address for offprints: S.Y. Khatib, Faculty of Medicine, Jordan University of Science and Technology, P.O. Box 3030, Irbid 22110, Jordan (E-mail: sykhatib@just.edu.jo)

conditions of perfusion [6]. Glibenclamide elicited sustained positive inotropic effect in isolated guinea pig heart, improved functional recovery of ischemic myocardium at 45 min reperfusion and accelerated the onset of contracture thus speeding up cardiac arrest in rat and guinea-pig heart [7–9]. Recently, glibenclamide was shown to increase intracellular calcium concentration in a dose dependent manner in cardiomyocytes [10]. Smallwood *et al.* [1] reported that glibenclamide attenuated action potential shortening (APD) during ischemia in a dose-related manner, but it did not effect on APD of the rabbit left ventricle. However, glyburide increased action potential duration at 95% repolarization in normoxic canine purkinje fibers.

The mechanism of action of the positive inotropic effect is not clear. Some reports suggested a direct effect of the drug [2, 3]. Other studies postulated an indirect effect mediated by cyclic AMP or catecholamines release [11]. However, the elevation of intracellular calcium and possible prolongation of action potential could suggest that the increase in intracellular Ca^{2+} responsible for the positive inotropic action to be attributed to increased slow inward calcium current, i_{Ca}.

The main aim of the present study was to investigate the effect of glyburide on i_{Ca} and on contractility in single ventricular myocytes isolated from guinea pig heart. We also looked into the effect of the drug on developed left ventricular pressure, coronary flow and heart rate in isolated Langendorff perfused rabbit heart.

Materials and methods

Isolated heart preparation

New Zealand white rabbits, weighing 2.5–3 kg, were lightly anaesthetised with diethylether then killed by a fast blow to the back of the neck to decerebrate the animal. The chest was immediately opened and the heart rapidly excised and washed in ice-cold Krebs solution. The heart was then cannulated within 2 min through the aorta on Langendorff apparatus consisting of water-jacketed double coils (Aimer Ltd., UK). Perfusion was started immediately with control Krebs solution running in one of the coils. The testing Krebs containing glyburide at 10 μM was placed in the other coil. The perfusing Krebs could be changed by turning a tap, which is part of the double coil, and placed, at the bottom of the apparatus. The control Krebs was continuously gassed with 95% O_2/5% CO_2 (pH 7.2–7.4) and consisted of (in mmol/L) NaCl, 100; KCl, 4; $MgSO_4$, 1.2; KH_2PO_4, 1.2; $CaCL_2$, 1.8; $NaHCO_3$, 25; CH_3COONa, 20; HEPES buffer 10 and glucose 10. The Krebs containing glyburide, running in the other coil, was also continuously gassed with 95% O_2/5% CO_2 and was identical to control Krebs in its constituents except for glyburide. The developed left ventricular pressure, DLVP, was recorded by a glass cannula inserted into the left ventricle through the left atrium and connected to a Harvard universal channel recorder via a pressure transducer PT 400 (Bell and Howell). For monitoring heart rate, HR, epicardial ECG was recorded by placing two pieces of cotton wool on the heart and connected to the channel recorder by AC/DC preamplifier. All Krebs solutions were filtered by in-line filters with 50 mm diameter and 8 mm pore size (Sartorius-Germany) and warmed to 37°C before perfusing the heart at 80 cm H_2O pressure. The heart was placed in jacketed glass chamber warmed to 37°C.

Experimental protocol

All hearts were initially perfused with normal Krebs until reaching steady state as indicated by constant, DLVP, HR and CF. This equilibrating period was reached after about 30 min of starting the perfusion. This equilibrating period was followed by 20 min of experimental perfusion. In the first 5 min the hearts were perfused with control Krebs before switching to that containing glyburide for an additional 15 min. A control group, n = 10, was run applying identical protocol except that the Krebs did not contain any glyburide at any stage.

Myocytes isolation

Guinea-pigs weighing 300–400 g were anaesthetized with chloroform. The heart was rapidly excised and dropped into a beaker containing cold HEPES-buffered physiological solution containing in mmol/L, Na^+, 130.4; Cl^-, 142.4; K^+, 5.4; HEPES, 5; glucose, 10; $H_2PO_4^-$, 0.4; Mg^{2+}, 3.5; taurine, 20; creatine, 0.75 and pH adjusted to 7.2 with NaOH. The washed heart was then cannulated on Langendorff apparatus and perfused at constant pressure of 30–40 mmHg and constant flow, set at 9 ml/min, with physiological solution containing 0.75 mmol/L Ca^{2+} at 37°C and equilibrated with 95% O_2/5% CO_2. The perfusion pressure was continuously monitored. After reaching stability, perfusion was switched to Ca^{2+} free solution for 5 min. The heart stopped beating while perfusion pressure increased slightly. Perfusion was then switched to re-circulated perfusion with physiological 'enzyme' solution containing collagenase II (1 mg/mL, Worthington), protease (type XIV, 0.1 mg/mL, Sigma) and Ca^{2+} (100 μM). The re-circulated perfusion continued for 9 min. During this perfusion the perfusion pressure increased sharply to about 90–100 mmHg from about 40 mmHg before steadily decreasing back to the initial value. At the end of enzyme perfusion the heart was cut down and the ventricles were cut free and incubated in Erlenmeyer flask containing physiological enzyme solution to which albumin was added at 2 g per 100 mL. The incubated heart was gently agitated in water bath at 37°C while being gassed with 95% O_2/5% CO_2 for 5 min. The solution was then filtered through gauze and the filtrate centrifuged a 400 rpm for 30 sec. The cell pellet was re-suspended in physiological solution containing 0.5 mmol/L Ca^{2+}. The remaining undigested ventricular tissues was re-incubated in enzyme solution and agitated again for 5 min followed by filtration. The process was repeated until all ventricular tissues were fully digested. The re-suspended cells were added together and kept at room temperature for at least 1 h before experimentation. All cells chosen for the study were large with as straight as possible edges, had clear striation and responded to stimulation with a rapid twitch.

Chamber perfusion

A drop of the cell suspension was placed in a bath mounted on a Nikon Diaphot inverted microscope. Perfusion of the path began after allowing cells to settle down for at least 5 min. The constituents of the solution used (mmol/L): Na^+, 135; K^+, 5; Mg^{2+}, 1; HCO_3^{2-}, 20; Cl^-, 102; SO_4^{2-}, 1; Ca^{2+}, 1; acetate, 20; glucose, 10; insulin, 5 U/L. This solution was equilibrated with 95% O_2/5% CO_2 to give pH 7.3. Five (250 mL) portions of solutions were prepared. One portion was used for control while Glyburide was added to the other four to give final concentrations of 0.1, 1, 10, and 100 μM. Up to four solutions were pumped by magnetic drive gear metering pumps (Micropump, USA) to solenoid valves (Lee products Ltd., UK). One of the solutions was directed to the bath by the solenoid valves and the remainder to waste. The bath had a volume of 70 μL and it was perfused at a rate of 2.4 mL/min (35 bath volumes per min). The solution was heated to 36°C by a heating coil wrapped around the glass inflow tubes before entering the chamber. The bath temperature was continuously monitored and regulated using a miniature bead thermistor mounted in the bath wall and feedback control of current flow through the heating coil. The temperature did not vary more than ± 0.2°C during the course of an experiment. The solution level and drainage from the bath were controlled by an electronic feedback system. Myocytes were stimulated by means of two platinum field electrodes placed on either side of the bath. After perfusion with normal Tyrode for 20–25 min the pump was switched to one of the Tyrode solutions containing glyburide starting with the lowest concentration for 15–20 min.

Drug preparation

Since glyburide is insoluble in water, a 10 mM stock solution was prepared in DMSO. The correct amount was used from this stock to prepare Tyrode or Krebs to the required concentration.

Recording and analysis of i_{Ca}

Rods shaped cells with clear striation were chosen. Membrane potential and currents were recorded using the GigaOhm-seal patch clamp technique in the whole-cell clamp configuration. Current amplification was accomplished with Axopatch 1-C (Axon Instruments, USA). The resistance of the patch pipettes ranged between 1–3 MigaOhm. For each seal, the pipette junction potential was offset to 0 mV with the pipette positioned near the membrane before initiating a seal. Membrane and seal capacitance was corrected electronically. A low-pass filter filtered the current with cut-off frequency of either 3 or 10 KHz [12]. During an experiment, membrane potential and current were monitored on a storage oscilloscope (Tektronix 5000 series) and a pen recorder (Gould 2400). The signals as well as timing pulses were also recorded on a videotape recorder (JVC, HR-158MS) using pulse code modulator (Neuro-corder, Neuro Data Instruments corporation, USA). The cells were constantly stimulated with pulses of 200 msec duration at 0.5 Hz which was chosen because increasing the frequencies above 0.5 Hz were shown by Boyett *et al.* [12] to decrease the i_{Ca}.. Applying voltage clamp pulses to 0 mV from a holding potential of –40 mV generated calcium current.

Measurements

Cardiac parameters
DLVP was measured as the difference between diastolic pressure and the maximum systolic pressure developed during the contraction i.e. the amplitude of the pressure trace. CF (mL/min) was determined from timed collection of effluent dripping from the heart. HR, (beat/min) was calculated from the ECG trace. DLVP, CF and HR at 15–20 min of perfusion with or without the Glyburide were expressed as % of the preceding values immediately before drug perfusion and compared with values for the control group.

I_{ca} measurement
The magnitude of i_{ca} was measured as the difference between the peak of i_{ca} and the current at the end of the 200 msec pulse. The magnitude of the current at the end of 15–20 min perfusion with the drug was calculated in nAmp and expressed as % of the immediately preceding current prior to drug perfusion.

Cell length (twitch shortening, TS)
To measure TS, cells were illuminated with long wavelength red light from 100 W halogen lamp (Nikon). The cell image was focused on to a linear-element photodiode array (Reticon). From the output of the array, the cell length was measured electronically as described in details by Boyett *et*

al. [13]. The twitch shortening was calculated as the amplitude of the cell length trace. The twitch shortening at the end of 15–20 min perfusion with the drug was calculated as percentage of the preceding twitch shortening before perfusion with the drug. The twitch shortening was considered as representative of cell contractility.

Statistical analysis

All values are expressed as mean ± S.D. For the rabbit experiments n = 6–10 and 10–14 for single myocytes. The significance was tested by Student's *t*-test and confirmed by non-parametric test 'Mann-Whitney test'. There was agreement between both tests, $p < 0.05$ was considered significant in both tests. The correlation coefficient was plotted by SPSS computer program.

Results

Effect of glyburide on DLVP, CF and HR

Glyburide at 10 μM increased DLVP to 130 ± 15% and CF to 136 ± 22% of control as shown (Fig. 1). Meanwhile the HR was not affected. The increase in both DLVP and CF was statistically significant, $p < 0.001$, when compared to the changes in the control group. The increase in DLVP and CF reached maximum within 2–3 min of starting glyburide perfusion and was sustained until the end of the experiment, i.e. 15–20 min. The control values refer to values of DVLP, CF and HR at the end of 20 min experimental perfusion without glyburide.

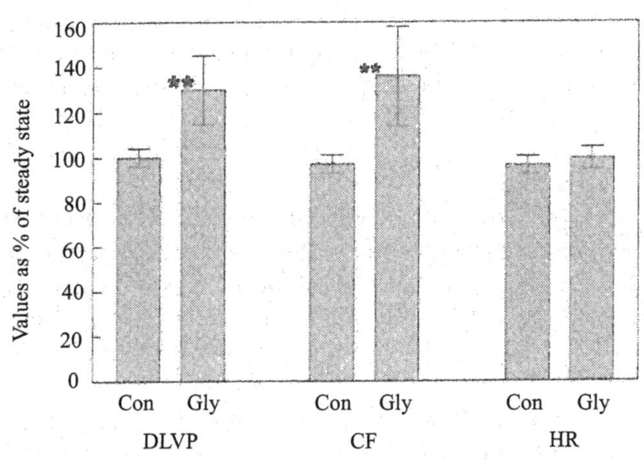

Fig. 1. Shows the effect of glyburide (Gly) 10 μM, on developed left ventricular pressure (DLVP), coronary flow (CF) and heart rate (HR) in the isolated rabbit heart. The results are expressed as % of preceding control values. The control group (Con) shows the results at the end of 20 min perfusion. All data are mean ± S.D., n = 10 for the control and n = 6 for the glyburide group.

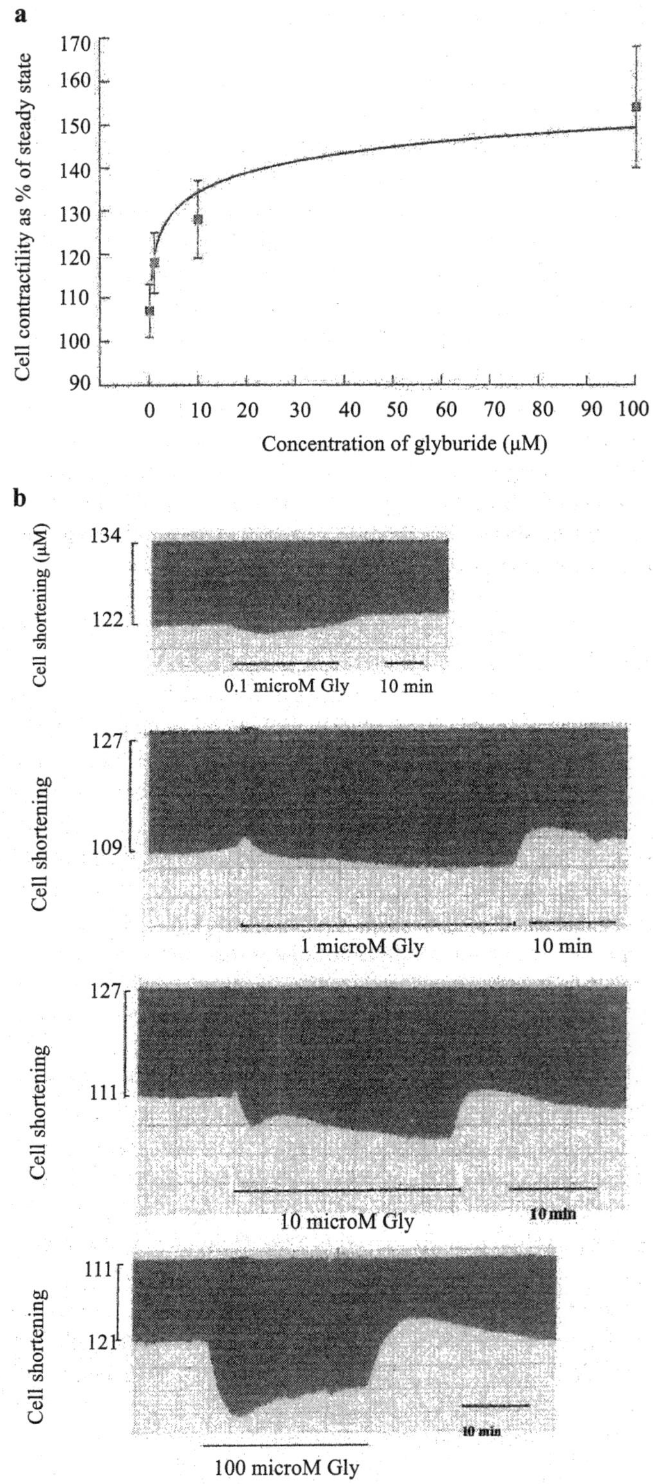

Fig. 2. (a) The effect of 0.1, 1, 10 and 100 μM of glyburide on cell contractility as represented by the twitch shortening expressed as % of the control twitch shortening immediately before glyburide perfusion in guinea pig myocytes. Each value represents the mean of 10–14 experiments, ± S.D., $p < 0.04$ for all values. (b) Shows slow time base of records showing the increase in cell shortening produced by different concentrations of glyburide. They represent different cells except those for 1 and 10 μM were obtained from the same cell.

Effect of glyburide on twitch shortening

Figure 2a shows the effect of different concentrations of glyburide (i.e. 0.1, 1, 10, 100 mM) on TS of guinea pig myocytes. TS was electronically recorded from the changes in cell length during contraction by optical method. Slow time base records showing the effect of the different concentrations of glyburide on TS are shown in Fig. 2b. Figure 2a shows that on increasing Glyburide concentration there was a significant increase in TS. The increase in TS as % of the preceding control was 7 ± 6, 18 ± 7, 28 ± 9 and 54 ± 14 produced by 0.1, 1, 10 and 100 μM respectively. The increase was significant for all concentrations, $p < 0.04$ for 0.1 μM and $p < 0.01$ for the other concentrations. The correlation coefficient (Fig. 3) shows a significant relationship between the drug concentration and TS in the range tested, $R = 0.692$, $p < 0.001$. Figure 2b also shows the increase in TS was reversible on going back to normal Tyrode.

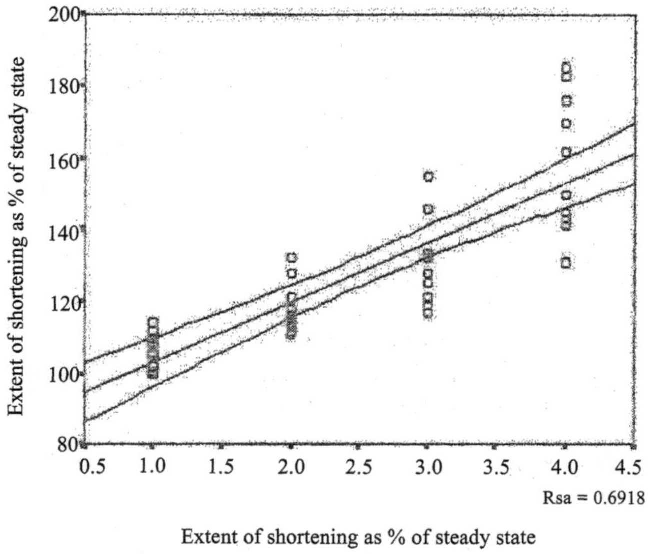

Fig. 3. The regression line plotted by SPSS program R = 0.692 suggesting a significant correlation between glyburide concentration and twitch shortening. $p < 0.001$, Pearson correlation, 1-tailed.

Effect of glyburide on i_{ca}

Table 1 shows the effect of two concentrations, 1 and 10 μM glyburide on i_{ca} in guinea pig myocytes. Current, mostly Ica, was recorded during 200 msec voltage clamp depolarizing pulses to 0 mV from a holding potential of –40 mV. The current was measured as the difference between peak inward current and the current at the end of the step depolarization. Because the absolute values of i_{ca} varies from cell to cell the depressed i_{ca} produced by the drug was calculated as % of the preceding control i_{ca} immediately before starting the drug perfusion. Superimposed i_{ca} at fast time base for control (i), 1 (ii) and 10 (iii) μM are shown in Fig. 4. The table shows a significant depression in i_{ca} produced by glyburide. This depression in the current seems to be concentration dependent. The percentage depression was 9 ± 6%, n = 10 and 19 ± 8%, n = 10, for 1 and 10 μM respectively. Only in 4 of these experiments the same cells were successfully tested for the effects of both concentrations. Glyburide at 0.1 μM did not have any measurable depression while 100 μM produced depression in current but was difficult to sustain for more than few contractions before cells ruptured, presumably due to the large increase in TS produced by such concentration.

Discussion

The results show glyburide caused a significant increase in DLVP, 30% and CF, 36% in isolated intact rabbit heart under normoxic conditions and significantly increased TS, that is a measure of contractility, while depressing i_{ca} in a dose dependent manner in single guinea-pig myocytes. The increase in TS was 7 ± 6, 18 ± 7, 28 ± 9 and 54 ± 14% produced by 0.1, 1, 10 and 100 μM respectively. The depression in i_{ca} was 9 ± 6% and 19 ± 8% for 1 and 10 μM.

The increase in CF can be explained by the increase in DLVP, which would be expected to increase the metabolic activity of cardiac cells. Schaffer *et al.* [14] found 3 μM of glyburide resulted in 45% increase in glucose utilization and lactate production in isolated perfused rat heart. More recently glyburide, 10 μM, resulted in enhanced 60% lactate production in pigs heart [15]. Hence, the increase in CF in our study could be explained by increased lactate production as a result on increased contractility [16].

The effects of glyburide on DVLP and TS suggest positive inotropic action in a dose dependent manner in intact

Table 1. Shows the effect of 1 and 10 μM glyburide on L-type slow calcium current at the end of 15–20 min superperfusion with Tyrode

Control i_{Ca} nA	1 μM glyburide	% Depression	Control i_{Ca} nA	10μM glyburide	% Depression
0.68 ± 0.15	0.62 ± 0.14*	9 ± 6	0.74 ± 0.17	0.6 ± 0.16*	19 ± 8

Shows the depressing effect of glyburide, 1 and 10 μM on the magnitude of L-type slow calcium current, i_{Ca} in myocytes isolated from guinea pig heart. The current at 15–20 min was expressed as % of the control current before Glyburide perfusion. (Mean ± S.D., n = 10–14, *$P < 0.001$ for both concentrations. Each group was compared to the respective control.

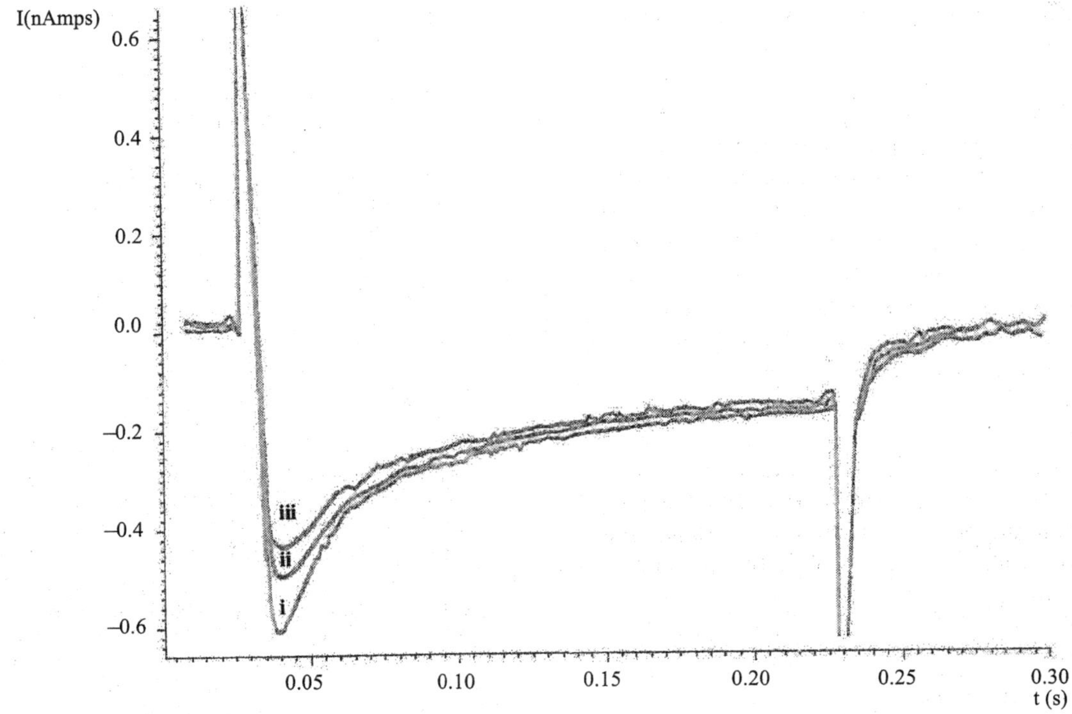

Fig. 4. Shows superimposed records of i_{Ca} from guinea pig cells at fast time base. Control (i), 1 μM (ii) and 10 μM (iii) glyburide. Two hundred msec duration pulses from 0 to –40 mV generated the current.

heart and in single myocytes. The correlation coefficient analyses suggest a significant correlation exist between contractility and glyburide concentration in the range of 0.1–100 μM. This effect of glyburide is in agreement with previous studies [2–6, 11]. In human volunteers glyburide produced a significant increase in ejection fraction and cardiac work [5]. In hypoxic myocardium glyburide, in a dose dependent manner (0.1–10 μM) attenuated the extent of a decrease in left ventricular pressure produce by hypoxia for 10 min in isolated Langendorff rabbit heart [6] while Tolbutamide reduced the severity of heart failure resulting from a period of ischemia [4]. In several studies glyburide reversed, reduced or antagonized the actions of potassium channel openers (e.g. Celikalim, Rilmakalim and Nicorandil) on intracellular calcium concentration and myocardial contractility. It reversed a negative inotropy induced by Celikalim, in feline left atrial papillary muscle [17] and antagonized a reduction in systolic intracellular free calcium and contractility that was induced by Rilmakalim in guinea-pig ventricular myocytes in a concentration dependent manner [18]. Glyburide also antagonized the action of Nicorandil which prevented an increase in intracellular Ca^{2+} of ventricular myocytes that had been exposed to high K^+ solution [19].

The mechanism of action of this positive inotropic effect is not clear. It is well documented that increased contractility is associated with increased intracellular Ca^{2+}. Frampton *et al.* [20] and Harrison *et al.* [21] using fluorescence technique found a direct relationship between TS and intracellular Ca^{2+}. Recently Kim *et al.* [10] reported, in single rat atrial myocytes, glibenclamide at 1, 10 and 100 μM increased intracellular Ca^{2+} from a basal value of 68 nM to 103, 166, and 289 nM respectively. It is, therefore, suggested that increased intracellular Ca^{2+} is responsible for the positive inotropic effect of glyburide. Such an increase in intracellular Ca^{2+} can not be explained by increased i_{ca} since it is depressed by glyburide as shown in this study. A possible increase in intracellular Na^+ can produce an increase in intracellular Ca^{2+} through the Na^+- Ca^{2+} exchange which was shown to play a significant role in cardiac contractile properties [22]. Increased intracellular Na^+ has been shown to increase intracellular Ca^{2+}, which was associated with increased contractility in guinea pig or rat myocytes [23]. An inhibition of Na^+/K^+ ATPase pump could lead to an increase in intracellular Na^+ and such inhibition of Na^+/K^+ ATPase pump by glyburide has been found in insulin secreting cells [24].

The possibility can not be ruled out that direct binding of glyburide to certain receptors on myocytes activates adenyl cyclase enzyme, which increases cAMP. Cyclic AMP is known to mediate the positive inotropic action of Catecholamines by increasing intracellular Ca^{2+}concentration [16].

The mechanism of i_{ca} depression is not clear. It could be due to decreased calcium influx as a result of decreased calcium concentration gradient or to direct depression effect on the L-type slow Ca^{2+} current, which needs further investigation.

In contradiction to the above, several studies reported a negative inotropic effect for glyburide and reduced recovery on reperfusion [25]. In such studies the period of ischemia was longer than 20 min. Schaffer *et al.* [4] found hearts subjected to periods of ischemia longer than 15 min were associated with a sever decline in mechanical recoverability. This could explain the contradictions in the effects of glyburide. In our studies and in those who found positive inotropy a short periods of ischemia or hypoxia (10–15 min) was used.

In conclusion, glyburide showed a positive inotropic action on intact rabbit heart and on single cardiomyocytes isolated from guinea-pig heart. This effect suggested increased intracellular Ca^{2+}, which seemed to be independent of the L-type slow calcium current since the latter was depressed. Such an increase in intracellular calcium could be accounted for by increased intracellular sodium.

Acknowledgements

The study was supported by the Deanship of Scientific Research at Jordan University of Science and Technology. The authors wish to thank the British Council for supporting S.Y. Khatib's visit to the UK. We also would like to thank Dr. Simon Harrison for his technical help and advice.

References

1. Smallwood JK, Ertel JP, Steinberg MI: Modification by glibenclamide of the electrophysiological consequences of myocardial ischemia in dog and rabbit. Arch Pharmacol 342: 214–220, 1990
2. Linden J, Brooker G: The positive inotropic action of sulfonylureas. Diabetes 27: 694–698, 1978
3. Pogatsa G, Dubeez E: The direct effect of hypoglycemic sulfonylureas on myocardial contractile force and arterial blood pressure. Diabetologia 13: 515–519, 1977
4. Schaffer SW, Poole WG, Lampson WG, Kramer JH: Effect of Tolbutamide on the mechanical function of the isolated rat heart subjected to global ischemia. J Mol Cell Cardiol 13: 341–345, 1981
5. Rothschid MA, Rothschild AH, Pfeifer MA: The inotropic action of Tolbutamide and Glyburide. Clin Pharmacol Ther 45: 642–649, 1989
6. Khatib SY, Al-Hader AA: Effect of glyburide on ATP, creatine phosphate and LVP in isolated rabbit heart. J Moll Cell Cardiol 24(suppl 1): O–07, 1993
7. Decking UK, Reffelman T, Schrader J, Kammermeier H: Hypoxia-induced activation of KATP channels limit energy depletion in the guinea-pig heart. Am J Physiol 269: H734–H742, 1995
8. Docherty JC, Gunter HT, Kuzio B, Shoemaker L, Yang L, Deslauriers R: Effects of Cromakalim and Glibenclamide on myocardial high-energy phosphates and intracellular pH during ischemia-reperfusion. J Mol Cell Cardiol 29: 1665–1673, 1997
9. Mitani A, Kinoshita K, Fukamachi K, Sakamoto M, Kurisu K, Tsuruhara Y, Fukamura F, Nakashima A, Tokunaga K: Effects of Glibenclamide and Nicorandil on cardiac function during ischemia and reperfusion in isolated perfused rat heart. Am J Physiol 261: H1864–H1871, 1991
10. Kim SH, Cho KW, Chang SH, Kim SZ, Chae SW: Glibenclamide suppresses stretch-activated ANP secretion: Involvement of K^+ ATP channel and L-type Ca^{2+} channel modulation. Pflügers Arch 434: 362–372, 1997
11. Dubach V, Burelschardt D, Raeder E, Forgo I *et al.*: Effect of intravenous Tolbutamide and Glibenclamide on myocardial contractility. Cardiology 64: 208–294, 1979
12. Boyett MR, Honjo H, Harrison S, Zang W, Kirby M: Ultra slow voltage-dependent inactivation of the calcium current in guineas-pig and ferret ventricular myocytes. Pflügers Arch 428: 39–50, 1994
13. Boyett MR, Moore M, Jewell BR, Montgomery RAP, Kirby MS, Orchard CH: An improved apparatus for the optical recording of contraction of single heart cells. Pflügers Arch 413: 197–205, 1988
14. Schaffer SW, Boen HT, Mahmood SM: Effect of Glyburide on myocardial metabolism and function. Am J Med 79(suppl 3B): 48–52, 1985
15. Wikstrom BG, Ronquist G, Waldenstrom A: Glyburide enhancement of lactate production in ischemic heart is modified by preconditioning: An *in vivo* experimental study in pigs by microdialysis technique. J Cardiovasc Pharmacol 27: 622–628, 1996
16. Ganong WF: Review of Medical Physiology. Appelton & Lange, Conecticut, 1991, pp 530, 574
17. Lodge NJ, Colatsky TJ, Cullinan CA, Follmer CH: Electromechanical effects of the putative potassium channel activator Celikalim on feline atrial and ventricular muscle. J Pharmacol Exp Ther 261: 1153–1159, 1992
18. Olbrich H, Muller M, Lindner S, Henke B *et al.*: Glimepiride inhibits the Relmakalim induced decrease in intracellular free calcium and contraction of isolated heart muscle cells from guinea pig to a lesser extent than Glibencalimde. Int J Cardiol 72: 53–63, 1999
19. Lopez JR, Jahangir R, Jahangir A, Shen WK, Terzie A: Potassium channel openers prevent potassium-induced calcium loading of cardiac cells: Possible implication in cardioplegia. J Thorac Cardiovasc Surg 112: 820–831, 1996
20. Frampton JE, Orchard CH, Boyett MR: Diastolic, systolic and sarcoplasmic [Ca^{2+}] during inotropic interventions in isolated rat myocytes. J Physiol 437: 351–375, 1991
21. Harrison S, Frampton J, McCall E, Boyett MR, Orchard CH: Contraction and intracellular Ca^{2+}, Na^+, and H^+ during acidosis in rat ventricular myocytes. Am J Physiol 262: C348–C357, 1992
22. Maxwell K, Scott J, Omelchenko A, Lukas A, Liyan L, Hnatowich M *et al.*: Functional role of regulation of Na^+/Ca^{2+} exchange assessed in transgenic mouse hearts. AJP – Heart Circ Physiol 277: H2212–H2221, 1999
23. Harrison SM, McCall E, Boyett MR: The relationship between contraction and intracellular sodium in rat and guinea pig ventricular myocytes. J Physiol 449: 517–550, 1992
24. Ribalet B, Mirrell CJ, Johnson DG, Levin SR: Sulfonylurea binding to a low-affinity site inhibits the Na/K-ATPase and the KATP channel in insulin-secreting cells. J Gen Physiol 107: 231–241, 1996
25. Billman GE, Avendano C, Halliwill JR, Burroughs JM: The effects of the ATP-dependent potassium channel antagonist, Glyburide, on coronary blood and susceptibility to ventricular fibrillation in unanesthetized dogs. J Cardiovasc Pharmacol 21: 197–204, 1993

Molecular and Cellular Biochemistry **242:** 89–99, 2003.

Metabolic phenotyping of the diseased rat heart using ^{13}C-substrates and *ex vivo* perfusion in the working mode

Geneviève Vincent,[1] Maya Khairallah,[1] Bertrand Bouchard[2] and Christine Des Rosiers[1,2]

Departments of [1]Biochemistry; [2]Nutrition, University of Montreal, Montreal, Quebec, Canada

Abstract

The objective of the present study was to compare energy substrate fluxes through metabolic pathways leading to mitochondrial citrate synthesis and release in normal and diseased rat hearts using ^{13}C-substrates and mass isotopomer analysis by gas chromatography-mass spectrometry (GCMS). This study was prompted by our previous finding of a modulated citrate release by perfused rat hearts and by the possibility that a dysregulated myocardial citrate release represents a specific chronic alteration of energy metabolism in cardiac patients. The 15-week-old spontaneously hypertensive rat (SHR) was chosen as our animal model of disease and the Wistar-Kyoto (WKY) rat as its matched control. *Ex vivo* work-performing hearts were perfused with a semi-recirculating buffer containing physiological concentrations of unlabeled (glucose) and ^{13}C-labeled ([U-$^{13}C_3$](lactate + pyruvate) and/or [1-^{13}C]oleate) substrates. In parallel to the continuous monitoring of indices of the heart's functional and physiological status, the following metabolic parameters were documented: (i) citrate release rates and citric acid cycle intermediate tissue levels, (ii) the contribution of fatty acids as well as pyruvate decarboxylation and carboxylation to citrate synthesis, and (iii) lactate and pyruvate uptake and efflux rates. Working hearts from both rat species showed a similar percent contribution of carbohydrates for citrate synthesis through decarboxylation (70%) and carboxylation (10%). SHR hearts showed the following metabolic alterations: a higher citrate release rate, which was associated with a parallel increase in its tissue level, a lower contribution of oleate β-oxidation to citrate synthesis, and an accelerated efflux rate of unlabeled lactate from glycolysis. These metabolic changes were not explained by differences in myocardial oxygen consumption, cardiac performance or efficiency, nor correlated with indices of tissue necrosis or ischemia. This study demonstrates how the alliance between *ex vivo* semi-recirculating working perfused rat hearts with ^{13}C-substrates and mass isotopomer analysis by GCMS, can provide an unprecedented insight into the metabolic phenotype of normal and diseased rat hearts. The clinical relevance of metabolic alterations herein documented in the SHR heart is suggested by its resemblance to those reported in cardiac patients. Taken altogether, our results raise the possibility that the increased citrate release of diseased hearts results from an imbalance between citrate synthesis and utilization rates, which becomes more apparent under conditions of substrate abundance. (Mol Cell Biochem **242**: 89–99, 2003)

Key words: SHR hearts, citrate release, citric acid cycle, energy metabolism, substrate flux parameters, isotopomer analysis

Introduction

The design of new treatments and optimization of current interventions to improve the statistics as well as the quality of life of patients suffering from cardiac diseases require a better understanding of the factors involved in such diseases. These factors include chronic alterations of myocardial energy metabolism such as a modified substrate oxidation profile with greater citrate and lactate release rates [1–7]. To better understand the causes and significance of these meta-

Address for offprints: C. Des Rosiers, Laboratoire du métabolisme intermédiaire, Y-3616, Centre hospitalier de l'Université de Montréal (CHUM) – Hôpital Notre-Dame, 1560 Sherbrooke E, Montréal, Québec, Canada H2L 4M1 (E-mail: christine.des.rosiers@umontreal.ca)

bolic alterations, the use of isotopically-labeled substrates, both *in vivo* and *ex vivo/in vitro*, has proven to be a very valuable approach (see ref. [8] for a recent review). Compared to measurements of metabolites, mRNA or protein levels, substrates labeled with radioactive (^{14}C, ^{3}H) or stable isotopes (^{13}C, ^{2}H) offer the possibility of evaluating the dynamic aspects of cardiac metabolism, namely substrate flux through pathways leading to energy production. Typically, tracing of energy substrate metabolism involves the determination of label incorporation into a metabolite and/or product reflecting glycolysis, glucose and fatty acid β-oxidation and/or the citric acid cycle (CAC). In recent years, studies with ^{13}C-substrates and either nuclear magnetic resonance (NMR) or gas chromatography–mass spectrometry (GCMS) expanded the field through the analysis of ^{13}C-isotopomers of relevant intermediary metabolites. One advantage of GCMS over NMR is its greater sensitivity that enables the determination of ^{13}C-enrichment of metabolites in the nanomolar range, which include CAC intermediates [9–14].

We developed a strategy, using ^{13}C-substrates and GCMS, to assess substrate fluxes through metabolic pathways leading to citrate synthesis in *ex vivo* Langendorff-perfused rat hearts [9–12] and *in situ* perfused pig hearts [13, 14]. As the first intermediate committed to the CAC, citrate plays a crucial role in cardiac energy metabolism. In addition, cytosolic citrate is proposed to play a role in substrate fuel partitioning by restricting glycolysis at the level of phosphofructokinase and/or long chain fatty acid (LCFA) β-oxidation following its conversion to malonyl-CoA, an inhibitor of carnitine palmitoyl transferase I [15, 16]. However, in the heart, the regulatory role of cytosolic citrate has been a subject of controversies because of the low activity of the mitochondrial tricarboxylate transporter [17, 18]. Interestingly, in one of the aformentioned studies, we found that hearts perfused with ^{13}C-substrates constantly release small quantities of citrate (nmol) whose ^{13}C-labeling pattern reflected that of tissue citrate, suggesting that citrate release reflected its mitochondrial efflux [9, 10]. In a subsequent study, citrate release was shown to be modulated by energy demand and substrate supply for citrate synthesis, namely, oxaloacetate (OAA) and acetyl-CoA, in agreement with the proposed role of cytosolic citrate in fuel partitioning [12]. These observations and the possibility that myocardial citrate release represents a specific chronic alteration of energy metabolism in patients suffering from cardiac diseases [1–3] prompted us to further examine this process, using ^{13}C-substrates and GCMS, in an animal model of cardiac pathology.

For this study, we chose a genetic model of hypertrophy, the 15-week-old spontaneously hypertensive rat (SHR). Age-matched Wistar-Kyoto (WKY) rats served as controls. SHR develop cardiac hypertrophy between 9 and 12 weeks of age; at 15 weeks of age, the cardiac hypertrophy appears to be well compensated [19, 20]. There is general agreement that the development of cardiac hypertrophy is associated with a metabolic shift from fatty acid to glucose utilization for energy production, referred to as the fetal metabolic phenotype (see for example refs [21–26]). However, it appears unclear whether increased glycolysis is associated with an increased lactate efflux and/or lower carbohydrate oxidation [22–24, 26] given the evidence for compartmentation of lactate metabolism in the heart [27]. In the light of these observations, we considered it important to design our ^{13}C-protocol to assess rates of lactate and pyruvate uptake and efflux [27] in parallel to fluxes through pathways leading to citrate synthesis and release.

Therefore, the objective of the present work was to quantitate substrate fluxes through metabolic pathways leading to citrate synthesis and release as well as fluxes through pathways affected by hypertrophy, in *ex vivo* work-performing SHR and WKY rat hearts. For this purpose, working hearts were perfused with buffer containing ^{13}C-substrates in a semi-recirculating mode to prevent recycling of unlabeled and ^{13}C-labeled metabolites, which is essential to the application of our ^{13}C-protocol [9, 10, 12]. In parallel to the continuous monitoring of indices of cardiac performance, the following metabolic parameters were documented: (i) citrate release rates and CAC intermediate tissue levels, (ii) the contribution of fatty acids as well as pyruvate decarboxylation and carboxylation to citrate synthesis, and (iii) lactate and pyruvate uptake and efflux rates. The results obtained demonstrated the following metabolic alterations in SHR hearts: higher citrate release rates and tissue levels, lower relative contribution of oleate β-oxidation to citrate synthesis as well as accelerated efflux of unlabeled lactate, presumably from glycolysis. These metabolic changes resembled those reported for cardiac patients.

Materials and methods

Materials

The sources of chemicals, biological products and ^{13}C-substrates have been identified previously [9–12]. Bovine serum albumin fatty acid-free (BSA fraction V, Intergen: 300 g of BSA in 1.5 L of modified Krebs-Henseleit buffer without glucose, with 0.1 mM EDTA) was dialyzed in membranes (molecular weight cut-off 6,000–8,000) at 4°C against 25 L of the same buffer for 28 h to reduce the background citrate concentration to the low μM range (1.2 ± 0.1 μM).

Heart perfusions in semi-recirculating working mode

Animal experimentation was approved by the local ethics committee in compliance with the guidelines of the Canadian

Council on Animal Care. Male SHR and WKY rats (15-week-old; 300–330 g; Charles River) were provided with food and water ad libitum. After anesthesia by intraperitoneal injection of sodium pentobarbital (65 mg/kg), hearts were cannulated rapidly and perfused retrogradely through the aorta at a constant pressure of 70 mmHg with semi-recirculating modified Krebs-Henseleit bicarbonate buffer containing (pH 7.4): 119 mM NaCl, 4.8 mM KCl, 2.5 mM $CaCl_2$, 1.2 mM KH_2PO_4, 1.2 mM $MgSO_4$, 25 mM $NaHCO_3$, 5.5 mM glucose, 8 nM insulin, 1 mM lactate and 0.2 mM pyruvate. A 10-min perfusion period was allowed for the insertion of a polyethylene catheter (PE-50) through the pulmonary vein into the left ventricle. The PE tubing, pulled through the ventricular wall and anchored in the apex of the heart by a fluted end, was connected to a pressure transducer for continuous monitoring of left ventricular functions. The left atrium was then cannulated through the pulmonary vein. Spontaneously beating hearts were switched to the anterograde work-performing mode and perfused via the left atrial cannula with semi-recirculating modified Krebs buffer (see 'Perfusion protocols' for details). Pressure of aortic afterload was set at 80 mmHg. Atrial preload was monitored continuously by a pressure transducer (Digi-Med Blood Pressure Analyzer, Micro-Med) and maintained at 11.5 mmHg. The present setup for working heart perfusions in the semi-recirculating mode, which is depicted in Fig. 1, is similar to that described for the perfusion of working hearts in the recirculating mode [8] except for the followings. The coronary effluent, which contains various metabolites released by the heart, is not recirculated but collected continuously. Thus, only aortic outflow is recirculated into the buffer reservoir. Consequently, the volume of buffer needed for a given perfusion experiment, 800 ml for 30 min, is greater than for similar heart perfusion in the recirculating mode. Proper oxygenation of 800 ml buffer containing fatty acid bound to albumin (pO_2: 500–550 mmHg) was achieved by its continuous pumping at a rate of 75 ml/min through a jacketed glass oxygenator consisting in part of 25-ft long silicone tubing rolled tightly in the gassed chamber [28]. Excess buffer was returned to the main reservoir through the overflow outlet of the oxygenator.

The following functional parameters were monitored continuously during the perfusion experiments: (i) atrial inflow and aortic outflow, using calibrated electromagnetic flow probes (Model FM501, Carolina Medical Electronics Inc.), (ii) temperature, with a thermocouple, and (iii) left ventricular contractile functions, namely, heart rate (HR), maximum left ventricular pressure (LVP_{max}), left ventricular end-diastolic pressure (LVEPD), and maximum value for the first derivative of LVP ($+dP/dt_{max}$), by pressure transducer (Digi-Med Heart Performance Analyzer, Micro-Med). The following physiological parameters were determined in influent and effluent perfusates collected at 20 min with a blood gas, elec-

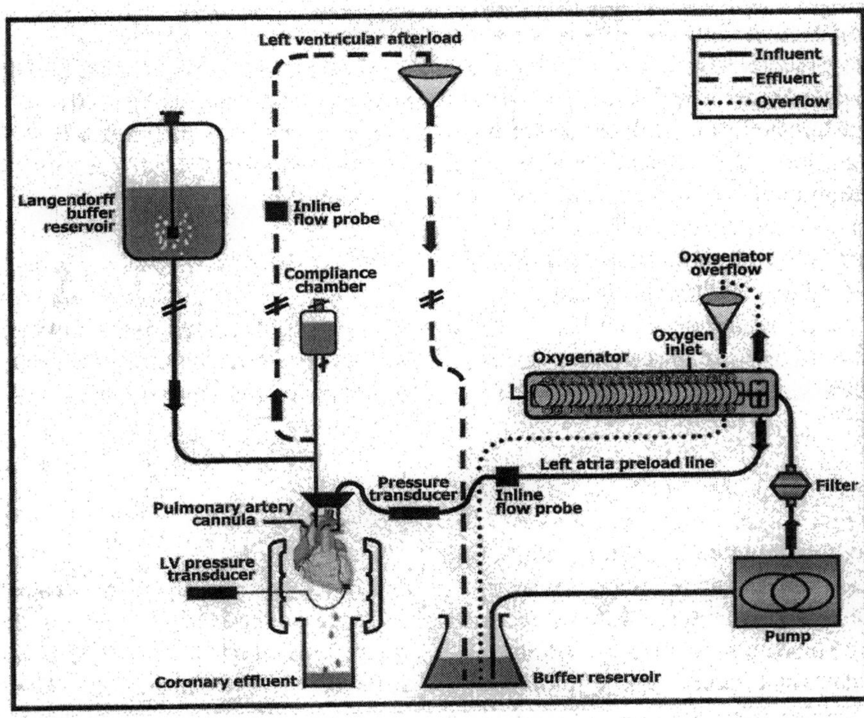

Fig. 1. Schematic overview of the semi-recirculating working heart setup. See 'Materials and methods' for details. LV pressure transducer: left ventricular pressure transducer.

trolytes and pH analyzer (ABL 77 series, Radiometer Copenhagen): pO_2, pCO_2, Ca^{2+} and other ion concentrations, and pH.

Perfusion protocols

Hearts from WKY rats and SHR were perfused for 30 min with semi-recirculating Krebs buffer containing physiological concentrations of glucose (5.5 mM), insulin (8 nM), carnitine (50 μM), [U-$^{13}C_3$]lactate (1 mM), and [U-$^{13}C_3$]pyruvate (0.2 mM), in the absence (n = 8) or presence of [1-^{13}C]oleate (0.4 mM) complexed to 3% dialyzed albumin, and 0.1 mM EDTA (n = 4–6). We also conducted a set of perfusion experiments with [1-^{13}C]oleate as the only ^{13}C-substrate (n = 4). The ionized calcium concentration of the albumin-containing buffer was determined to be 1.8 mM. The substrate concentrations in the perfusion buffer mimic those of plasma. Lactate and pyruvate were added in a physiological ratio to clamp the redox state and hence minimize perturbations of the cytosolic redox state that occur when lactate is supplied alone. The addition of carnitine compensates for its potential loss during heart isolation [29]. We verified that substrate concentrations and ^{13}C-enrichments in the buffer reservoir remained unchanged throughout the perfusion. A near isotopic steady-state was reached after 20–25 min of heart perfusion in the working mode with ^{13}C-substrates. Therefore, we chose 30 min as the duration of our perfusion experiments, after which the hearts were freeze-clamped with aluminium tongs chilled in liquid N_2. Under all conditions, citrate release rates (25–30 min), lactate and pyruvate uptake and efflux rates (25–30 min), and lactate dehydrogenase release rates (LDH; 5, 15 and 25 min) were quantitated from influent and effluent perfusate samples collected at the times indicated, while ^{13}C-enrichment and concentration of CAC intermediates (citrate, α-ketoglutarate (α-KG), succinate, fumarate, malate and OAA) as well as flux ratios relevant to substrate selection for citrate synthesis, namely, pyruvate and fatty acids from ^{13}C-enrichment of the acetyl (carbons 4 and 5: C_{4+5}) and OAA ($C_{1+2+3+6}$) moiety of citrate were measured in freeze-clamped powdered tissues.

Analytical procedures

Citrate release rates were quantitated by isotope dilution GCMS and flow rate measurements, as described previously [12]. Levels of CAC intermediates were determined in 100-mg tissue samples spiked with 10 nmol [1,5-$^{13}C_2$]citrate, 25 nmol [1,4-$^{13}C_2$]succinate, and 5 nmol [U-$^{13}C_4$]fumarate. Quantification was achieved using standard curves. GCMS assays of ^{13}C-mass isotopomer distribution (MID) of CAC intermediates and related metabolites in heart tissue samples (citrate and its OAA moiety, α-KG, succinate, fumarate, malate and pyruvate) have been described previously in detail [9–12]. Perfusate lactate concentration and LDH activity were determined by enzymatic assay with a Roche Cobas Fara spectrophotometer [11, 12, 30]. The ^{13}C-MID of lactate and pyruvate in influent and effluent perfusates was determined after treatment with 1 M NaB^2H_4 (see 'Calculations' below for details).

Calculations

Myocardial oxygen consumption (MVO_2: μmol/min) was calculated from the product of O_2 concentration (mM) difference between influent and effluent perfusates coming out of the oxygenator and pulmonary artery, respectively, and the coronary flow rate (ml/min). A value of 1.06 mM was taken as the concentration of dissolved O_2 at 100% saturation [31]. Intracellular pH value was calculated using venous CO_2 pressure (pCO_2, mmHg), as described by Bünger *et al.* [32]: $pHi = 7.524e^{(-0.0008786*pCO2)}$. Rate pressure product (RPP; mmHg·beats·min^{-1}) was calculated from: [(LVP_{max} (mmHg) – LVEPD (mmHg)) * HR (beats/min)]. Cardiac power [CP (mWatts) = cardiac output (m^3/sec) * LVP_{dev} (Pascals) * 10^3] and cardiac efficiency [CE (mWatts/μmol·min^{-1}) = CP (mWatts)/MVO_2 (μmol/min)] were calculated with a conversion factor of 133.32 Pascals per mmHg. To express tissue metabolite concentrations per g dry weight, the conversion factor for 1 g of freeze-clamped powdered heart tissue was determined to be 8.9 ± 0.2 (n = 29).

Mass isotopomers of metabolites containing 1 to n ^{13}C-atoms were identified as Mi with $i = 1, 2, \ldots n$. The absolute molar percent enrichment (MPE) of individual ^{13}C-labeled mass isotopomers (Mi) of a given metabolite was calculated as follows:

$$MPE(Mi) = \%\,A_{Mi}/[A_M + \Sigma A_{Mi}], \quad (1)$$

where A_M and A_{Mi} represent the peak areas from ion chromatograms corrected for natural abundance, corresponding to unlabeled (M) and ^{13}C-labeled (Mi) mass isotopomers, respectively.

Flux parameters

The development of equations to calculate flux ratios relevant to citrate synthesis in hearts perfused with [U-$^{13}C_3$](lactate + pyruvate) and/or [1-^{13}C]oleate has been described previously in detail [9, 10, 12]. In brief, flux ratios were calculated from the measured MID of the following tissue metabolites: citrate and its OAA moiety (OAA^{CIT}), from which we extrapolated the acetyl moiety of citrate (AC^{CIT}), pyruvate and succinate.

In this study, we reported the following flux rates, expressed relative to that of citrate synthase (CS): (i) oleate oxidation: OLE/CS = 9 * M1 AC^{CIT}/M1 oleate (Eq. 3 of Ref. [12]), (ii) pyruvate decarboxylation: PDC/CS = M2 AC^{CIT}/M3 pyruvate (Eq. 5 of ref. [9]), (iii) pyruvate carboxylation: PC/CS = OAA^{CIT}/M3 pyruvate (Eq. 4 of ref. [9]), and (iv) the contribution of other substrates (OS), such as endogenous fatty acids and/or amino acids, to the formation of acetyl-CoA: OS/CS = 1 – (PDC/CS + OLE/CS). The measured MPE M3 OAA^{CIT} was corrected for the fraction of M3 OAA molecules coming from citrate isotopomers metabolized in the CAC, as described in detail in ref. [9] (Eqs. 8–10).

To extrapolate the MPE of AC^{CIT} from the measured MID of citrate and of its OAA moiety, we used different mathematical approaches that depended on the choice of ^{13}C-substrate(s). For hearts perfused with [1-^{13}C]oleate as the sole ^{13}C-substrate, we employed Eq. 18 of ref. [9]. For hearts perfused with [U-$^{13}C_3$](lactate + pyruvate) in the absence or presence of [1-^{13}C]oleate, we used Eqs. 2 and 2a, respectively:

$$\text{MPE M2 AC}^{CIT} = 0.5 * [\Sigma \text{ MPE } (\text{M}i * i)(\text{citrate} - \text{OAA}^{CIT})] \quad (2)$$

$$\text{MPE M2 AC}^{CIT} = 0.5 * [\Sigma \text{ MPE } (\text{M}i * i)(\text{citrate} - \text{OAA}^{CIT}) - \text{MPE M1 AC}^{CIT}] \quad (2a)$$

Equation 2 provides a simple mean to extrapolate the MPE AC^{CIT} from the measured MID of citrate and OAA^{CIT} in hearts perfused with [U-$^{13}C_3$](lactate + pyruvate) as the only ^{13}C-substrates. In Eq. 2a, the MPE M1 of AC^{CIT} resulting from oleate oxidation was determined in separate experiments in which [1-^{13}C]oleate was the only ^{13}C-substrate. These equations were used because the MPE values of AC^{CIT} calculated with Eqs. 18 and 19 of ref. [9] were imprecise. This is explained by the fact that tissue citrate was predominantly enriched in isotopomers of higher masses (M3–M6), while Eqs. 18 and 19 are based solely on the MPE M1 and M2 of citrate and OAA^{CIT} (see ref. [9] for details).

Lactate and pyruvate uptake and efflux rates

These rates were determined by a modification of the NMR approach described recently [27]. In brief, in hearts perfused with a non-recirculating buffer containing unlabeled glucose and [U-$^{13}C_3$](lactate + pyruvate), lactate produced by glycolysis from exogenous glucose and endogenous glycogen is unlabeled (M) and can be distinguished by GCMS from the [U-$^{13}C_3$]lactate (M3) added to the buffer. The uptake of $^{13}C_3$-labeled lactate is quantified from the difference between its influent and effluent perfusate concentration. In practice, the MID of perfusate lactate and pyruvate was determined by GCMS in samples treated with 1 M NaB^2H_4. This treatment reduces pyruvate into lactate, and the 4 mass isotopomers of lactate obtained can be distinguished by GCMS: ^{12}C-lactate → ^{12}C-lactate (M), ^{13}C-lactate → ^{13}C-lactate (M3), ^{12}C-pyruvate → ^{12}C-lactate deuterated (M1), and ^{13}C-pyruvate → ^{13}C-lactate deuterated (M4). The concentrations of unlabeled (M, M1) and ^{13}C-labeled (M3, M4) lactate and pyruvate were calculated from their corresponding MIDs and enzymatically-determined perfusate lactate concentrations, which included both unlabeled (M) and ^{13}C-labeled lactate (M3). Efflux and uptake rates of lactate and pyruvate (μmol/min) were obtained by multiplying their perfusate concentrations in unlabeled (M, M1) and ^{13}C-labeled (M3, M4) (μmol/ml) isotopomers, respectively, by the coronary flow rate (ml/min).

Statistical analysis

The data are expressed as means ± S.E. of n = 4–8 heart perfusions. The unpaired *t*-test was applied for statistical evaluation of the data. A probability of $p < 0.05$ was considered to be significant.

Results

As a whole, the metabolic parameters measured in hearts perfused under normoxia for 30 min with a non-recirculating buffer containing 5.5 mM glucose, 8 nM insulin, 50 μM carnitine, 1 mM [U-$^{13}C_3$]lactate, 0.2 mM [U-$^{13}C_3$]pyruvate, in the absence or presence of 0.4 mM [1-^{13}C]oleate/albumin, showed a similar trend. Furthermore, for most parameters, differences observed between WKY and SHR hearts were independent of the presence of oleate. Therefore, due to space limitation, we chose to present in greater detail the results obtained with WKY and SHR hearts perfused in the presence of the LCFA oleate since this condition better mimics the physiological setting. However, we will report in the text any differences in values for functional and metabolic parameters measured in WKY and SHR hearts perfused in the absence of oleate.

Functional and physiological parameters

The values of the various functional and physiological indices measured throughout the perfusion experiments in the presence of oleate are summarized in Table 1. All measured values, in both WKY and SHR hearts, were comparable to those reported in the literature [22, 33]. Thus, except for slightly higher aortic flow, SHR hearts showed values similar to that of WKY hearts for all parameters measured, including MVO_2, cardiac flows, CP and CE. In the absence of oleate, SHR hearts showed values for RPP, LVP_{dev}, cardiac flows, CP

Table 1. Functional and physiological parameters of perfused working WKY and SHR hearts

Parameters	WKY	SHR
HR (beats/min)	225 ± 12	237 ± 9
RPP (mmHg·beats·min^{-1}·10^{-3})	34.6 ± 1.1	34.9 ± 1.0
LVP_{dev} (mmHg)	155 ± 11	148 ± 6
Cardiac output (ml/min)	48 ± 2	52 ± 2
Aortic flow (ml/min)	30 ± 2	35 ± 1*
Coronary flow (ml/min)	18 ± 2	17 ± 1
pH	7.35 ± 0.02	7.35 ± 0.01
MVO_2 (μmol/min)	8.9 ± 0.5	9.5 ± 0.5
CP (mWatts)	16 ± 1	18 ± 1
CE (mWatts/μmol·min^{-1})	1.8 ± 0.1	1.9 ± 0.1

Data are means ± S.E. of 4–6 heart perfusion experiments with 5.5 mM glucose, 8 nM insulin, 50 μM carnitine, 1 mM [U-$^{13}C_3$]lactate, 0.2 mM [U-$^{13}C_3$]pyruvate and 0.4 mM [1-^{13}C]oleate/albumin. Values of heart rate (HR), left ventricular pressure developed (LVP_{dev}) and cardiac flows, monitored throughout the perfusion experiments, were averaged for the entire 0–30-min period. MVO_2 and pH_i were calculated from pO_2 and pCO_2 values determined in influent and effluent perfusates collected at 20 min. RPP – rate pressure product; CP – cardiac power; CE – cardiac efficiency. *$p < 0.05$: SHR vs. WKY.

and CE that were similar to those observed in the presence of oleate (data not shown). However, they were significantly lower than those measured in WKY hearts (RPP = 43.7 ± 2.2 mmHg·beats·min^{-1}·10^{-3}; LVP_{dev} = 203 ± 8 mmHg; cardiac output = 66 ± 3 ml/min; aortic flow = 48 ± 2 ml/min; CP = 30 ± 2 mWatts, and CE = 3.4 ± 0.2 mWatts/μmol·min^{-1}; $p < 0.05$). Under all conditions examined, LDH release rates were low (< 125 mU/min) and similar for both rat species.

Citrate release rates and CAC intermediate tissue concentrations

Similar to Langendorff-perfused hearts [9, 10, 12], hearts perfused in the working mode with non-recirculating buffer containing ^{13}C-substrates constantly released a small amount of citrate, whose ^{13}C-MID reflected that of tissue citrate (data not shown). Furthermore, citrate release was not correlated with that of LDH, an index of necrosis, or succinate, an index of ischemia (between 10 and 13 nmol/min) [10]. The citrate release rates of WKY and SHR hearts perfused in the absence or presence of oleate ranged between 5 and 15 nmol/min. Figure 2A presents the citrate release rates for WKY and SHR hearts perfused in the presence of oleate. The results showed a significantly higher release rate for SHR hearts. In the absence of oleate, citrate release rates of WKY and SHR hearts revealed a similar pattern but only when the data were expressed relative to their respective CP (data not shown).

Figure 2B displays the tissue concentrations of citrate and other CAC intermediates measured in hearts perfused in the presence of oleate. Comparison of Figs 2A and 2B reveals

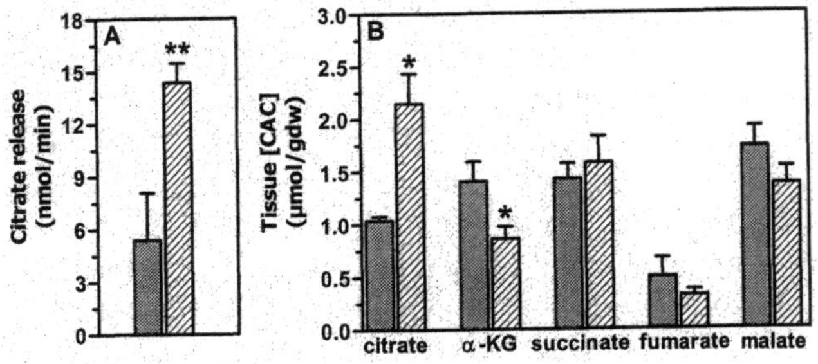

Fig. 2. Citrate release rates (A) and CAC intermediate tissue concentrations (B) in WKY (filled bars) and SHR (hatched bars) hearts. Data are means ± S.E. of 4–6 heart perfusion experiments with 5.5 mM glucose, 8 nM insulin, 50 μM carnitine, 1 mM [U-$^{13}C_3$]lactate, 0.2 mM [U-$^{13}C_3$]pyruvate, and 0.4 mM [1-^{13}C]oleate/albumin. Citrate release rates were quantitated in effluent perfusate samples, collected between 25–30 min, by isotope dilution GCMS and flow rate measurements. CAC intermediates were quantitated by GCMS in tissue homogenates spiked with standards. *$p < 0.05$ and **$p < 0.01$: SHR vs. WKY.

that in SHR hearts, the citrate release rate and tissue level increased in parallel. This increment was specific for citrate. Indeed, the α-KG tissue level was decreased, while that of other CAC intermediates as well as total CAC pool size were similar for both rat species. For hearts perfused in the absence of oleate, the total CAC pool size was 2-fold lower, and there was no difference in tissue citrate or α-KG levels between rat species, but SHR hearts showed a significantly greater tissue succinate levels (1.4 ± 0.1 vs. 0.9 ± 0.1 μmol/g dry weight; $p < 0.05$). Under all conditions, tissue levels of isocitrate represented less than 3% those of citrate and were similar for both rat species (data not shown).

^{13}C-enrichment and flux data

Table 2 reports the MIDs of CAC intermediates isolated from WKY and SHR hearts perfused with [U-$^{13}C_3$](lactate + pyruvate) and [1-^{13}C]oleate. Note that these data are representative of those obtained in hearts perfused in the absence of oleate. The ^{13}C-MID data for the various CAC intermediates were similar in both rat species. From these ^{13}C-enrichment data, we would like to emphasize the following points. First, tissue citrate was highly enriched in M4, M5 and M6 isotopomers, which is typical for hearts perfused with [U-$^{13}C_3$](lactate + pyruvate) under conditions of substantial pyruvate decarboxylation [10]. There was little ^{13}C-labeling of CAC intermediates resulting from [1-^{13}C]oleate oxidation. In fact, when WKY and SHR hearts were perfused with [1-^{13}C]oleate as the sole ^{13}C-substrate, we detected only M1 isotopomers for citrate and its OAA moeity (MPE citrate: 2.8 ± 0.4 and 1.2 ± 0.1; OAA^{CIT}: 1.3 ± 0.2 and 0.8 ± 0.1, respectively; $p < 0.01$; SHR vs. WKY). Second, as observed previously for Langendorff-perfused rat hearts [9–12] and *in situ*

Table 2. ^{13}C-Labeling of CAC intermediates isolated from working WKY and SHR hearts perfused with [U-$^{13}C_3$]lactate, [U-$^{13}C_3$]pyruvate and [1-^{13}C]oleate

Metabolites	Rat species	MPE (%)						
		M1	M2	M3	M4	M5	M6	ΣM*i*
Citrate	WKY	11.9 ± 0.3	18.4 ± 0.5	20.1 ± 0.5	15.7 ± 0.6	11.0 ± 0.6	4.3 ± 0.3	81.4 ± 2.0
	SHR	11.4 ± 0.5	18.9 ± 0.3	20.2 ± 0.1	16.0 ± 0.4	11.2 ± 0.5	4.6 ± 0.3	82.3 ± 0.5
α-KG	WKY	13.0 ± 0.3	21.6 ± 0.2	19.8 ± 0.5	15.1 ± 0.3	11.4 ± 0.4		80.9 ± 1.0
	SHR	12.1 ± 0.4	21.7 ± 0.4	20.2 ± 0.2	15.1 ± 0.4	11.3 ± 0.7		80.5 ± 0.6
Succinate	WKY	17.5 ± 1.0	19.7 ± 0.9	18.4 ± 1.0	11.5 ± 0.7			67.1 ± 3.5
	SHR	18.6 ± 0.4	20.5 ± 0.3	18.3 ± 0.5	11.0 ± 0.8			68.3 ± 1.4
Fumarate	WKY	19.1 ± 0.4	21.4 ± 0.3	21.5 ± 0.6	13.2 ± 0.5			75.3 ± 1.4
	SHR	19.1 ± 0.6	21.3 ± 0.3	21.0 ± 0.3	13.6 ± 0.5			75.0 ± 0.4
Malate	WKY	19.6 ± 0.5	21.3 ± 0.4	20.6 ± 0.5	12.4 ± 0.4			73.9 ± 1.5
	SHR	19.7 ± 0.5	21.1 ± 0.2	20.4 ± 0.2	12.1 ± 0.5			73.3 ± 0.4
OAA^{CIT}	WKY	18.1 ± 0.7	23.8 ± 0.7	22.0 ± 1.4	7.3 ± 0.6			71.1 ± 2.3
	SHR	18.9 ± 1.3	24.6 ± 0.7	23.0 ± 1.0	8.0 ± 0.7			74.5 ± 1.0

Data are means ± S.E. of 4–6 heart perfusion experiments with 5.5 mM glucose, 8 nM insulin, 50 μM carnitine, 1 mM [U-$^{13}C_3$]lactate, 0.2 mM [U-$^{13}C_3$]pyruvate and 0.4 mM [1-^{13}C]oleate/albumin. The MID of CAC intermediates was determined in tissue homogenates by GCMS.

perfused pig hearts [13], most of the ^{13}C-dilution occurred between α-KG and succinate, suggesting a small entry of unlabeled carbon through anaplerosis ($p < 0.01$ vs. citrate, n = 4–6). The MIDs of malate and fumarate were identical, reflecting rapid equilibration by fumarase. Finally, the MPE in M3 of fumarate, malate and the OAA moiety of citrate was higher than that of succinate, indicating the entry of M3 isotopomers by pyruvate carboxylation.

Table 3 reports the MPE values for acetyl-CoA of citrate, the corrected OAA of citrate and the pyruvate as well as the flux parameters for WKY and SHR hearts perfused in the presence of oleate. From this table, it is apparent that the relative rate of pyruvate decarboxylation (PDC/CS), which was similar for WKY and SHR hearts, was the predominant source of acetyl-CoA for citrate synthesis in these hearts. From the MPE of tissue pyruvate in M3 isotopomers, we conclude that ca. 60% of tissue pyruvate arose from exogenously supplied pyruvate and/or lactate; the remaining 40% arose from exogenously supplied glucose and/or endogenous glycogen. The relative rate of oleate β-oxidation (OLE/CS) was below 20%, and was significantly greater for WKY than SHR hearts. Consequently, the relative contribution of other substrates to acetyl-CoA formation, such as endogenous fatty acids and/or amino acids, was also greater for SHR than WKY hearts (OS/CS = 0.25 ± 0.02 and 0.09 ± 0.01%, respectively; $p < 0.05$).

We also report, for the first time in the perfused working rat heart, the relative rates of pyruvate carboxylation (PC/CS), an anaplerotic reaction. From the PC/CS values, which were similar for both rat species, we conclude that this anaplerotic reaction generates ca. 10% of OAA for citrate synthesis. Thus, flux through pyruvate carboxylation was 6-fold lower than through pyruvate decarboxylation. Note that the precision of estimates for PC/CS was lower than for PDC/CS flux ratios. This was due to a high degree of ^{13}C-recycling in the CAC, as reflected by the 4-fold difference between the measured and corrected MPE M3 OAA moiety of citrate (ca. 20 and 5%, respectively, from Tables 2 and 3). In WKY and SHR

Table 3. ^{13}C-enrichment and flux values determined in perfused working WKY and SHR hearts

MPE and flux values	WKY	SHR
MPE AC^{CIT}: (M1)	1.53 ± 0.31	0.65 ± 0.11*
MPE AC^{CIT}: (M2)	45.9 ± 1.3	42.1 ± 2.7
MPE OAA$^{CIT}_{corrected}$: (M3)	5.1 ± 1.5	7.0 ± 1.1
MPE pyruvate: (M3)	62.1 ± 0.7	59.8 ± 1.5
OLE/CS	0.17 ± 0.03	0.07 ± 0.01*
PDC/CS	0.74 ± 0.03	0.70 ± 0.04
PC/CS	0.11 ± 0.03	0.12 ± 0.02
PC/PDC	0.14 ± 0.04	0.17 ± 0.03

Data are means ± S.E. of 4–6 heart perfusion experiments. Flux ratios are calculated from the measured MID of the following tissue metabolites: (i) citrate and of its OAA moiety ($C_{1+2+3+6}$; OAA^{CIT}), from which we extrapolate the acetyl moiety of citrate (C_{4+5}; AC^{CIT}), (ii) pyruvate, and (iii) succinate. Flux parameters are expressed relative to citrate synthase (CS). The relative rate of oleate oxidation (OLE/CS = 9 * M1 AC^{CIT}/M1 oleate) was determined in hearts perfused with [1-^{13}C]oleate as the sole ^{13}C-substrate. The relative rates of pyruvate decarboxylation (PDC/CS = M2 AC^{CIT}/M3 pyruvate) and pyruvate carboxylation (PC/CS = OAA^{CIT}/M3 pyruvate) were determined in hearts perfused with [U-$^{13}C_3$](lactate + pyruvate) and [1-^{13}C]oleate as ^{13}C-substrates. The MPE M3 of OAA^{CIT} was corrected for the formation of M3 OAA from CAC metabolism of citrate isotopomers (see 'Materials and methods'). *$p < 0.05$: SHR vs. WKY.

hearts perfused in the absence of oleate, estimates of PC/CS (0.12 ± 0.01 and 0.13 ± 0.01, respectively) and PC/PDC (0.17 ± 0.03 and 0.21 ± 0.02, respectively) were similar to those obtained in the presence of oleate.

Lactate and pyruvate uptake and efflux rates

Figure 3A depicts the concentrations of unlabeled and ^{13}C-labeled lactate and pyruvate in influent and effluent perfusates of WKY and SHR hearts perfused in the presence of oleate. Figure 3A is also representative of data obtained in hearts perfused in the absence of oleate. We found that under all conditions examined, the relative proportion of ^{13}C-lactate in the effluent perfusates was significantly lower than that of the influent perfusate, indicating the release of unlabeled lactate (Fig. 3A, left panel). However, at least for WKY hearts, total lactate concentration, that is, unlabeled plus labeled (^{12}C + ^{13}C), in influent and effluent perfusates was similar. This is because the increase in the concentration of ^{12}C-lactate in effluent perfusate was compensated by a similar decrease in the concentration of ^{13}C-lactate, reflecting its uptake. The situation differed for SHR hearts where total lactate concentration in the effluent perfusate was significantly greater than in the influent perfusate due to a proportionally greater increase in ^{12}C-lactate concentration in the effluent perfusate. In contrast to lactate, there was no increase in the proportion of unlabeled pyruvate in the effluent perfusate, although there was a decrease in that of ^{13}C-pyruvate (Fig. 3A, right panel; $p < 0.001$).

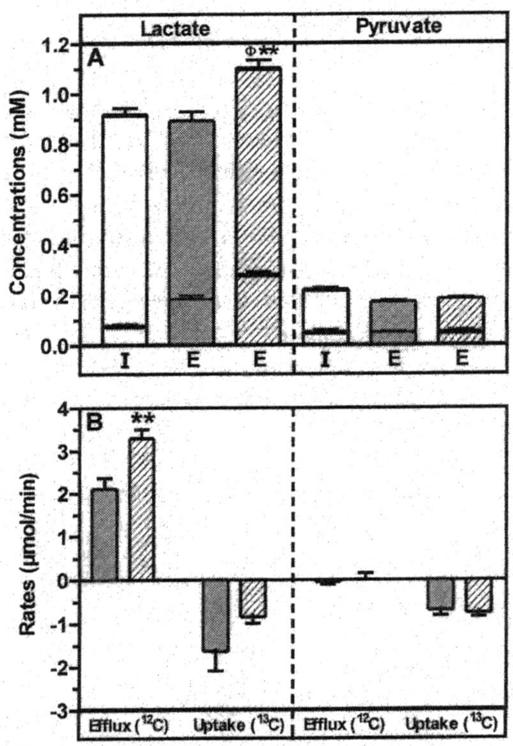

Fig. 3. Total perfusate lactate and pyruvate concentrations (A) and lactate and pyruvate uptake and efflux rates (B) in WKY (filled bars) and SHR (hatched bars) hearts. Data are means ± S.E. of 4–6 heart perfusion experiments with 5.5 mM glucose, 8 nM insulin, 50 µM carnitine, 1 mM [U-$^{13}C_3$]lactate, 0.2 mM [U-$^{13}C_3$]pyruvate and 0.4 mM [1-^{13}C]oleate/albumin. Stacked bars depict the concentrations of unlabeled (^{12}C: lower part) and labeled (^{13}C: upper part) lactate and pyruvate in influent (I; open bars) and effluent (E; hatched and filled bars) perfusates collected between 25 and 30 min and treated with NaB^2H_4. Lactate and pyruvate efflux (^{12}C) and uptake (^{13}C) rates were calculated from the product of their concentration differences in influent and effluent perfusates and coronary flow rates. **$p < 0.01$: SHR vs. WKY; and Φ$p < 0.001$: effluent vs. influent perfusate.

The results of Fig. 3A and coronary flow rate measurements allow the calculation of lactate and pyruvate efflux and uptake rates, which are depicted in Fig. 3B. These data demonstrates that SHR hearts showed higher efflux rates of unlabeled lactate compared to WKY hearts, indicating increased glycolytic activity. However, the rates of uptake of ^{13}C-labeled lactate and pyruvate did not differ significantly between rat species. Interestingly, while lactate and pyruvate are supplied at the physiological ratio of 5, lactate and pyruvate uptake rates differ only by a factor of 2, indicating preferential pyruvate uptake.

Discussion

This study reports a detailed comparison of substrate fluxes through metabolic pathways linked to energy production in normal WKY and diseased SHR hearts, using ^{13}C-substrates and mass isotopomer analysis by GCMS. Specifically, our ^{13}C-protocol was designed on the basis of previous studies in Langendorff-perfused rat hearts, to quantitate substrate fluxes through metabolic pathways leading to citrate synthesis and release as well as fluxes through pathways affected by hypertrophy [21–26]. For the present study, our ^{13}C-protocol was transposed to the *ex vivo* work-performing rat heart in view of its greater relevance to the *in vivo* situation in terms of energy demand (workload) and energy supply (substrates). Given the cost of ^{13}C-substrates, perfusion of working hearts with semi-recirculating buffer represents a valid compromise to meet the criteria for isotopic studies. The use of a recirculated buffer would impact on the capacity of the heart perfusion system to reach isotopic steady-state conditions and, hence, on the precision of estimated values of the flux parameters. Indeed, since the heart releases unlabeled and/or labeled metabolites, the ^{13}C-enrichment of supplied substrates changes with time. This point is best illustrated with lactate for which we report a higher concentration of unlabeled isotopomer, resulting in a lower percentage of ^{13}C-lactate in the effluent than influent perfusate (Fig. 3A).

The methodological limitations encountered in assessing flux parameters relevant to citrate synthesis in semi-recirculating working rat hearts are similar to those reported previously in *ex vivo* Langendorff-perfused rat hearts and *in situ* perfused pig hearts [9–14]. These include potential slight underestimation of the relative contributions of pyruvate carboxylation and decarboxylation, since these flux ratios are calculated from whole-tissue pyruvate M3 enrichment rather than intracellular enrichment. In addition, under the conditions of the present study, we were unable to estimate the contribution of β-oxidation from 0.4 mM [1-^{13}C]oleate in hearts perfused with [U-$^{13}C_3$](lactate + pyruvate), as described previously [12], since tissue citrate was predominantly ^{13}C-labeled from the metabolism of [U-$^{13}C_3$](lactate + pyruvate). Therefore, the low ^{13}C-enrichment of citrate in M1 isotopomers resulting from [1-^{13}C]oleate oxidation (< 3%) was determined in separate perfusion experiments with [1-^{13}C]oleate as the only ^{13}C-substrate. An alternative approach would be to perfuse hearts with [3-^{13}C](lactate + pyruvate) and [U-^{13}C]oleate, although the use of this ^{13}C-substrate mixture restricts flux measurements to the contributions of pyruvate and fatty acids to acetyl-CoA formation.

As a whole, hearts from control WKY rats perfused in the semi-recirculating work-performing mode showed values for indices of cardiac performance and MVO_2 and a pattern of substrate selection for energy production that are consistent with those reported by others for working normal rat hearts perfused under similar conditions [22, 33]. In brief, we estimated that more than 70% of acetyl-CoA for citrate synthesis is supplied by carbohydrates through pyruvate decarboxylation, and in the presence of 0.4 mM oleate, less than 20% arises from the β-oxidation of this exogenous LCFA. We also report, for the first time in the working heart, the following metabolic characteristics, which resembled those documented previously in *ex vivo* Langendorff-perfused rat hearts and *in situ* pefused pig hearts using ^{13}C-substrates and mass isotopomer analysis by GCMS [9–14]. First, working hearts constantly release a small amount of citrate (5–15 nmol/min) at a rate representing at most 0.5% of the CAC flux rate. Second, the relative contribution of anaplerotic pyruvate carboxylation to OAA for citrate synthesis is estimated to be ca. 10% and compensates largely for the mitochondrial citrate efflux [34]. Finally, working hearts simultaneously release unlabeled lactate and take up ^{13}C-labeled lactate and pyruvate for mitochondrial oxidation. Since lactate released by the heart was unlabeled, it is likely to reflect its production by glycolysis from exogenous glucose or endogenous glycogen. Based on measured lactate efflux rates, the percent contribution of this glycolytically-derived ATP is estimated to be ca. 5% of total mitochondrial ATP production resulting from carbohydrate and lipid oxidation (from MVO_2 values). It is noteworthy that our lactate uptake/efflux data are consistent with the proposed compartmentation of lactate metabolism in the heart, in which glycolytically-derived lactate production and oxidation of exogenous lactate operate as separate metabolic pathways [27]. They are also consistent with the concept of an intracellular lactate shuttle [35].

Let us now examine functional and metabolic differences between working WKY and SHR hearts. As a whole, working SHR hearts perfused with a physiological substrate mixture maintained normal values for indices of contractile and physiological function. They showed a similar percent contribution of carbohydrates for citrate synthesis through decarboxylation (70%) and carboxylation (10%). However, they showed the following metabolic alterations: a 2.7-fold elevation in citrate release, a 4.3-fold lower contribution of β-oxidation of exogenous oleate to acetyl-CoA formation for citrate synthesis, and a 1.6-fold increase in the release of unlabeled lactate from glycolysis. These metabolic changes of working SHR hearts are consistent with those reported previously in this and other animal models of cardiac hypertrophy (e.g. aortic banding) and/or in cardiac patients, using other techniques [1–3, 21–26]. In brief, there is general agreement that the hypertrophied heart returns to the fetal metabolic phenotype. This implicates changes in gene expression for proteins involved in glucose, lactate and LCFA uptake and metabolism that result in a shift in the heart's substrate preference for energy production from fatty acids to carbohydrates. For example, the higher glycolytic rate probably results from a shift in LDH isoforms favouring glycolytic lactate efflux over oxidation. Although the percent of glycolytically derived ATP is only increased from 5 to at most 10%, this may be crucial for ion pump function [36]. In addition, the lower contribution of β-oxidation from exogenous oleate to acetyl-CoA in SHR hearts is consistent with the reported defect in myocardial fatty acid uptake and β-oxidation [25].

This investigation reports also additional unique information relevant to our understanding of the metabolic causes underlying dysregulated citrate release in the diseased heart. In fact, we found that the greater citrate release by SHR hearts was not correlated with changes in total CAC pool size, indices of cellular integrity (LDH release) or ischemia (succinate release and pH_i). Based on measured MVO_2 values, one would presume that the CAC flux rate was not altered in SHR hearts. However, the observed differences in tissue levels of citrate (increased) and α-KG (decreased) in SHR hearts perfused in the presence of oleate, but not in the absence of oleate, suggest a potential imbalance in citrate synthesis (citrate synthase) and utilization (aconitase) rates, which becomes more apparent under conditions of mitochondrial substrate abundance. This interpretation would be consistent with our previous finding of modulation of mitochondrial citrate efflux by substrate supply [12], and with the reported lowering of myocardial citrate release in cardiac patients fed a low-fat low-calorie diet [2]. Of potential significance, aconitase, whose total activity in heart homogenates is 10-fold lower than that of citrate synthase, is subject to inactivation by a number of free radicals (superoxide anions, nitric oxide and peroxynitrite) [37–39], which are among many factors involved in the pathophysiology of heart diseases [40]. Restricted mitochondrial citrate utilization at the level of aconitase could impact on the contractile reserve of the hypertrophied heart. Further investigations are however needed to examine these aspects.

We recognize that the following considerations should be kept in mind regarding the results obtained in this study. First, although SHR hearts show many metabolic characteristics similar to hypertrophied rat hearts after aortic banding [22], the SHR is a genetic model of cardiac diseases that also exhibit distinct features that are age-dependent [19, 41]. Future studies are therefore warranted in SHR hearts to correlate metabolic changes in citrate synthesis and release with age and with a recognized indicator of the severity of disease, such as the expression of the atrial and/or brain natriuretic peptides [42]. Second, its matched control, the WKY rat, has repeatedly been reported to present high biological variability since it may not constitute a singly inbred strain [43, 44]. In this regard, we noticed and excluded some WKY rats with atypical and enlarged hearts. Finally, we chose to compare the metabolic phenotype of WKY and SHR hearts perfused at the same afterload, although SHR hearts are subject *in vivo* to a higher afterload. Nevertheless, working SHR hearts perfused at 80 mmHg showed values for cardiac flows and performance, MVO_2 and pH_i that resembled those of working WKY hearts. As a whole, the working SHR heart revealed alterations in carbohydrate and lipid metabolism that are consistent with those reported for age-matched SHR, using other techniques [22]. In view of our previous finding of modulation of citrate release by energy demand in Langendorff-perfused rat hearts, one might wonder, however, what would be the impact of a higher workload on the citrate release rate. This is, in fact, difficult to predict for SHR hearts because of the possibility of an imbalance between citrate synthesis and utilization rates. Furthermore, it remains to be demonstrated whether citrate release is modulated by energy demand in the more physiological *ex vivo* work-performing rat heart. Therefore, additional series of perfusion experiments are needed to document the effect of different workloads and fatty acid concentrations on metabolic pathways linked to citrate synthesis and release, both in WKY and SHR working hearts.

In conclusion, this study demonstrates how the alliance between *ex vivo* semi-recirculating working perfused rat hearts with ^{13}C-substrates and mass isotopomer analysis by GCMS, can provide an unprecedented insight into the metabolic phenotype of normal and diseased rat hearts. Given the high sensitivity of GCMS, we foresee no major difficulties in transposing our ^{13}C-protocol to the *ex vivo* working mice heart model, as it was done with radioactive substrates [45]. The use of our ^{13}C-protocol in this study has enabled the demonstration of metabolic changes in SHR hearts, which include an accelerated efflux of citrate and unlabeled lactate, as well as lower β-oxidation of exogenously-supplied LCFA. Similar metabolic alterations have been reported previously for cardiac patients, supporting the relevance of findings in our study model to the clinical situation [1–3]. Interestingly, taken altogether, the results from this work raise the possibility that the increased citrate release by SHR hearts may result from an imbalance between citrate synthesis and utilization rates, which becomes more apparent under conditions of mitochondrial substrate abundance. In this regard, further investigations are warranted to specifically examine if changes in the activity and/or expression of CAC enzymes are part of the mitochondrial alterations associated with the progression of cardiac diseases.

Acknowledgements

The authors thank Drs. John Chatham, Blandine Comte and Alain Comtois for their assistance in setting up the working heart preparation. Thanks are also due to Ovid Da Silva of the Research Support Office, CHUM Research Center, for editing this text. This work was presented at the XVII World Heart Congress of the International Society for Heart Research (ISHR) held in Winnipeg, and at its satellite meeting held in Banff in July 2001. This study was supported by the Canadian Institutes of Health Research (CIHR Grants # 9575 and 10920 to C.D.R and a studentship to G.V.).

References

1. Thomassen AR, Nielsen TT, Bagger JP, Henningsen P: Myocardial exchanges of glutamate, alanine and citrate in controls and patients with coronary artery disease. Clin Sci 64: 33–40, 1983
2. Thuesen L, Nielsen TT, Thomassen A, Bagger JP, Henningsen P: Beneficial effect of a low-fat low-calorie diet on myocardial energy metabolism in patients with angina pectoris. Lancet 2: 59–62, 1984
3. Thomassen AR, Nielsen TT, Bagger JP, Henningsen P: Cardiac metabolic and hemodynamic effects of insulin in patients with coronary artery disease. Diabetes 38: 1175–1180, 1989
4. Opie LH: Cardiac metabolism – emergence, decline, and resurgence. Part I. Cardiovasc Res 26: 721–733, 1992
5. Stanley WC, Lopaschuk GD, Hall JL, McCormack JG: Regulation of myocardial carbohydrate metabolism under normal and ischemic conditions. Potential for pharmacological interventions. Cardiovasc Res 33: 243–257, 1997
6. Taegtmeyer H, Goodwin GW, Doenst T, Frazier OH: Substrate metabolism as a determinant for postischemic functional recovery of the heart. Am J Cardiol 80: 3A–10A, 1997
7. Lopaschuk GD, Rebeyka IM, Allard MF: Metabolic modulation: A means to mend a broken heart. Circulation 105: 140–142, 2002
8. Barr RL, Lopaschuk GD: Methodology for measuring *in vitro/ex vivo* cardiac energy metabolism. J Pharmacol Toxicol Meth 43: 141–152, 2000
9. Comte B, Vincent G, Bouchard B, Des Rosiers C: Probing the origin of acetyl-CoA and oxaloacetate entering the citric acid cycle from the ^{13}C labeling of citrate released by perfused rat heart. J Biol Chem 272: 26117–26124, 1997

10. Comte B, Vincent G, Bouchard B, Des Rosiers C: A ^{13}C mass isotopomer study of anaplerotic pyruvate carboxylation in perfused rat hearts. J Biol Chem 272: 26125–26131, 1997
11. Laplante A, Vincent G, Poirier M, Des Rosiers C: Effects and metabolism of fumarate in the perfused rat heart. A ^{13}C mass isotopomer study. Am J Physiol 272: E74–E82, 1997
12. Vincent G, Comte B, Poirier M, Bouchard B, Des Rosiers C: Citrate release by perfused rat hearts: A window on mitochondrial cataplerosis. Am J Physiol 278: E846–E856, 2000
13. Panchal AR, Comte B, Huang H, Kerwin T, Darvish A, Des Rosiers C, Brunengraber H, Stanley WC: Partitioning of pyruvate between oxidation and anaplerosis in swine hearts. Am J Physiol 279: H2390–H2398, 2000
14. Panchal AR, Comte B, Huang H, Dudar B, Roth B, Chandler M, Des Rosiers C, Brunengraber H, Stanley WC: Acute hibernation decreases myocardial pyruvate carboxylation and citrate release. Am J Physiol 281: H1613–H1620, 2001
15. Depré C, Rider MH, Hue L: Mechanisms of control of heart glycolysis. Eur J Biochem 258: 277–290, 1998
16. Ruderman NB, Saha AK, Vavvas D, Witters LE: Malonyl-CoA, fuel sensing, and insulin resistance. Am J Physiol 276: E1–E18, 1999
17. Cheema-Dhadli S, Robinson BH, Halperin ML: Properties of the citrate transporter in rat heart: Implications for regulation of glycolysis by cytosolic citrate. Can J Biochem 54: 561–565, 1976
18. Sluse FE, Meijer AJ, Tager JM: Anion translocators in rat-heart mitochondria. FEBS Lett 18: 149–153, 1971
19. Kinney Lapier TL, Rodnick KJ: Changes in cardiac energy metabolism during early development of female SHR. Am J Hypertens 13: 1074–1081, 2000
20. Haneda T, Ichihara K, Abiko Y, Onodera S: Functional and metabolic responses to ischemia in the perfused heart isolated from normotensive and spontaneously hypertensive rats. Jpn Circ J 50: 607–613, 1986
21. Barger PM, Kelly DP: Fatty acid utilization in the hypertrophied and failing heart: Molecular regulatory mechanisms. Am J Med Sci 318: 36–42, 1999
22. Allard MF, Schönekess BO, Henning SL, English DR, Lopaschuk GD: Contribution of oxidative metabolism and glycolysis to ATP production in hypertrophied hearts. Am J Physiol 267: H742–H750, 1994
23. Christe ME, Rodgers RL: Altered glucose and fatty acid oxidation in hearts of the spontaneously hypertensive rat. J Mol Cell Cardiol 26: 1371–1375, 1994
24. Doenst T, Goodwin GW, Cedars AM, Wang M, Stepkowski S, Taegtmeyer H: Load-induced changes *in vivo* alter substrate fluxes and insulin responsiveness of rat heart *in vitro*. Metabolism 50: 1083–1090, 2001
25. Hajri T, Ibrahimi A, Coburn CT, Knapp FF Jr, Kurtz T, Pravenec M, Abumrad NA: Defective fatty acid uptake in the spontaneously hypertensive rat is a primary determinant of altered glucose metabolism, hyperinsulinemia, and myocardial hypertrophy. J Biol Chem 276: 23661–23666, 2001
26. Seymour A-M L, Chatham JC: The effects of hypertrophy and diabetes on cardiac pyruvate dehydrogenase activity. J Mol Cell Cardiol 29: 2771–2778, 1997
27. Chatham JC, Des Rosiers C, Forder JR: Evidence of separate pathways for lactate uptake and release by the perfused rat heart. Am J Physiol 281: E794–E802, 2001
28. Gamcsik MP, Forder JR, Millis KK, McGovern KA: A versatile oxygenator and perfusion system for magnetic resonance studies. Biotech Bioeng 49: 348–354, 1996
29. Rodgers RL, Christe ME, Tremblay GC, Babson JR, Daniels T: Insulin-like effects of a physiologic concentration of carnitine on cardiac metabolism. Mol Cell Biochem 226: 97–105, 2001
30. Bergmeyer HU: In: Methods of Enzymatic Analysis, vol. 3, 2nd edn. Verlag Chemie International, Academic Press, New York, 1974
31. Starnes JW, Wilson DF, Erecinska M: Substrate dependence of metabolic state and coronary flow in perfused rat heart. Am J Physiol 249: H799–806, 1985
32. Bünger R, Mallet RT, Hartman DA: Pyruvate-enhanced phosphorylation potential and inotropism in normoxic and postischemic isolated working heart. Near-complete prevention of reperfusion contractile failure. Eur J Biochem 180: 221–233, 1989
33. Goodwin GW, Cohen DM, Taegtmeyer H: [5-^{3}H]glucose overestimates glycolytic flux in isolated working rat heart: Role of pentose phosphate pathway. Am J Physiol 280: E502–E508, 2001
34. Gibala MJ, Young ME, Taegtmeyer H: Anaplerosis of the citric acid cycle: Role in energy metabolism of heart and skeletal muscle. Acta Physiol Scand 168: 657–665, 2000
35. Brooks GA: Intra- and extra-cellular lactate shuttles. Med Sci Sports Exerc 32: 790–799, 2000
36. Weiss JN, Venkatesh N: Metabolic regulation of cardiac ATP-sensitive K^+ channels. Cardiovasc Drugs Ther 7: 499–505, 1993
37. Gardner PR, Fridovich I: Superoxide sensitivity of the *Escherichi coli* aconitase. J Biol Chem 266: 19328–19333, 1991
38. Janero DR, Hreniuk D: Suppression of TCA cycle activity in the cardiac muscle cell by hydroperoxide-induced oxidant stress. Am J Physiol 270: C1735–C1742, 1996
39. Nulton-Persson AC, Szweda LI: Modulation of mitochondrial function by hydrogen peroxide. J Biol Chem 276: 23357–23361, 2001
40. Dhalla AK, Hill MF, Singal PK: Role of oxidative stress in transition of hypertrophy to heart failure. J Am Coll Cardiol 28: 506–514, 1996
41. Rupp H, Schulze W, Vetter R: Dietary medium-chain triglycerides can prevent changes in myosin and SR due to CPT-1 inhibition by etomoxir. Am J Physiol 269: R630–R640, 1995
42. Boomsma F, van den Meiracker AH : Plasma A- and B-type natriuretic peptides: Physiology, methodology and clinical use. Cardiovasc Res 51 : 442–449, 2001
43. H'Doubler PB Jr, Petersen M, Shek W, Auchincloss H, Abbott WM, Orkin RW: Spontaneously hypertensive and Wistar-Kyoto rats are genetically disparate. Lab Anim Sci 41: 471–473, 1991
44. Kurtz TW, Morris RC Jr: Biological variability in Wistar-Kyoto rats. Hypertension 10: 127–131, 1987
45. Belke DD, Larsen TS, Lopaschuk, GD, Severson DL: Glucose and fatty acid metabolism in the isolated working mouse heart. Am J Physiol 277: R1210–R1217, 1999

Molecular and Cellular Biochemistry **242:** 101–107, 2003.

Increased glycolysis as protective adaptation of energy depleted, degenerating human hibernating myocardium

Achim M. Vogt,[3] Albrecht Elsässer,[1] Holger Nef,[1] Christoph Bode,[1] Wolfgang Kübler[3] and Jutta Schaper[2]

[1]*Department of Cardiology, University of Freiburg/Br., Freiburg;* [2]*Department of Experimental Cardiology, Max-Planck-Institute for Physiological and Clinical Research (W.G. Kerckhoff-Institute), Bad Nauheim;* [3]*Department of Cardiology, University of Heidelberg, Germany*

Abstract

In the current study on human hibernating myocardium (HHM), we tested the hypothesis that increased glycolysis might exert a positive effect during a supply-demand balance situation by augmentation of myocardial energy formation. In 14 patients HHM was preoperatively detected by clinical methods and validated by the recovery of contractile function three months following revascularization. During open-heart surgery, transmural biopsies were removed from the hibernating areas and analyzed using biochemical and morphologic methods. Metabolite contents were normalized for the degree of fibrosis (control: 9.8 ± 0.5%, HHM 28.1 ± 3.0%; $p < 0.05$), providing values for cardiomyocytes only. In energy depleted HHM, severe intracellular degeneration, glycogen accumulation and myocyte loss were found. Elevated lactate levels (2.22 ± 0.26 vs. 25.38 ± 3.53 μmol/wet wt, $p < 0.001$) were indicative of an increased anaerobic glycolytic flux. In conclusion the presence of abundant intracellular glycogen and an increased anaerobic glycolysis in HHM is indicative of a protective adaptation of this myocardium, which might balance energy deficit and may limit structural damage. (Mol Cell Biochem **242**: 101–107, 2003)

Key words: hibernating myocardium, energy metabolism, tissue degeneration, apoptosis, glycolysis

Introduction

Although precise pathophysiologic mechanisms of myocardial hibernation still remain elusive, there are numerous reports showing that this myocardium is jeopardized by continuous disintegration. Intracellular degeneration and fibrosis as well as cell death could be demonstrated, as underlying mechanisms [1, 2].

Recently, severe energy depletion in human hibernating myocardium was shown [3]. As all processes responsible for the structural degeneration of HHM are known to be controlled by the cellular energy status [4–10], the hypothesis might be brought forward that metabolic alterations may contribute to the structural disintegration of HHM, thereby representing a common and basic pathomechanism.

Clinically, human hibernating myocardium (HHM) exhibits an increased glucose uptake. This characteristic feature, which is exploited as diagnostic procedure [11], is accompanied by an amplified sarcolemmal abundance of glucose transporters [12] and an increased cellular glycogen content [1]. In myocardial hypoperfusion, the beneficial effects of an augmented ATP generation by glycolysis have been repeatedly shown in clinical as well as in experimental studies [13, 14]. If this also holds true for HHM, the increase in glycolytic flux in HHM may not only be useful in diagnostic procedures but it may also represent an effective protective adaptation of HHM against supply-demand imbalance situations by limiting energy loss and thus structural damage.

It was the aim of the current study to test this hypothesis of HHM's pathophysiology by investigating myocardial

Address for offprints: A. Elsässer, Department of Cardiology, University of Freiburg, Hugstetterstrasse 3, D-79106 Freiburg, Germany (E-mail: elsaesser@med1.ukl.uni-freiburg.de)

biopsies taken from HHM and normal myocardium using microscopic and biochemical techniques. In this study, we addressed the question whether there is evidence for an augmented glycolytic energy formation, indicating a protective adaptation of deenergized and degenerating HHM in order to balance energetic and thus structural deterioration.

Materials and methods

Fourteen patients with angiographically documented coronary heart disease and a reduced left ventricular function resulting in the indication for coronary bypass surgery were studied. Informed written consent from each patient for every investigation and approval of the institutional hospital review board of the University of Freiburg had been obtained.

To detect HHM preoperatively, established clinical methods such as low-dose dobutamine echocardiography, thallium-201 scintigraphy using a stress-redistribution-reinjection-protocol, radionuclide ventriculography and coronary angiography including left ventriculography were used as described in detail previously [15]. All preoperative clinical investigations were repeated three months after revascularization in order to document in these regions the extent of functional recovery after restoration of adequate perfusion, thereby validating the preoperative diagnosis 'hibernating myocardium'.

The results obtained in HHM were compared with left ventricular biopsies from normal human myocardium from patients undergoing operative correction of atrial septal defects (n = 3) and from donor hearts not used for transplantation (n = 4).

Transmural Tru-cut® needle biopsies were removed during open-heart surgery from the beating heart before establishment of extracorporeal circulation and cardioplegic arrest. These biopsies were immediately shock frozen in liquid nitrogen for high performance liquid chromatography (HPLC) and immunohistochemistry or immersed in 3% glutaraldehyde buffered with 0.1 M Na cacodylate (at pH 7.4, 440 mosmol/l) for microscopy. Shock-frozen samples were kept in liquid nitrogen until further use. Following clinical validation of the myocardial regions as hibernating myocardium (3 months postoperatively), the specific examinations of the biopsies were started.

Myocardial metabolite contents were determined by HPLC and are given here in µmol/g wet weight [16]. Considering the error arising from varying degrees of myocardial fibrosis in control and hibernating myocardium, the total myocardial metabolite content were normalized to the cellular myocardial fraction (CMF), which results in values for myocellular content:

$$content_{CMF} = \frac{100}{(100 - fibrosis\ [\%])} \cdot content_{total}.$$

Small tissue samples were embedded in Epon following routine procedures. Ultrathin sections were stained with uranyl acetate and lead citrate and viewed and photographed in a Philips CM 10 electron microscope.

TUNEL was performed on cryosections with an ApopTag kit (Boehringer, Mannheim, Germany) following the supplier's instructions. TUNEL positive cells were detected by labeling with FITC in a light microscope equipped for fluorescence (DM, Leica, Germany).

Epon embedded semi-thin sections were stained in Epon with periodic acid-Schiff's reagent (PAS) for glycogen and evaluated by light microscopy.

The fibrotic content of HHM was determined by immunohistochemical labeling of fibronectin. Three different tissue sections, which were obtained from varying biopsy levels, were evaluated by confocal microscopy. Five different fields of vision were randomly chosen and fibrosis was quantified employing the point counting method following stereological principles [1, 17]. The fibrotic area of the tissue samples was expressed as % of the total myocardial area.

Statistical analysis

Data are expressed as mean values ± S.E. To evaluate the clinical data, the Friedman and Dunn test, ANOVA and Scheffé test, the paired *t*-test and the Mann-Whitney-rank-sum test were used. Fibrosis and the metabolite contents were analyzed using ANOVA and Scheffé testing. A p value of < 0.05 was considered as significant difference.

Results

All patients exhibited a recovery of regional contractile performance (radionuclide ventriculography: 24.5 ± 2.0% preoperatively to 55.6 ± 3.2% postoperatively, $p < 0.0001$) and a significant improvement of the global function (radionuclide ventriculography : 26.4 ± 2.3% preoperatively to 46.7 ± 3.2% postoperatively, $p < 0.0001$) three months after revascularization, thereby unequivocally validating our clinical preoperative diagnosis 'hibernating myocardium' (Table 1).

In human hibernating myocardium, increased levels of lactate (control: 2.21 ± 0.26, HHM: 25.37 ± 3.52 µmol/g wet wt/myocyte fraction) were found, which are indicative of an increased anaerobic glycolytic flux. Furthermore, an augmentation of ATP provisional stores by substrate chain phosphorylation was present (Table 1).

Numerous myocytes in HHM showed a lack of contractile material as well as alterations of mitochondria and occurrence of non-specified cytoplasm. These areas were frequently filled

Table 1. Lactate content, clinical and morphological parameters

	Normal		Hibernation		
	Mean	S.E.M.	Mean	S.E.M.	p
Lactate					
Myocardial content (μmol/g wet wt)	2.000	0.240	17.389	2.313	0.0002
Myocellular content (μmol/g wet wt CMF)	2.217	0.261	25.377	3.526	0.0002
Clinical and morphological parameters					
Global EF (%)	26.4	2.3	46.7	3.2	< 0.0001
Regional EF (%)	24.5	2.0	55.6	3.2	< 0.0001
Fibrosis (%)	9.857	0.508	28.062	3.012	0.010

Upper part: lactate content given as total myocardial as well as myocellular content (CMF – cellular myocardial fraction). Lower part: clinical and morphological parameters. EF – ejection fraction. P-values as indicated (ANOVA).

with degenerative material, i.e. with completely empty or lipid-containing vacuoles, lipofuscin and glycogen (Fig. 1a).

The intercellular distances between myocytes were widened. In addition, substantial amounts of fibrotic material – including collagen fibrils, lipofuscin-containing macrophages and electron-dense 'ground substance' (representing fibronectin, laminin and the proteoglycans) – were present (Fig. 1b). These degenerative changes were not observed in myocardium from normal human left ventricles.

Ultrastructural evidence of apoptosis was found in a few myocytes. Typical morphological signs of apoptosis were: intact mitochondria and an intact sarcolemma, nuclei of varying size and shape with distinct chromatin clumping and formation of apoptotic bodies indicating nuclear disassembly. Due to the sequestration of cellular particles from myocytes, the sizes and shapes of apoptotic cells varied greatly. This cellular debris was embedded in the extracellular matrix or it was engulfed by macrophages.

Scattered TUNEL positive myocytes were rarely found in HHM (Fig. 2). Because of the limited size of the biopsy samples, percentage values were not considered to be representative and therefore they were not calculated. In control myocardium, apoptosis was not detected.

PAS staining revealed large intracytoplasmic areas filled with glycogen in HHM (Fig. 3).

In HHM, an increased degree of fibrosis was observed, which was randomly located in the endocardial, midmyo-

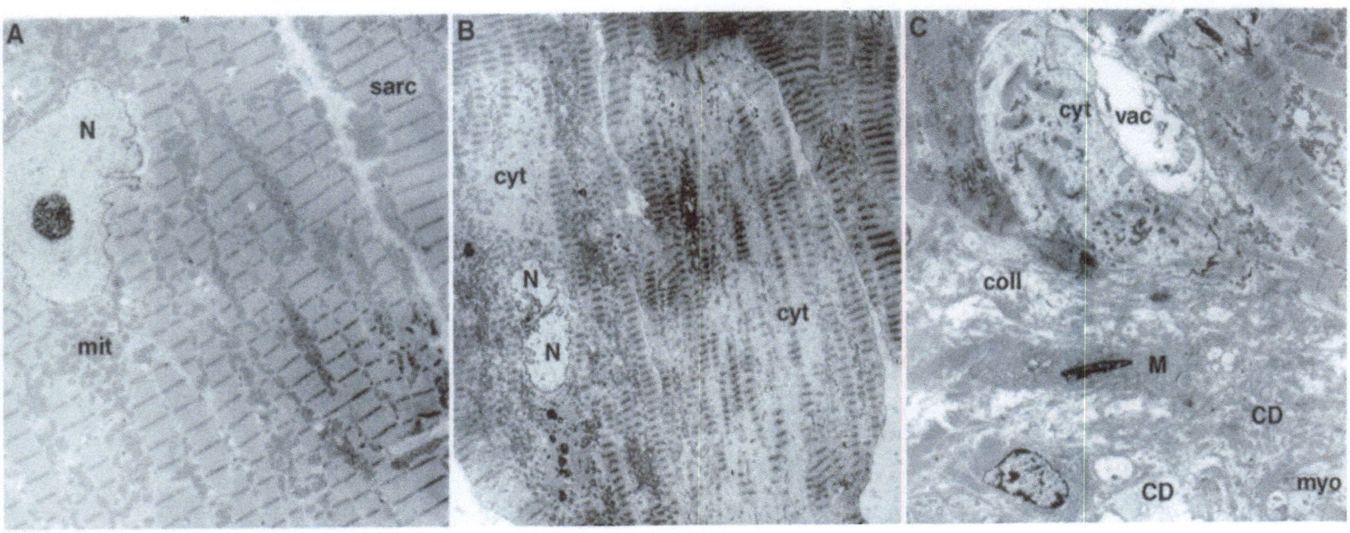

Fig. 1. Electron microscopic appearance of cellular degeneration in human hibernating myocardium. (A) Control myocardium with nucleus (N) and prominent nucleolus, numerous mitochondria (mit) and regularly arranged sarcomeres (sarc). Magnification ×3200. (B) Severe degeneration characterized by loss of myofilaments, small dark mitochondria and large areas of nonspecific cytoplasm (cyt) with glycogen accumulation and lipofuscin. The nucleus (N) is extremely lobulated. Magnification ×1600. (C) Degenerated myocytes (myo) with cytoplasmic areas (cyt) and vacuoles (vac) surround a widened extracellular space. The interstitium contains collagen fibrils (coll), macrophages (M), fibroblasts (lower left corner) and cellular debris (CD). Magnification ×2000.

cardial and epicardial layers (control: 9.8 ± 0.5%, HHM 28.1 ± 3.01%, $p < 0.05$).

Discussion

Tissue disintegration by energy depletion in human hibernating myocardium

Recently, an in-depth analysis of human hibernating myocardium's energy metabolism substantiated that HHM is severely energy depleted myocardium [3]. In this tissue, intra- and extracellular structural degeneration as well as myocyte loss by apoptosis could be observed. As all of these mechanisms are known to be controlled by the cellular energy status [4–10], our data, taken together, support the hypothesis that the supply-demand imbalance observed in human hibernating myocardium profoundly contributes to the continuous disintegration of myocardial integrity.

Degeneration of myocytes is a frequent pathomorphological finding in HHM. Furthermore, reduced contents in mRNAs encoding contractile, cytoskeletal and adhesive proteins [1] are indicative of diminished rates of synthesis. Both, impaired protein synthesis and predominating protein degradation finally result in cellular atrophy by sequestration of cellular particles into the extracellular space and the initiation of repair mechanisms with the development of replacement fibrosis [1, 17]. Consequently, a vicious cycle of continuous myocardial deterioration is commenced [2] which is not only of theoretical interest but also of practical interest because the degree of fibrosis represents the main cut-off point determining the extent of functional recovery after restoration of an adequate perfusion [1, 2, 17].

As apoptosis requires metabolic energy, mild to moderate decreases in ATP levels (reductions of up to 70%) may determine the fate of myocardial cells to die a suicidal cell death. If ATP contents drop to lower values, however, oncotic cell death occurs [6–9]. Moreover, graded decreases in ATP-levels (between 25 and 70%) or increases in the ADP/ATP ratio (0.2 or greater) not only play an important role in maintaining the apoptotic cascade, but may, additionally, suffice to initialize programmed cell death by apoptosis [18–20]. Since the levels for ATP in HHM were decreased but not completely exhausted and an increase in the ADP/ATP ratio was observed (see above and [3]), our data are in good accordance with both current hypotheses. Therefore, at the energy level observed in HHM, apoptosis most probably is initiated and maintained

Fig. 2. Detection of apoptosis using the TUNEL method. One nucleus of a myocyte shows positive labeling (green) indicating fragmentation of DNA. Myocytes are red and other nuclei are blue.

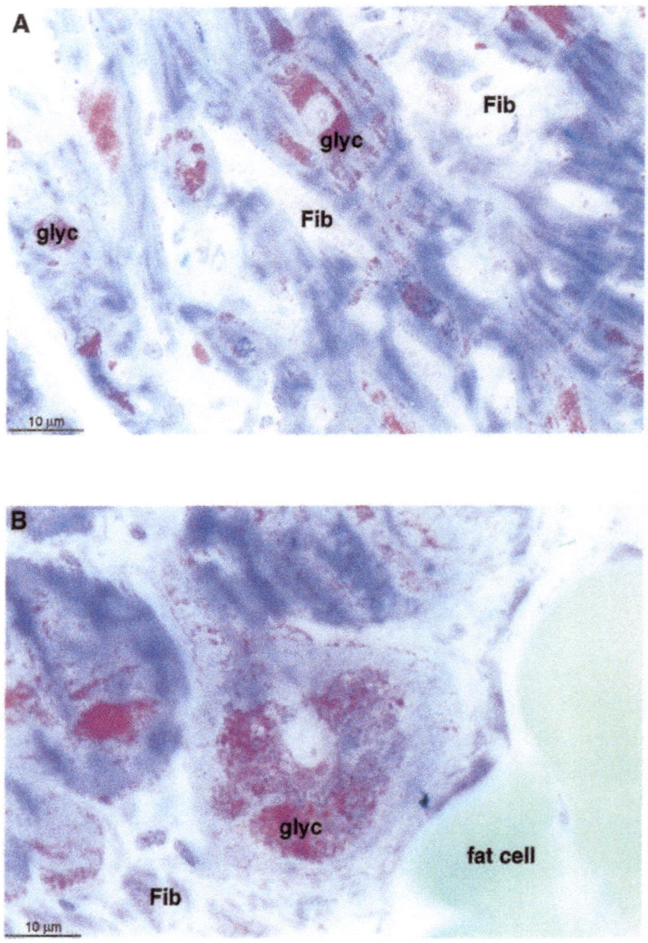

Fig. 3. PAS staining of human hibernating myocardium. (A) An accumulation of glycogen (glyc) in many myocytes is evident at a small magnification. (B) Shows several myocytes at higher magnification that contain much glycogen (glyc) in the perinuclear area. Fib= fibrosis.

unless an adequate energy metabolism is restored by revascularization. Otherwise, an increasing loss of high energy phosphates will result in a switch from apoptotic to oncotic cell death [6, 7, 10].

Increased glycolysis in human hibernating myocardium: Impact on augmented ischemic tolerance

In spite of varying concepts concerning the pathophysiology of HHM, its increased uptake of radiolabelled glucose tracers (^{18}-fluoro-deoxyglucose) represents the gold standard for the clinical diagnosis [11]. In addition to an augmented glucose uptake, most probably due to an elevated sarcolemmal abundance of glucose transporters [12], an enlarged content in glycolytic substrate, as evidenced by prominent cellular glycogen stores, is a constant finding of hibernating myocytes [1]. However, this increased uptake and storage of substrate(s) of the Embden-Meyerhof pathway may not only be of diagnostic value [21], but it may also contribute to the metabolic adaptation of HHM providing an increased resistance of this energy depleted myocardium to situations of severe supply-demand imbalance [22].

Myocardial tissue survival in situations of perfusion deficits critically depends on the cellular energy status [23]. To limit the energy deficit in conditions of energy shortage, as observed during hypoxia and ischemia, myocardial ATP formation may be amplified by an increase glycolytic substrate chain phosphorylation. By this means, the efficiency of ATP-synthesis is elevated when reduced oxygen-availability becomes the limiting factor [24–27]. This mechanism was repeatedly shown to be active providing myocardial cell survival in various states of myocardial hypoperfusion [13, 14]. In isolated cardiac myocytes subjected to hypoxia, increased glucose uptake and accelerated glycolysis have been shown to decrease the occurrence of apoptosis [22].

In human hibernating myocardium, a vast abundance of glycogen was seen. Moreover, an increased myocardial lactate content was observed, indicating – against the background of an increased sarcolemmal availability of glucose transporters [12] – an elevated anaerobic glycolytic flux. As any increase in glycolytic flux indicates a corresponding increment in ATP formation, our data suggest that an amplified flux through glycolysis in HHM may represent a powerful metabolic adaptation to balance the energetic deficit observed in HHM and thus to better tolerate this supply-demand imbalance.

Proposed pathomechanism of myocardial hibernation and its clinical implications

As described earlier [3], the supply-demand imbalance of HHM induces disturbances of myocardial energy metabolism, resulting in energy depletion (left side in Fig. 4). The consecutive drop in high-energy phosphate levels and an alteration in the ADP/ATP ratio initiate and maintain tissue degeneration and cell loss, which finally ends in structural disintegration. At the same time, the supply-demand imbalance initiates an increase in flux through glycolysis (right side in Fig. 4). This adaptive mechanism provides the endangered myocardium with additional ATP stores, which in turn will contribute to balance the myocardial energy shortage and to maintain the ADP/ATP ratio at a low level. When the glycolytic capacity is sufficiently high to compensate the energy deficit, structural disintegration will be arrested (upper part in Fig. 5). According to the concept of Camici and Dutka [28], such a condition represents 'functional hibernation'. If the severity of energy depletion exceeds the capacity of glycolytic ATP formation (lower part in Fig. 5), more and more cardiomyocytes degenerate and undergo apoptotic cell death, a situation that initiates the transition to 'structural hibernation [28]. However, even under these circumstances, an increase in glycolytic flux might partially compensate for the energetic deterioration and might slow down this detrimental process.

Finally, only revascularization will restore this impairment in energy metabolism and initiate myocardial recovery and structural restitution. However, if revascularization is not performed in time, an increasing degree of replacement fi-

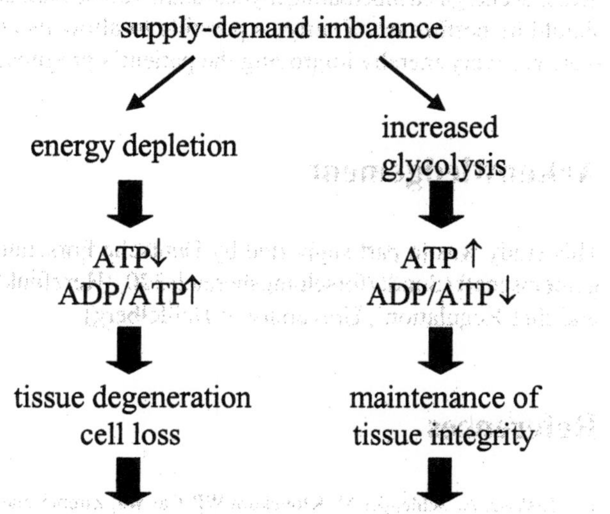

Fig. 4. Proposed pathophysiology of structural disintegration of HHM and protective adaptation by increased glycolysis. The supply-demand imbalance in HHM leads to energy depletion, as reflected by decreased ATP levels and an increased ADP/ATP ratio. This initiates tissue degeneration and cell loss which finally causes structural disintegration (left). Simultaneously, the supply-demand imbalance results in an increased glycolytic flux, which keeps ATP levels high and the ADP/ATP ratio low, thereby protecting HHM by maintaining tissue integrity.

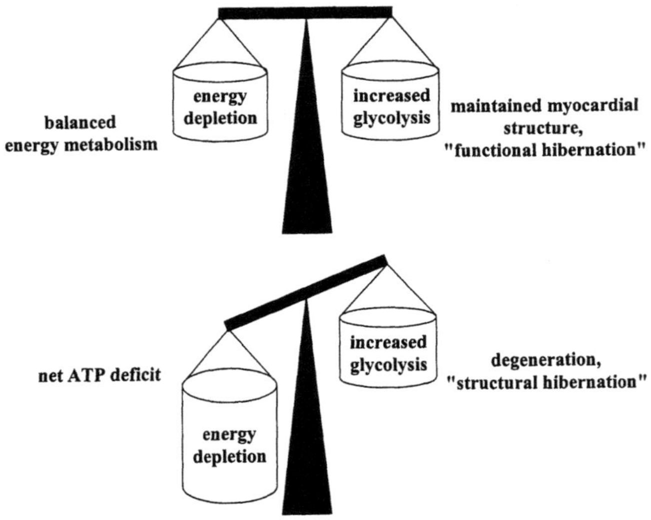

Fig. 5. Different energetic situations in HHM. At less critical states of supply-demand imbalance, increased glycolysis suffices to compensate for energy shortage, maintaining myocardial structure (upper part, 'functional hibernation') [28]. If the severity of energy depletion exceeds the capacity of glycolytic ATP formation, however, a net ATP deficit occurs initiating and maintaining myocardial degeneration (lower, 'structural hibernation) [28].

brosis will prevent the complete structural (and thus functional) restoration despite the presence of adequate revascularization. Loss of myocardial structural and functional integrity will thus be a continuous process. Therefore, to save this highly endangered, deenergized hibernating myocardium, revascularization should be performed as early as possible to allow its complete recovery thereby improving the patient's prognosis.

Acknowledgement

This study was in part supported by Deutsche Forschungsgemeinschaft (Sonderforschungsbereich 320, 'Herzfunktion und ihre Regulation', University of Heidelberg).

References

1. Elsässer A, Schlepper M, Klovekorn WP, Cai WJ, Zimmermann R, Muller KD, Strasser R, Kostin S, Gagel C, Munkel B, Schaper W, Schaper J: Hibernating myocardium: an incomplete adaptation to ischemia. Circulation 96: 2920–2931, 1997
2. Elsässer A, Decker E, Kostin S, Hein S, Skwara W, Müller KD: A self-perpetuating vicious cycle of tissue damage in human hibernating myocardium. Mol Cell Biochem 213: 17–28, 2000
3. Elsässer A, Müller K, Skwara W, Bode C, Kübler W, Vogt A: Severe energy deprivation of human hibernating myocardium as possible pathomechanism of contractile dysfunction, structural degeneration and cell death. J Am Coll Cardiol 39: 1189–1198, 2002
4. Kao R, Rannels DE, Morgan HE: Effects of anoxia and ischemia on protein synthesis in perfused rat hearts. Circ Res 38: I124–I130, 1976
5. Williams EH, Kao RL, Morgan HE: Protein degradation and synthesis during recovery from myocardial ischemia. Am J Physiol 240: E268–E273, 1981
6. Leist M, Single B, Castoldi AF, Kuhnle S, Nicotera P: Intracellular adenosine triphosphate (ATP) concentration: A switch in the decision between apoptosis and necrosis. J Exp Med 185: 1481–1486, 1997
7. Richter C, Schweizer M, Cossarizza A, Franceschi C: Control of apoptosis by the cellular ATP level. FEBS lett 378: 107–110, 1996
8. Eguchi Y, Shimizu S, Tsujimoto Y: Intracellular ATP levels determine cell death fate by apoptosis or necrosis. Cancer Res 57: 1835–1840, 1997
9. Depre C, Taegtmeyer H: Metabolic aspects of programmed cell survival and cell death in the heart. Cardiovasc Res 45: 538–548, 2000
10. Nicotera P, Leist M, Ferrando-May E: Intracellular ATP, a switch in the decision between apoptosis and necrosis. Toxicol Lett 102–103: 139–142, 1998
11. Camici PG, Wijns W, Borgers M, De Silva R, Ferrari R, Knuuti J, Lammertsma AA, Liedtke AJ, Paternostro G, Vatner SF: Pathophysiological mechanisms of chronic reversible left ventricular dysfunction due to coronary artery disease (hibernating myocardium). Circulation 96: 3205–3214, 1997
12. Brosius FC, Nguyen N, Egert S, Lin Z, Deeb GM, Haas F, Schwaiger M, Sun D: Increased sarcolemmal glucose transporter abundance in myocardial ischemia. Am J Cardiol 80: 77a–84a, 1997
13. Depre C, Vanoverschelde J-L, Taegtmeyer H: Glucose for the heart. Circulation 99: 578–588, 1999
14. King L, Opie L: Glucose and glycogen utilization in myocardial ischemia – changes in metabolism and consequences for the myocyte. Mol Cell Biochem 180: 3–26, 1998
15. Elsässer A, Müller K, Vogt A, Strasser R, Gagel C, Schlepper M, Klövekorn W: Assessment of myocardial viability: Dobutamine echocardiography and thallium-201 single-photon emission computed tomographic imaging predict the postoperative improvement of left ventricular function after bypass surgery. Am Heart J 135: 463–475, 1998
16. Vogt AM, Ackermann C, Noe T, Jensen D, Kübler W: Simultaneous detection of high energy phosphates and metabolites of glycolysis and the Krebs cycle by HPLC. Biochem Biophys Res Commun 248: 527–532, 1998
17. Elsässer A, Schlepper M, Zimmermann R, Muller KD, Strasser R, Klovekorn WP, Schaper J: The extracellular matrix in hibernating myocardium — a significant factor causing structural defects and cardiac dysfunction. Mol Cell Biochem 186: 147–158, 1998
18. Lieberthal W, Menza SA, Levine JS: Graded ATP depletion can cause necrosis or apoptosis of cultured mouse proximal tubular cells. Am J Physiol 274: F315–F327, 1998
19. Nagy PV, Feher T, Morga S, Matko J: Apoptosis of murine thymocytes induced by extracellular ATP is dose- and cytosolic pH-dependent. Immunol Lett 72: 23–30, 2000
20. Chow S, Kass G, Orrenius S: Purines and their roles in apoptosis. Neuropharmacology 36: 1149–1156, 1997
21. Uren NG, Camici PG: Hibernation and myocardial ischemia: clinical detection by positron emission tomography. Cardiovasc Drugs Ther 6: 273–279, 1992
22. Malthora R, Brosius FI: Glucose uptake and glycolysis reduce hypoxia-induced apoptosis in cultured neonatal rat cardiac myocytes. J Biol Chem 274: 12567–12575, 1999
23. Jennings R, Reimer K: The cell biology of acute myocardial ischemia. Annu Rev Med 42: 225–246, 1991
24. Bing R: Coronary circulation and cardiac metabolism. In: A. Fishman, D. Richards (ed). Circulation of the Blood: Men and Ideals. Oxford University Press, New York, 1964

25. Bing R: The metabolism of the heart. In: Harvey Lecture Series. Academic Press, New York, 1954–55, pp 27–70
26. Vary T, Reibel D, Neely J: Control of energy metabolism in heart muscle. Ann Rev Physiol 43: 419–430, 1981
27. Neely J, Morgan H: Relationship between carbohydrate and lipid metabolism and the energy balance of heart muscle. Annu Rev Physiol 36: 413–459, 1974
28. Camici P, Dutka D: Repetitive stunning, hibernation, and heart failure: contribution of PET to establish a link. Am J Physiol 280: H929–H936, 2001

Molecular and Cellular Biochemistry **242:** 109–114, 2003.

Optimal conditions for heart cell cryopreservation for transplantation

Hiroki Yokomuro, Donald A.G. Mickle, Richard D. Weisel and Ren-Ke Li

Divisions of Cardiovascular Surgery, Toronto General Hospital, Toronto General Research Institute, University Health Network; Department of Surgery, University of Toronto, Toronto, Ontario, Canada

Abstract

Cultured myocyte transplantation into an infarcted myocardium has been shown to improve contractile function. Cryopreservation of cultured muscle cells or heart tissue will be important for the technology to be practical. This study, using fetal cardiomyocytes, evaluated the optimal conditions for muscle cell cryopreservation. Study 1: Fetal rat cardiomyocytes were isolated and cultured. The freshly isolated and passage 1, 2, 3 and 4 cells were cryopreserved in a solution containing 70% IMDM, 20% FBS and 10% DMSO and stored in –196°C for 1, 2, 4, 8, 12 and 24 weeks. The cells were thawed and cultured. Cell number and contractility were evaluated at 0, 2, 4, 6, 8 and 10 days of culture. Study 2: Rat myocardium was cryopreserved in sizes of 0.2, 2 and 6 mm^3 for 1 week. The tissue was thawed and cells were isolated. Cell growth and contractility were evaluated. (1) Cardiomyocytes grew and contracted after cryopreservation. Storage time did not affect cell survival rate, beating cell numbers and beating rates. Increasing cell passage prior to cryopreservation decreased the percentage of beating cells. (2) Cells isolated from cryopreserved tissue grew *in vitro* and contracted normally. Cell yield decreased with increased cryopreserved tissue size. Fetal rat cardiomyocytes survived and functioned after *in vitro* cryopreservation. Viable cells can be isolated from cryopreserved myocardium and cultured. Cryopreservation of small pieces of myocardium is preferred for maximal cell yields. (Mol Cell Biochem **242**: 109–114, 2003)

Key words: cell transplantation, cryopreservation, cardiomyocyte

Introduction

Various cell types have been transplanted experimentally into the myocardium, myocardial scar tissue and subcutaneous connective tissue of animals [1, 2]. We have demonstrated that primary cultures of fetal cardiomyocytes transplanted into the connective tissue of adult rats survived and formed contractile cardiac tissue [3]. The transplanted tissue enlarged and contracted regularly and spontaneously. In experimental models of left ventricular cryonecrosis and coronary artery occlusion, cultured fetal heart cells [4], adult heart cells [5], adult smooth muscle cells [6] and skeletal muscle cells [7, 8] transplanted into the myocardial scar have been shown to limit scar expansion and improve regional and global heart function. Recently, cultured skeletal muscle cells have been implanted into the dysfunctional myocardium of patients with congestive heart failure [9]. Preliminary clinical data and experimental results have indicated that myoblast transplantation could be used clinically to improve the contractile function of myocardial infarct patients.

Cryopreservation of muscle cells could make this technique more suitable for the practical clinical application of cell transplantation. The timing of the transplantation will depend on the clinical status of the patient, which may vary. The rate of growth of the cells may also vary which could alter the optimal time for transplantation. These problems would be overcome by the cryopreservation of the patient's tissue and cultured cells so that they are available when the patient is ready to receive the transplantation. In addition, repeat transplantations may be required. We have demonstrated that

Address for offprints: R.K. Li, Toronto General Hospital, CCRW 1-815, 200 Elizabeth Street, Toronto, Ontario, Canada, M5G 2C4 (E-mail: RenKeLi@uhnres.utoronto.ca)

fetal rat cardiomyocytes can be cryopreserved [10]. However, the optimal conditions for heart cell cryopreservation have not been determined. In this study we evaluated the effect of cryopreservation time and cell passage on *in vitro* survival and function of fetal cardiomyocytes. Myocardial tissues were cryopreserved to investigate the possibility of these tissues being used for future cell isolation and culture.

Materials and methods

Animals

All procedures performed on animals were approved by the Animal Care Committee of the University Health Network. The experimental animals used were Sprague-Dawley rats (Charles River Canada Inc., Quebec, Canada). Rat hearts, obtained from 18–20 day gestational rat, were used for cell isolation. All experiments were performed in accordance with the 'Guide for the Care and Use of Laboratory Animals' published by the National Academy Press (1996).

Cell isolation

Fetal rat cardiomyocytes were isolated using an enzymatic digestion method [3, 4]. Fetal rat hearts were washed with phosphate buffered saline (PBS) and the ventricular tissue was minced and incubated in a PBS solution containing trypsin (0.2%), collagenase (0.1%), and glucose (0.02%) for 30 min at 37°C. The cardiac cells were then isolated by repetitive pipetting of the digested myocardial tissue. The cells in the supernatant were transferred into a tube containing 15 ml of cell culture medium (Iscove's Modified Dulbecco's Medium (IMDM) with 10% fetal bovine serum (FBS), penicillin G (100 U/ml), streptomycin (100 μg/ml)). The tube was centrifuged at 600 g for 5 min at room temperature. The cardiomyocyte purity after the first 24 h of culturing was 94 ± 3.5% (N = 8). The cell pellet was resuspended in the cell culture medium (freshly isolated cells) and cultured (passage 1) at 37°C in 5% CO_2/95% air. When the cells reached 90% confluence, the cells were trypsinized (0.05% trypsin in PBS) and separated into 3 dishes (passage 2). This process was repeated for passages 3 and 4. Cells at each passage were detached from culture dish and suspended in freezing medium contained 70% IMDM, 20% FBS, and 10% dimethylsulfoxide (DMSO).

Cell cryopreservation and thawing

The cell suspension in the freezing medium was transferred to cryovials (NALGENE), which were placed into a freezing chamber (NALGENE) and stored at –80°C overnight. The cells were cooled at a rate of –1°C/min until they reached –80°C and the cryovials were then stored in liquid nitrogen.

The samples were rapidly thawed in a water bath at 37°C. The freezing medium was diluted in four steps (with 10, 5, 2.5, and 0% DMSO) by centrifuging and removing the supernatant at 5 min intervals to extract the intracellular DMSO. The cell pellet was resuspended in culture medium with 2% FBS for the survival study or with 10% FBS for the growth study. The cells were cultured at 37°C in 5% CO_2/95% air.

Cell counting

Each culture dish was washed 3 times with 5 mL phosphate-buffered saline to remove culture medium and all non-viable cells. The viable cells that remained attached to the culture dish were detached using a 0.05% trypsin solution and counted using an electronic cell counter.

Survival rate

The cryopreserved cells, after thawing, were cultured for 24 h in medium with 2% FBS. The cells were then detached from the culture dish using trypsin solution and the cell number was determined using an electronic cell counter.

Cell growth curve

The cells, cryopreserved for different time periods, were thawed and cultured for 0, 1, 2, 4, 6, 8 and 10 days. At each time point, the cells were detached from the culture dish with the trypsin solution and counted.

Percentage of beating cells and beating rates

After cryopreservation the cells were cultured for 24 h and the number of beating cells and beating rates were determined with a microscope within the 5 randomly selected fields. The number of beating and non-beating cells was counted by group-blinded observer.

Contractility of fetal rat cardiomyocytes were evaluated at pre-cryopreservation. The cells were cultured for 24 h. Isoproterenol or propanolol was added into culture medium at final concentrations of 0, 0.25, 0.5 and 1.0 μg/ml for isoproterenol and 0, 1.2, 2.4 and 4.8 μg/ml for propanolol. The beating rates were determined with a microscope. Five microscopic fields were randomly selected and cell beats per minute was recorded.

Tissue cryopreservation and cell isolation

Fresh myocardium was collected from fetal rat hearts and cut into 0.2, 2 and 6 mm^3 sizes (N = 6 for each group). The tissue was suspended in freezing medium, cryopreserved for 1 week and thawed as described in the 'Cell cryopreservation and thawing' section. The 2 and 6 mm^3-cryopreserved tissues of were minced to 1 mm^3 in size. The tissue from the three groups was digested using trypsin and collagenase as described in the 'Cell isolation' section. The isolated cells were cultured for 24 h and the cell number, percentage of beating cells and cell beating rate were recorded as described above. The cell numbers were also determined on 4, 8 and 14 days of culture.

Data analysis

All data were expressed as the mean ± S.D. Analysis of variance was used to test the difference between groups.

Results

Cell survival after cryopreservation

The survival rates of passage 1 cardiomyocytes cryopreserved for 1, 2, 4, 8, 12 and 24 weeks were similar among the groups (N = 8 for each group) (Fig. 1A). Cell passaging did not affect the cell survival rate (Fig. 2A). Cell growth rates over 10 days following cryopreservation showed that storage time and cell passage had no significantly affect on cell growth (Figs 1B and 2B).

Contractility of cryopreserved fetal cardiomyocytes in vitro

The fetal rat cardiomyocytes contracted regularly and spontaneously for 24 h following isolation from the myocardium. The cells had some characteristics of myocardial tissue (Fig. 3). Cell contractility increased with the addition of isoproterenol to the culture and decreased with the addition of propanolol (Fig. 3). The fetal cardiomyocytes cultured after cryopreservation contracted regularly and spontaneously as was observed prior to cryopreservation. The percentage of beating cells was not affected by the duration of cryopreservation (N = 8 for each group) (Fig. 1C). However, the percentage of beating cells decreased significantly with increasing passage number prior to cryopreservation (Fig. 2C). The number of beating cells decreased significantly ($p < 0.05$) with increasing culture time, as the cells began to form a tissue *in vitro*. The contractile rate of cryopreserved cells was not affected by duration of cryopreservation or passaging prior to preservation.

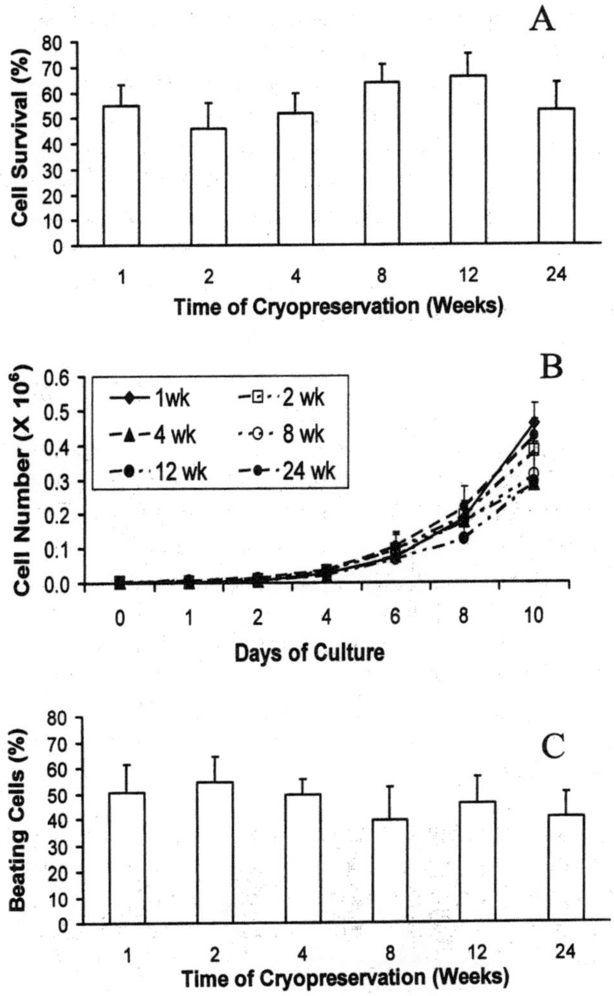

Fig. 1. Effects of cryopreservation time (from 1–24 weeks) on cellular survival (A), cellular growth (B) and contractility (C) after thawing were evaluated on cultured fetal rat cardiomyocytes (passage 1). The data was expressed as mean ± 1 S.D. (N = 8 for each group). The time of cryopreservation did not affect the cardiomyocyte survival, growth and contractility after thawing.

Cells isolated from cryopreserved myocardium

Cardiomyocytes were successfully isolated and cultured from fetal myocardium, which was cryopreserved in liquid nitrogen for 1 week in sizes of 0.2, 2.0 and 6.0 mm^3 (N = 6 for each group) (Fig. 4). Fewer cells were isolated for culture with increasing tissue size ($p < 0.05$ for all groups). By 24 h after isolation from the myocardial pieces, the cells proliferated in culture, contracted regularly and spontaneously.

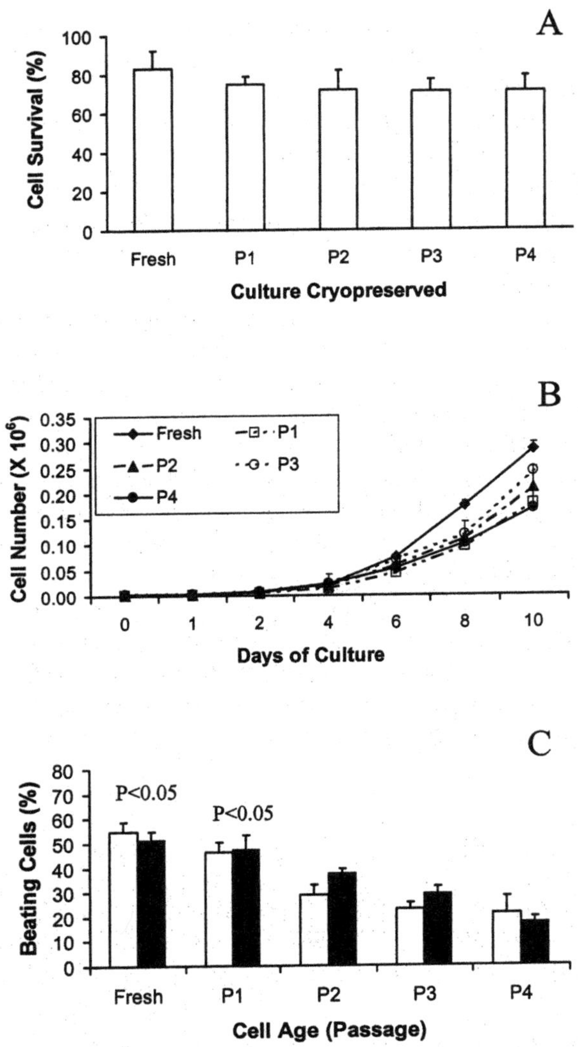

Fig. 2. Effect of cell age (freshly isolated (fresh) and passage 1 (P1), 2 (P2), 3 (P3) and 4 (P4)) on cellular survival (A), cellular growth (B) and contractility (C) after 4 weeks of cryopreservation were evaluated using cultured fetal rat cardiomyocytes. The data was expressed as mean ± 1 S.D. (N = 8 for each group). Only cell age affected the percentage of beating cells. At each passage there was no difference in the number of beating cells between the cryopreserved (open bar) and the non-cryopreserved (closed bar) cells. However, there is a decrease ($p < 0.05, < 0.05$) in the numbers of beating cells in the fresh and passage 1 cells compared to the passage 2, 3 and 4 cells.

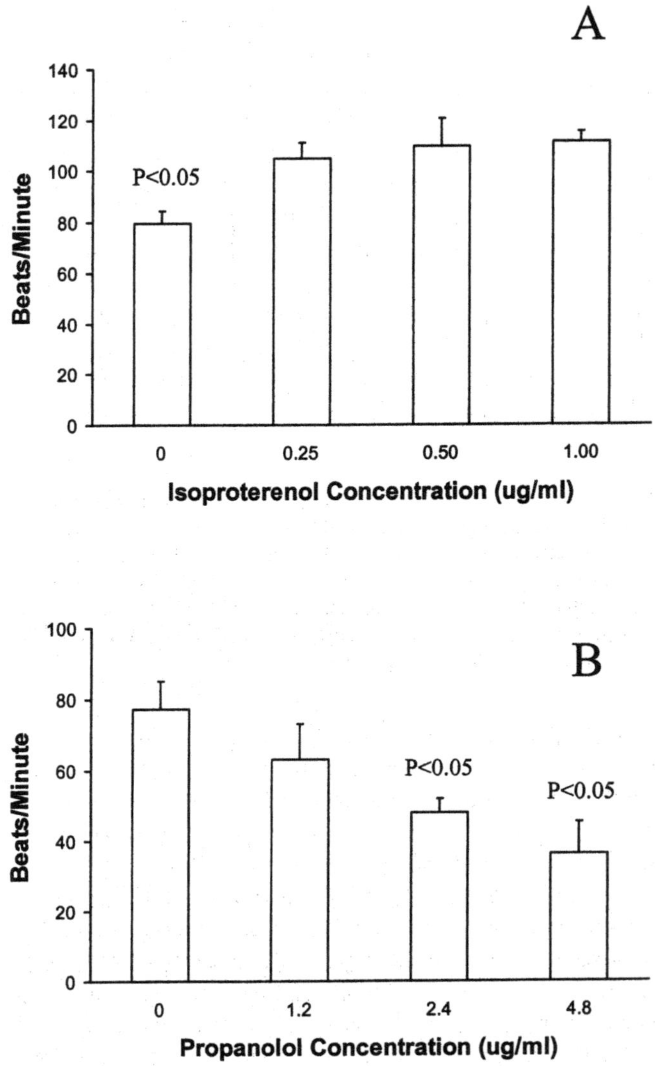

Fig. 3. Effect of isoproterenol and propanolol on the contractile rates of cultured fetal rat cardiomyocytes (passage 1) (N = 6 for each group). Increasing concentrations of isoproterenol and propanolol increased and decreased cardiomyocyte contractile rates, respectively.

Discussion

Cell transplantation has been shown to be a potential new technology to treat patients with heart disease. Several cell types, such as heart cells, skeletal muscle cells, smooth muscle cells and bone marrow stromal cells, have been transplanted into the left ventricular scar of adult rats, rabbits and pigs [4–8, 11]. The transplanted cells survived in the transplanted area and formed a tissue. The transplant tissue prevented scar tissue thinning and ventricular dilatation, which improved heart function. The data has lead to the autotransplantation of cultured skeletal muscle cells into a myocardial infarct scar of a patient having aortocoronary bypass surgery [9]. Although the benefits of the transplant and the bypass surgery could not be definitively separated, the clinical results were supportive of the cell transplant improving myocardial function.

Cell growth *in vitro* is a continuous process of cell division until confluence is reached in the culture dish. Cells are detached from the culture surface with trypsin, the cell suspension is aliquoted into a number of culture dishes (passaging) and the cells are again grown to confluence. With each cell division, cultured primary cells lose characteristics of the original tissue. Eventually the dedifferentiated cultured cells lose their capacity to differentiate into their original pheno-

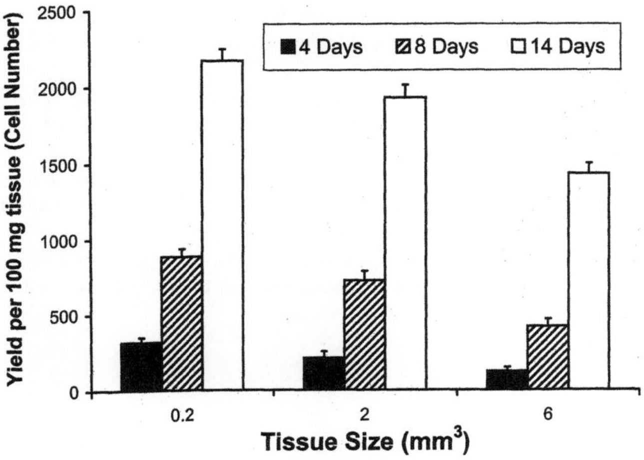

Fig. 4. Fetal myocardium, 0.2, 2 and 6 mm^3 in size, was cryopreserved for 1 week. The cells were isolated and cultured from the cryopreserved tissue. The cells were counted on day 4, 8 and 14 of culture. The data are expressed as the mean ± 1 S.D. (N = 6 for each group). The cell numbers on each day of the 0.2 mm^3 group were significantly greater ($p < 0.05$ for all time points) than those of the 2 mm^3 group, which were greater ($p < 0.05$ for all time points) than those of the 6 mm^3 group.

type. The current study showed that the percentage of beating cells decreased with subculturing indicating that the dedifferentiated cardiomyocytes lose their ability to re-form their contractile apparati. However, cryopreservation did not affect cell dedifferentiation.

Clinically, cell transplantation should not be performed until the inflammatory response of an acute ischemic episode has subsided [12]. Therefore, transplantation may need to be delayed until the patient has adequately recovered from an ischemic event. The patient may not be clinically ready to be transplanted when the culture is ready. Retransplantation may be required at a later date, for example, patients with a repeat infarction may require additional cell transplantation. Cultures from different patients may grow at different rates. In all these situations, the culture may need to be stored until transplantation becomes possible and cryopreservation allows a delay in surgery until the patient is able to receive the transplant.

Liquid nitrogen can be used to store tissue or cells [13]. After the cells were cryopreserved in the liquid nitrogen, the time of storage did not affect the cell survival and beating rates. There were no differences in cell survival rate, percentages of beating cells and rates of cell contractility among the cell populations cryopreserved for 1 week and for 24 weeks. After cryopreservation the fetal cardiomyocytes contracted regularly and spontaneously, as did the non-cryopreserved cells. The small number of passages did not affect cell survival and function. The cryopreserved cells grew similarly to the freshly prepared non-cryopreserved cells.

Small pieces of myocardium can be cryopreserved and cells, normal in appearance and function, can be isolated and cultured. Tissue size affected the viable cell yield. The larger the piece of myocardial tissue cryopreserved, the lower the cell yield. The optimal tissue size to be cryopreserved for maximal cell yields, should be less than 2.0 mm^3 in size. Our data indicate that a myocardial biopsy can be cryopreserved and transported frozen in liquid nitrogen to a cell isolation and culture facility. Some of the tissue could be banked for future use and the remainder processed for transplantation.

The cryopreservation technique needs to be optimized. Although some of the cryopreserved cells survived, the survival rate was 60–80%. Since the storage time and culturing prior to cryopreservation did not affect the survival rate, changes to the cryopreservation solution and technique need to be made. The optimal DMSO concentration has not been determined. The use of glycerol and high concentrations of fetal bovine serum to improve cell survival rate after cryopreservation should also be investigated. Also the optimal conditions for cryopreserving rat fetal cells might not be the same as the conditions best suited for adult cells.

In summary, fetal cardiomyocytes can be cryopreserved and thawed with a 60–80% yield. The time of cryopreservation had no effect on cell survival, growth and beating rates in culture. Cryopreservation of small pieces of myocardium will permit optimal use of the patient's tissue.

Acknowledgements

Dr. Ren-Ke Li is a Career Investigator of the Heart and Stroke Foundation of Canada. This research was supported by RKL's research grants from the Heart and Stroke Foundation of Ontario (NA4603 and NA4829) and Canadian Institute of Health Research (MOP14795).

References

1. Li RK, Yau TM, Sakai T, Mickle DA, Weisel RD: Cell therapy to repair broken hearts. Can J Cardiol 14: 735–744, 1998
2. El Oakley RM, Ooi OC, Bongso A, Yacoub MH: Myocyte transplantation for myocardial repair: A few good cells can mend a broken heart. Ann Thorac Surg 71: 1724–1733, 2001
3. Li RK, Mickle DA, Weisel RD, Zhang J, Mohabeer MK: *In vivo* survival and function of transplanted rat cardiomyocytes. Circ Res 78: 283–288, 1996
4. Li RK, Jia ZQ, Weisel RD, Mickle DA, Zhang J, Mohabeer MK, Rao V, Ivanov J: Cardiomyocyte transplantation improves heart function. Ann Thorac Surg 62: 654–660, 1996
5. Li RK, Weisel RD, Mickle DA, Jia ZQ, Kim EJ, Sakai T, Tomita S, Schwartz L, Iwanochko M, Husain M, Cusimano RJ, Burns RJ, Yau TM: Autologous porcine heart cell transplantation improved heart function after a myocardial infarction. J Thorac Cardiovasc Surg 119: 62–68, 2000
6. Li RK, Jia ZQ, Weisel RD, Merante F, Mickle DA. Smooth muscle cell transplantation into myocardial scar tissue improves heart function. J Mol Cell Cardiol 31: 513–522, 1999

7. Taylor DA, Atkins BZ, Hungspreugs P, Jones TR, Reedy MC, Hutcheson KA, Glower DD, Kraus WE: Regenerating functional myocardium: Improved performance after skeletal myoblast transplantation. Nat Med 4: 929–933, 1998
8. Scorsin M, Hagege A, Vilquin JT, Fiszman M, Marotte F, Samuel JL, Rappaport L, Schwartz K, Menasche P: Comparison of the effects of fetal cardiomyocyte and skeletal myoblast transplantation on postinfarction left ventricular function. J Thorac Cardiovasc Surg 119: 1169–1175, 2000
9. Menasche P, Hagege AA, Scorsin M, Pouzet B, Desnos M, Duboc D, Schwartz K, Vilquin JT, Marolleau JP: Myoblast transplantation for heart failure. Lancet 357: 279–280, 2001
10. Yokomuro H, Li RK, Mickle DA, Weisel RD, Verma S, Yau TM: Transplantation of cryopreserved cardiomyocytes. J Thorac Cardiovasc Surg 121: 98–107, 2001
11. Tomita S, Li RK, Weisel RD, Mickle DA, Kim EJ, Sakai T, Jia ZQ: Autologous transplantation of bone marrow cells improves damaged heart function. Circulation 100: II247–II256, 1999
12. Li RK, Mickle DA, Weisel RD, Rao V, Jia ZQ: Optimal time for cardiomyocyte transplantation to maximize myocardial function after left ventricular injury. Ann Thorac Surg 72: 1957–1963, 2001
13. Keung YK, Lau S, Elkayam U, Chen SC, Douer D: Cardiac arrhythmia after infusion of cryopreserved stem cells. Bone Marrow Transplant 14: 363–367, 1994

Molecular and Cellular Biochemistry **242**: 115–120, 2003.

Different pathways for sodium entry in cardiac cells during ischemia and early reperfusion

Delphine Baetz,[1] Monique Bernard,[2] Caroline Pinet,[1] Sophie Tamareille,[1] Soad Chattou,[1] Houda El Banani,[1] Alain Coulombe[1] and Danielle Feuvray[1]

[1]*CNRS UMR 8078, Hôpital Marie Lannelongue, Université Paris-Sud XI;* [2]*CRMBM, Faculté de Médecine Timone, Marseille, France*

Abstract

A number of data are consistent with the hypothesis that increases in intracellular Na^+ concentration (Na^+_i) during ischemia and early reperfusion lead to calcium overload and exacerbation of myocardial injury. However, the mechanisms underlying the increased Na^+_i remain unclear. ^{23}Na nuclear magnetic resonance spectroscopy was used to monitor Na^+_i in isolated rat hearts perfused with a high concentration of fatty acid as can occur under some pathological conditions. Whole-cell patch-clamp experiments were also performed on isolated cardiomyocytes in order to investigate the role of voltage-gated sodium channels. Na^+_i increased to substantially above control levels during no-flow ischemia. The results show that a pharmacological reduction of Na^+_i increase by cariporide (1 µmol/L, a Na^+/H^+ exchange blocker) is not the only protection against ischemia-reperfusion damage, but that such protection may also be brought about by metabolic action aimed at reducing fatty acid utilization by myocardial cells. This action was obtained in the presence of etomoxir (0.1 µmol/L), an inhibitor of carnitine palmitoyltransferase-1 (the key enzyme involved in fatty acid uptake by the mitochondria) which also decreases long-chain acyl carnitine accumulation. The possibility of Na^+ channels participating in Na^+_i increase as a consequence of alterations in cardiac metabolism was studied in isolated cells. Sustained I_{Na} was stimulated by the presence of lysophosphatidylcholine (LPC, 10 µmol/L) whose accumulation during ischemia is, at least partly, dependent on increased long-chain acyl carnitine. Current activation was particularly significant in the range of potentials between –60 and –20 mV. This may have particular relevance in ischemia. The quantity of charge carried by sustained I_{Na} was reduced by 24% in the presence of 1 µmol/L cariporide. Therefore, limitation of long-chain fatty acid metabolism, and consequent limitation of ischemia-induced long-chain acyl carnitine accumulation, may contribute to reducing intracellular Na^+ increase during ischemia-reperfusion. (Mol Cell Biochem **242**: 115–120, 2003)

Key words: intracellular sodium, ischemia-reperfusion, Na-NMR spectroscopy, isolated rat hearts, cardiac cells, Na^+ current, fatty acid metabolism

Introduction

Many pathophysiological processes in cardiac ischemia and reperfusion are associated with cellular ionic disturbances. In particular, a number of results are consistent with the hypothesis that increases in intracellular free Na^+ concentration (Na^+_i) during ischemia may ultimately lead to calcium overload and exacerbation of myocardial injury [1–5]. However, the mechanisms underlying the increased Na^+_i early in ischemia and during reperfusion remain unclear. The most plausible underlying mechanisms are failure to extrude Na^+ via the Na^+/K^+-ATPase and Na^+ influx via Na^+/H^+ exchange, Na^+-HCO_3^- cotransport and the voltage-gated Na^+ channel. Na^+ efflux via the Na^+/K^+-ATPase is attenuated during ischemia [6], therefore activation of other mechanisms leads to increased intracellular Na^+. As to the Na^+/H^+ exchanger,

Address for offprints: D. Feuvray, Laboratoire de Physiologie Cellulaire, Bât 443, Université Paris-Sud XI, 91405 Orsay cedex, France (E-mail: danielle.feuvray@ibaic.u-psud.fr)

controversy exists over its possible role in mediating Na^+ influx during ischemia, particularly because both pH_i and pH_o decline during ischemia so that extracellular acidosis may inhibit Na^+/H^+ exchange [7]. However, several studies have identified this transporter as mediating Na^+ influx in ischemic hearts [8, 9], whereas a recent work suggested that it may be inhibited during total ischemia [10]. In addition to the Na^+/H^+ exchanger, activation of a Na^+-HCO_3^- cotransporter (NBC) might also contribute to Na^+ influx [11, 12]. In this respect, recent data indicated that administration of anti-NBC antibody to ischemic-reperfused rat hearts has beneficial effects on systolic contractility and also on diastolic function in both zero- and low-flow global ischemia [13]. Whether or not this cotransporter is involved in Na^+_i changes during ischemia has, so far, not been investigated. On the other hand, several reports favour the idea that a sustained component of the Na^+ current plays a role in ischemic Na^+ rise [14–17]. However, the relative contribution of these mechanisms is debatable and may largely depend on the ischemic conditions. In this respect, the role of myocardial metabolic modulation during ischemia and reperfusion may be of importance. Indeed, long-chain fatty acids, which are the major substrates for the heart, can be detrimental to the ischemic and reperfused heart. This is because they require more oxygen than does glucose to produce an equivalent amount of ATP [18], and also because metabolic products accumulate in ischemic myocardium [19, 20]. Thus, shifting the energy substrate preference away from fatty acid oxidation and toward glucose oxidation may be important during an episode of severe ischemia, and upon reperfusion [21].

We examined Na^+_i in isolated rat hearts perfused with high concentrations of fatty acids as can occur under some pathological conditions [22]. Results were compared with those obtained in either the absence of exogenous fatty acid or in the presence of etomoxir, an inhibitor of carnitine palmitoyltransferase-1, the key enzyme involved in fatty acid uptake by the mitochondria. We compared results obtained under these conditions with those obtained in the presence of cariporide, a known Na^+/H^+ exchange blocker [23]. We also performed additional experiments on isolated cardiomyocytes in order to investigate the role of amphipathic metabolites which accumulate in the sarcolemma during ischemia [24] on the development of a sustained component of the sodium current.

Material and methods

Animals

All procedures were in accordance with the regulations laid down by the Ministère de l'Agriculture, France, for the care and use of laboratory animals.

Isolated heart perfusions and Na-NMR spectroscopy

Male Wistar rats weighing 300–350 g were anesthetized with thiopentone (5 mg/100 g body wt, i.p.) and the hearts were quickly removed and then Langendorff perfused. Hearts were initially perfused at constant pressure (60 mmHg; 10-min stabilization period) with a modified Krebs-Henseleit bicarbonate perfusate, consisting of (mM): NaCl 118, KCl 5.9, $MgSO_4$ 1.2, $CaCl_2$ 1.25, $NaHCO_3$ 25, glucose 11, gassed with 95% O_2–5% CO_2 (pH 7.4, 37°C). The range of values for coronary flow was 15.6 ± 0.6 mL/min. Thereafter, and in order to exclude any effect of the different experimental conditions on coronary flow rate, constant-flow perfusion was initiated by setting the flow rate to the level attained at the end of the stabilization period and continued (control perfusion) until inducing ischemia. All perfusion solutions were filtered (0.8 µm; Millipore) prior to use. Perfusion pressure was monitored with a Statham 23 ID pressure transducer connected by polyethylene tubing to a side arm on the aortic cannula. The experimental protocol consisted of 20 min of control perfusion, followed by 30 min of no-flow global ischemia, and 30 min of reperfusion at a constant pre-ischemic coronary flow rate. Two different substrate combinations were used in separate heart groups: 5.5 mM glucose and 1.2 mM palmitate prebound to 3% bovine serum albumin (BSA), or 5.5 mM glucose in the presence of 3% BSA. In a second group, 10^{-6} M cariporide (a gift from Hoechst AG, Germany) was added to the solution containing fatty acid, 10 min before inducing ischemia and was present throughout. In a third group, 10^{-7} M etomoxir (2-[6-(4-chlorophenoxy)-hexyl]-oxirane-2-carboxylate ; Projekt-Entwicklung GmbH, Germany) was added to the solution containing fatty acid 10 min before inducing ischemia and was present throughout.

^{23}Na-NMR spectra were acquired in a Bruker-Nicolet WP-200 spectrometer at 52.9 MHz. Each ^{23}Na-NMR spectrum was (1-min time-resolved) obtained by accumulation of 256 consecutive free induction decays using a pulse of 90° and a 0.2 sec recycle time using 1 K data points and 2.5 KHz spectral width. The shift agent $TmDOTP^{5-}$ was used to distinguish the Na^+_o and Na^+_i resonance of the spectra, and Na^+ signals were quantified using the Na^+ resonance of a standard solution in a glass capillary. Spectra were processed for the quantitative analysis of the intracellular component (Na^+_i) and the reference in two steps. First, removal of the overlapping spectral extracellular component was carried out by the HLSVD method [25], and then the resulting reference and Na^+_i peaks were quantified with a time domain fitting routine (AMARES) [26].

Isolated myocytes and current recordings

For cardiac myocytes experiments, rats (Wistar, 250–300 g body wt) or guinea-pigs (SPF Hartley, 250–300 g) were anesthetized and the heart was rapidly removed. Myocytes were isolated, as previously described [17, 27]. Rod-shaped myocytes were used on the day of isolation.

Whole-cell, voltage-clamp configuration of the patch-clamp technique was used to record sodium currents as we have described previously [17]. All experiments were carried out at room temperature (20–22°C). I_{Na} was elicited by 2 sec (or 3 sec when necessary) membrane depolarizations from a holding potential of –100 mV applied with a stimulation frequency of 0.2 Hz. The amplitude of fast inactivating Na^+ current (I_{Na}) was measured as peak inward current with reference to the current measured at the end of test pulse. I_{NaL} is referred to as the Na^+ current activated by lysophosphatidylcholine (L-α-lysophosphatidylcholine, palmitoyl, Sigma) (LPC). At a given test pulse, I_{NaL} was characterized in the following way. The quantity of charge transferred during a 2 sec pulse was evaluated by calculating the time integral of the difference current obtained by subtracting the control trace from that recorded in the presence of LPC. The quantity of charge was then normalized to unit of cell membrane capacitance, Cm (pC/pF). Cm was obtained at the beginning of each experiment by measuring the resistance in series and the time constant of the experimental decay of the capacitive transient elicited by a 10 mV depolarizing pulse.

Data analysis

Data are presented as means ± S.E. Statistical significance was estimated by Student's *t*-test or variance analysis followed by Student-Newman-Keuls test to locate differences between groups. Significance was set at $p < 0.05$.

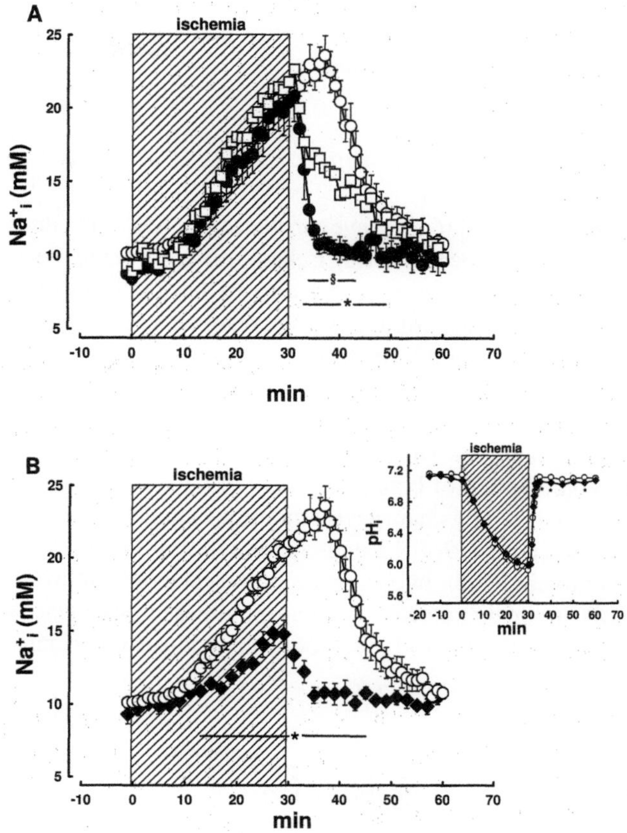

Fig. 1. Changes in Na^+_i during 30 min of no-flow ischemia and reperfusion. (A) for hearts receiving (–O–; n = 8) or not receiving (–□–; n = 6) fatty acid and for fatty-acid-perfused hearts that received 10^{-7} M etomoxir (–●–; n = 8). (B) for fatty-acid-perfused hearts that received (–◆–; n = 7) or not (–O–; n = 8) 10^{-6} M cariporide. Changes in pH_i in hearts receiving or not cariporide are also shown (see inset). *–●– and –◆– vs. –O–; §–□– vs. –O–. Reproduced in part from El Banani *et al.* [4] with permission.

Results and discussion

Changes in Na^+_i in ischemic and reperfused hearts

Figure 1A shows the mean values of Na^+_i obtained from the ^{23}Na-NMR spectra. In both groups of hearts receiving or not receiving fatty acid, Na^+_i similarly increased to substantially above control levels during no-flow ischemia (204 and 230% increase, respectively, at end ischemia; $p = 0.56$). A similar increase of Na^+_i (229%) was also observed in fatty acid-perfused hearts that received etomoxir, a carnitine palmitoyltransferase-1 inhibitor, which also decreases long-chain acyl carnitines [28]. Upon reperfusion, Na^+_i continued to increase in untreated hearts receiving palmitate during the first few minutes of reperfusion, whereas Na^+_i decreased immediately upon reperfusion in the etomoxir-treated hearts as well as in the group that did not receive palmitate. We may therefore hypothesize that some ischemia-induced alterations in fatty acid metabolism may be responsible for the additional Na^+_i increase observed at reperfusion, and which could be attenuated or even suppressed (present results) by etomoxir.

In an attempt to elucidate which of the Na^+_i regulating mechanisms might contribute to ischemia-induced Na^+_i increase, we performed experiments in which fat-perfused hearts received cariporide, a known Na^+/H^+ exchange blocker [23]. In the presence of cariporide, Na^+_i increase during ischemia was markedly attenuated (Fig. 1B). It is worth noting that both inhibitors, etomoxir which did not change Na^+_i during ischemia but significantly reduced Na^+_i upon reperfusion, or cariporide which noticeably reduced Na^+_i during ischemia and consequently upon reperfusion, induced a significant improvement in ventricular function recovery during reperfusion (Fig. 2). Nevertheless, Na^+/H^+ exchange blockade was not accompanied by any significant change in pH_i decrease during the ischemic period, and a significant effect of cariporide on pH_i was only observed after 5 min of reperfusion (see inset in Fig. 1B). In the absence of a pH_i effect the role of Na^+/H^+ exchange activity during no-flow ischemia may be questioned.

Na^+ current in cardiac cells

Recent studies [16, 29] have concluded that Na^+ influx through Na^+ channels occurs during ischemia after myocytes become inexcitable. We therefore investigated the possibility of these channels participating in Na^+_i increase as a consequence of alterations in cardiac metabolism. Lysophosphatidylcholine (LPC) accumulates during ischemia. LPC accumulation reflects either increased production via PLA_2-catalyzed phosphatidylcholine hydrolysis or decreased LPC catabolism, or a combination of both processes [24]. We have previously shown that long-chain acyl carnitine accumulates in ischemic myocardium [19, 20]. Long-chain acyl carnitine acts to competitively inhibit two enzymes of the catabolic pathways of LPC (namely cytosolic lysophospholipase transacylase and lysophospholipase) [24]. Moreover, acidosis contributes to inhibition of microsomal lysophospholipase [24]. Thus, during ischemia long-chain acyl carnitine and acidosis contribute to LPC accumulation. Figure 3 shows the effect of LPC on Na^+ current in both rat ventricular and guinea-pig atrial myocytes. In this example of Na^+ current recording in a rat ventricular myocyte (Fig. 3A), the action of LPC on the activation of tetrodotoxin (TTX, 30 μmol/L)-sensitive I_{NaL} was assessed in response to a depolarization to – 20 mV from a holding potential of –100 mV. The LPC-induced current showed very little inactivation over the pulse duration. Figure 3C shows the relationship between the quantity of charge carried by LPC-induced I_{NaL}, normalized for each cell by taking into account the membrane capacitance, and test potentials. Current activation was particularly important in the range of test potentials between –60 and –20 mV, which may have particular relevance for the ischemic situation. Similar effects of LPC were obtained from guinea-pig atrial myocytes (Fig. 3B). In this respect, it is worth mentioning that local generation of thrombin secondary to intracoronary arterial thrombi [30] may stimulate G-protein coupled receptors on cardiomyocytes, resulting in increased production of LPC

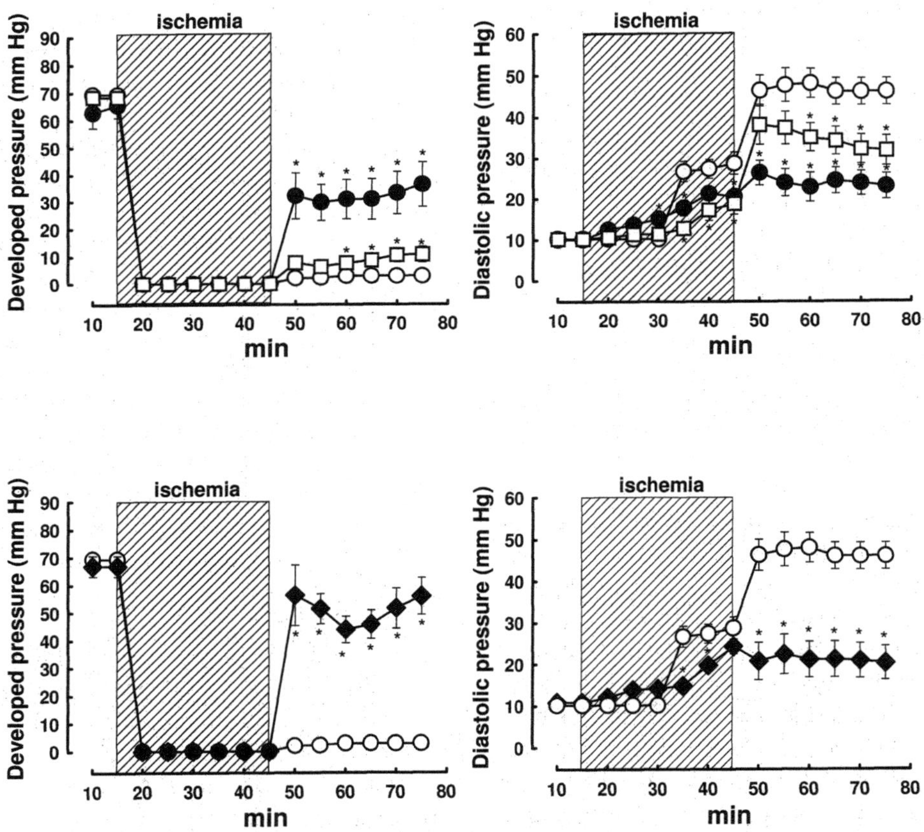

Fig. 2. Cardioprotective effects of etomoxir (10^{-7} M, upper panel) and cariporide (10^{-6} M; lower panel) during and after no-flow ischemia. Changes in diastolic pressure and left ventricular developed pressure in hearts receiving (–○–; n = 8) or not receiving (–□–; n = 6) fatty acid and for fatty-acid-perfused hearts that received etomoxir (–●–; n = 8) or cariporide (–◆–; n = 7). Both etomoxir and cariporide were added to the perfusion solution 10 min before ischemia and maintained throughout the reperfusion. *vs. –○–.

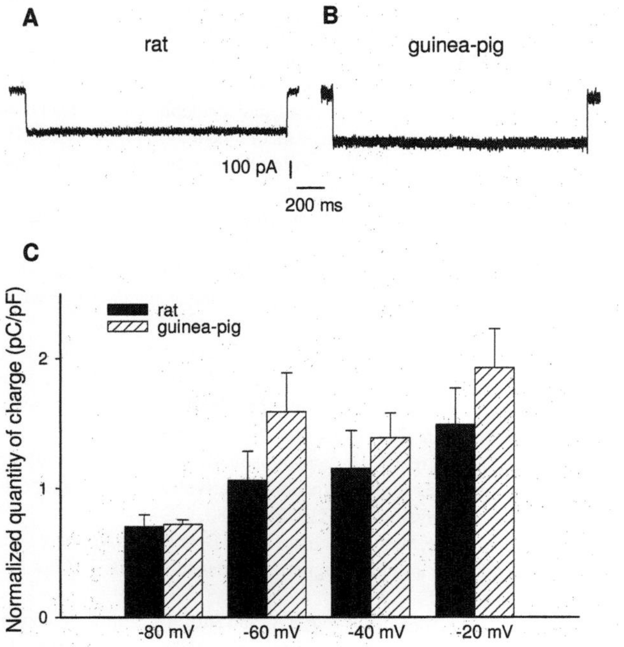

Fig. 3. Typical LPC-induced I_{NaL} recordings in rat ventricular (A) and guinea-pig atrial (B) myocytes. Currents were recorded after 2 min exposure to 10 μmol/L LPC extracellularly. (C) The quantity of charge carried by I_{NaL} over a 2 sec pulse normalized as pC/pF is plotted against test potentials for rat ventricular (black bar; n = 7) and guinea-pig atrial (hatched bar; n = 4) myocytes.

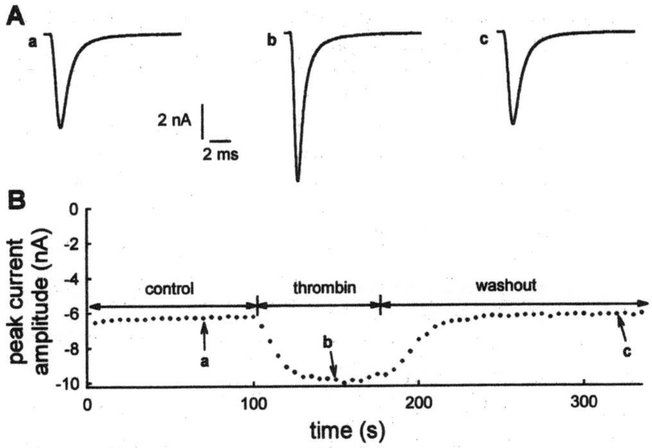

Fig. 4. (A) Typical recording of the effect of thrombin (32 IU/mL) on peak I_{Na} amplitude in a rat ventricular myocyte. Current was induced by a test-pulse to –40 mV from a holding potential of –100 mV. (B) Time-course of the effect of thrombin. The average calculated current density was 137.6 ± 4.8 pA/pF vs. 92.8 ± 2.9 pA/pF under control conditions. $p < 0.001$; n = 6 myocytes in each group.

[24]. Furthermore, preliminary data indicate that thrombin itself is a potent activator of voltage-gated Na^+ channels ([31] and Fig. 4).

Several different studies have suggested that various Na^+/H^+ exchange inhibitors may, in addition, affect the slowly inactivating sodium current [14, 32]. We therefore investigated the effects of cariporide and EIPA (an amiloride derivative), each at the concentration of 1 μmol/L, on LPC-induced I_{NaL}. The quantity of charge carried by I_{NaL} over a 2 sec pulse (in response to a depolarizing step to –20 mV) was reduced by 34 and 24% in the presence of EIPA and cariporide, respectively. Thus, the observation of lower ischemic Na^+_i levels in the cariporide-treated hearts (see Fig. 1B) may have partly resulted from some degree of cariporide-induced inhibition of I_{NaL}.

Therefore, not only a pharmacological reduction of Na^+_i increase may afford protection against ischemia-reperfusion damage, but such protection, through a reduction of the ionic imbalance and associated deleterious effects on cardiac function, may also be brought about by a metabolic action aimed at reducing fatty acid utilization by myocardial cells. In this context, it has also been shown that long-chain acyl carnitine which, alongside LPC, accumulates in the sarcolemma during ischemia, markedly increases the slowly inactivating sodium current [33]. In addition and interestingly, the Na^+/K^+-ATPase is reversibly inhibited by palmitoyl carnitine [34]. Therefore, limitation of long-chain fatty acid metabolism, and consequent limitation of ischemia-induced long-chain carnitine accumulation, may limit Na^+_i increase during ischemia-reperfusion via slowly inactivating Na^+ channels and/or a relative increase in Na^+/K^+-pump function.

Although this remains to be further investigated, the sodium-bicarbonate cotransporter (NBC) could also contribute to the control of Na^+_i. From our earlier results [12] it was clear that an HCO_3^--dependent mechanism, most likely the NBC, contributed to pH_i recovery after ischemia in perfused rat hearts. It was recently shown that administration of anti-NBC antibody to ischemic-reperfused rat hearts produces marked protective effects since the time of onset and amplitude of ischemia- and reperfusion-induced contracture are significantly reduced compared to the non-treated heart [13]. Furthermore, our preliminary results [35] suggest that stimulation of NBC activity may be particularly important in the presence of a cardioactive hormone such as angiotensin II.

Acknowledgement

Part of this work was supported by grants from the BQR (Université Paris-Sud XI) and Fondation de France. D. Baetz was supported by a GRRC grant and S. Tamareille by a grant from Ministère de la Recherche.

References

1. Tani M, Neely JR: Role of intracellular Na^+ in Ca^{2+} overload and depressed recovery of ventricular function of reperfused ischemic rat hearts. Circ Res 65: 1045–1056, 1989

2. Pierce GN, Czubryt MP: The contribution of ionic imbalance to ischemia/reperfusion-induced injury. J Mol Cell Cardiol 27: 53–63, 1995
3. Murphy E, Cross H, Steenbergen C: Sodium regulation during ischemia vs. reperfusion and its role in injury. Circ Res 84: 1469–1470, 1999
4. El Banani H, Bernard M, Baetz D, Cabanes E, Cozzone P, Lucien A, Feuvray D: Changes in intracellular sodium and pH during ischaemia-reperfusion are attenuated by trimetazidine. Comparison between low- and zero-flow ischaemia. Cardiovasc Res 47: 688–696, 2000
5. Cross HR, Lu L, Steenbergen C, Philipson KD, Murphy E: Overexpression of the cardiac Na^+/Ca^{2+} exchanger increases susceptibility to ischemia/reperfusion injury in male, but not female, transgenic mice. Circ Res 83: 1215–1223, 1998
6. Cross HR, Radda GK, Clarke K: The role of Na^+/K^+ ATPase activity during low flow ischemia in preventing myocardial injury: A ^{31}P, ^{23}Na and ^{87}Rb NMR spectroscopic study. Magn Res Med 34: 673–685, 1995
7. Vaughan-Jones RD, Wu ML: Extracellular H^+ inactivation of Na^+-H^+ exchange in the sheep cardiac Purkinje fibre. J Physiol (Lond) 428: 441–466, 1990
8. Pike MM, Luo CS, Clark MD, Kirk KA, Kitakaze M, Madden MC, Cragoe EJ, Pohost GM: NMR measurements of Na^+ and cellular energy in ischemic rat heart: Role of Na^+-H^+ exchange. Am J Physiol 265: H2017–H2026, 1993
9. Imahashi K, Hashimoto K, Yamagushi H, Nishimura T, Kusuoka H: Alteration of intracellular Na^+ during ischemia in diabetic rat hearts: the role of reduced activity in Na^+/H^+ exchange against stunning. J Mol Cell Cardiol 30: 509–517, 1998
10. Park CO, Xiao XH, Allen DG: Changes in intracellular Na^+ and pH in rat heart during ischemia: Role of Na^+/H^+ exchanger. Am J Physiol 276: H1581–H1590, 1999
11. Le Prigent K, Lagadic-Gossmann D, Mongodin E, Feuvray D: HCO_3^--dependent alkalinizing transporter in adult rat ventricular myocytes: Characterization and modulation. Am J Physiol 273: H2596–H2603, 1997
12. Khandoudi N, Bernard M, Cozzone P, Feuvray D: Mechanisms of intracellular pH regulation during postischemic reperfusion of diabetic rat hearts. Diabetes 44: 196–202, 1995
13. Khandoudi N, Albadine J, Robert P, Krief S, Berrebi-Bertrand I, Martin X, Bevensee MO, Boron WF, Bril A: Inhibition of the cardiac electrogenic sodium bicarbonate cotransporter reduces ischemic injury. Cardiovasc Res 52: 387–396, 2001
14. Haigney MCP, Lakatta EG, Stern MD, Silverman HS: Sodium channel blockade reduces hypoxic sodium loading and sodium-dependent calcium loading. Circulation 90: 391–399, 1994
15. Le Grand B, Vie B, Talmant JM, Coraboeuf E, John GW: Alleviation of contractile dysfunction in ischemic hearts by slowly inactivating Na^+ current blockers. Am J Physiol 269: H533–H540, 1995
16. Xiao XH, Allen DG: Role of Na^+/H^+ exchanger during ischemia and preconditioning in the isolated rat heart. Circ Res 85: 723–730, 1999
17. Chattou S, Coulombe A, Diacono J, Le Grand B, John G, Feuvray D: Slowly inactivating component of sodium current in ventricular myocytes is decreased by diabetes and partially inhibited by known Na^+-H^+ exchange blockers. J Mol Cell Cardiol 32: 1181–1192, 2000
18. Liu B, El Alaoui-Talibi Z, Clanachan AS, Schulz R, Lopaschuk GD: Uncoupling of contractile function from mitochondrial TCA cycle activity and O_2 consumption during reperfusion of ischemic rat hearts. Am J Physiol 270: H72–H80, 1996
19. Feuvray D: Structural, functional, and metabolic correlates in ischemic hearts: Effects of substrates. Am J Physiol 240: H391–H398, 1981
20. Neely JR, Feuvray D: Metabolic products and myocardial ischemia. Am J Pathol 102: 282–291, 1981
21. Lewandowski ED: Metabolic mechanisms associated with antianginal therapy. Circ Res 86: 487–489, 2000
22. Lopaschuk GD, Collins-Nakai R, Olley PM *et al.*: Plasma fatty acid levels in infants and adults after myocardial ischemia. Am Heart J 128: 61–67, 1994
23. Scholz W, Albus U, Counillon L, Gögelein H, Lang HJ, Linz W, Weichert A, Schölkens BA: Protective effects of HOE 642, a selective sodium-hydrogen exchange subtype 1 inhibitor, on cardiac ischaemia and reperfusion. Cardiovasc Res 29: 260–268, 1995
24. McHowat J, Creer MH: Lysophosphatidylcholine accumulation in cardiomyocytes requires thrombin activation of Ca^{2+}-independent PLA_2. Am J Physiol 272: H1972–H1980, 1997
25. Pijnappel WWF, Van Der Boogart A, De Beer R, Van Ormondt D: SVD-based quantification of magnetic resonance signals. J Magn Res 97: 122–134, 1992
26. Vanhamme L, Van Den Boogart A, Van Huffel S: Improved method for accurate and efficient quantification of MRS data with use of prior knowledge. J Magn Reson 129: 35–43, 1997
27. Antoine S, Lefèvre T, Coraboeuf E, Nottin R, Coulombe A: B-type Ca^{2+} channels activated by chlorpromazine and free radicals in membrane of human atrial myocytes. J Mol Cell Cardiol 30: 2623–2636, 1998
28. Kantor PF, Dyck JR, Lopaschuk GD: Fatty acid oxidation in the reperfused ischemic heart. Am J Med Sci 318: 3–14, 1999
29. Van Emous JG, Nederhoff MGJ, Ruigrok TJC, Van Echteld CJA: The role of the Na^+ channel in the accumulation of intracellular Na^+ during myocardial ischemia: Consequences for post-ischemic recovery. J Mol Cell Cardiol 29: 85–96, 1997
30. Davies MJ, Thomas A: Thrombosis and acute coronary artery lesions in sudden cardiac ischemic death. N Engl J Med 310: 1137–1140, 1984
31. Pinet C, Le Grand B, John GW, Coulombe A: Thrombin is a potent activator of voltage-gated sodium channels in human atrial myocytes. J Mol Cell Cardiol 33: (abstr) A94, 2001
32. Le Grand B, Marty A, Talmant JM, John GW: HOE 694 affords protection versus veratrine contractures in rat atria by Na^+ channel blockade. Fundam Clin Pharmacol 10: 467–473, 1996
33. Wu J, Corr PB: Palmitoyl carnitine modifies sodium currents and induces transient inward current in ventricular myocytes. Am J Physiol 266: H1034–H1046, 1994
34. Tanaka M, Gilbert J, Pappano AJ: Inhibition of sodium pump by L-palmitoylcarnitine in single guinea-pig ventricular myocytes. J Mol Cell Cardiol 24: 711–719, 1992
35. Baetz D, Feuvray D: Modulation of Na^+-HCO_3^- cotransport (NBC) activity by intracellular signals in adult rat ventricular myocytes. J Mol Cell Cardiol 33: (abstr) A7, 2001

Molecular and Cellular Biochemistry **242**: 121–127, 2003.

Comparison of connexin expression patterns in the developing mouse heart and human foetal heart

Steven R. Coppen,[1] Riyaz A. Kaba,[1] Deborah Halliday,[1] Emmanuel Dupont,[1] Jeremy N. Skepper,[2] Suzy Elneil[3] and Nicholas J. Severs[1]

[1]*Cardiac Medicine, National Heart and Lung Institute, Faculty of Medicine, Imperial College, Royal Brompton Hospital, London;* [2]*University of Cambridge, Cambridge;* [3]*Addenbrooke's Hospital, Cambridge, UK*

Abstract

Heart muscle cells are electrically coupled by gap junctions, clusters of low-resistance transmembrane channels composed of connexins (Cx). The expression of the three major connexins (Cx43, Cx40 and Cx45) present in cardiac myocytes is known to be developmentally regulated but it is not clear how the patterns in the human heart compare with those found in the mouse. This issue is of importance given the wide use of transgenic mice to investigate gene function with the aim of extrapolating the results to human. In the present study we applied immunoconfocal microscopy to investigate the spatial distribution of the three connexins in the developing mouse heart and foetal human heart. Although Cx45 labelling was present at low levels throughout the developing mouse heart and human foetal (9-week) heart, it was most prominent in the conduction tissues. In the developing mouse heart, Cx40 was widely expressed at embryonic day 12.5 (E12.5) but at E17.5 expression was restricted to the conduction tissues and atria. In the 9-week human foetal heart, the Cx40 labelling pattern was similar to the E15 mouse heart, being far more abundant in conduction tissues (bundle branches to Purkinje fibres) and atria than in the ventricular muscle. Cx43 labelling became more apparent in the ventricular myocardium as development of the mouse heart progressed but was virtually undetectable in the central conduction system. In the human foetal heart Cx43 was virtually undetectable in the atria but was the predominant connexin in the ventricles. We conclude that, at least in some key aspects, the pattern of connexin expression in the developing mouse heart parallels that found in the human embryonic heart. (Mol Cell Biochem **242**: 121–127, 2002)

Key words: gap junctions; connexins, intercellular communication, embryonic development

Introduction

Gap junctions are clusters of transmembrane channels which directly link the cytoplasmic compartments of neighbouring cells. These intercellular channels allow transfer of small signalling molecules and ions, a function necessary for tissue homeostasis, development and differentiation [1–3]. In the heart, a key role of gap junctions is electrical coupling of individual myocytes, thereby mediating the orderly spread of the impulse throughout the heart [4]. Each gap junction channel is composed of a pair of connexons (hemichannels), one contributed by each of the connected cells. Each connexon consists of six protein sub-units corresponding to connexin molecules. The connexins are a multigene family of conserved proteins. To date, 19 different connexin genes have been identified in the mouse and 20 in human [5]. Most tissues, including heart, express more than one connexin isotype [6] and, from *in vitro* experiments, the connexin composition of gap junctions has been shown to affect such channel properties as unitary conductance, voltage sensitivity and molecular permeability [7]. Hence, the variations in the pattern of expression of the major connexins expressed in the heart, connexins 43, 40 and 45, are hypothesised to be key determinants of electrophysiological specialisation in different cardiac tissues [6].

In the mouse heart, Cx45 is the first detectable connexin in the developing cardiac conduction system [8] and continues to be expressed in all parts of the conduction system in

Address for offprints: N.J. Severs, Cardiac Medicine, National Heart and Lung Institute, Faculty of Medicine, Imperial College, Royal Brompton Hospital, Sydney Street, London, SW3 6NP, UK (E-mail: n.severs@ic.ac.uk)

the adult [9, 10]. Cx40 is also present in parts of the conduction system where it is confined to a central core within the connexin45-expressing tissues [9]. In addition, Cx40 is expressed in the atrium in most mammalian species where it is co-expressed with Cx43. Cx43 is the most abundant connexin in the working ventricular myocardium but is barely detectable in the central parts of the mouse cardiac conduction system (AV node, His bundle and bundle branches). The spatiotemporal patterns of expression of the three cardiac connexins have been shown to be regulated during heart development [11–14]. With the ever-wider application of transgenic mice to analyse the role of connexins [15] and the desire to extrapolate the results to human, it has become important to determine how far the developmental patterns of expression in the human resemble those in the mouse. The current study set out to compare the spatiotemporal patterns of connexin expression in the normal developing mouse heart and human foetal heart as a first step towards determining the extent to which such extrapolations are valid.

Materials and methods

Tissue collection

Foetal hearts (embryonic day 12.5, 15.5 and 17.5) were obtained from timed pregnant Balb-c mice. Three human foetal hearts at 9 weeks gestation were obtained from surgical termination of pregnancy. The hearts were washed in phosphate-buffered saline (PBS), placed in plastic moulds orientated so that frontal sections could be made with respect to the major axis of the heart, and covered with tissue mounting medium (Cryo-M-Bed). Hearts from adult Sprague Dawley rats were treated the same way except that the heart chambers were injected with a 50% solution of mounting medium in PBS prior to placing in the moulds. The hearts were snap frozen by floating the moulds on isopentane cooled by liquid nitrogen. Animal procedures were conducted according to the Animals (Scientific Procedures) Act, 1986 (UK), and the collection of human material was approved by the local ethics committees.

Cells and culture conditions

Non-transfected HeLa cells and HeLa cells transfected to express Cx40 and Cx43 were provided by Professor Klaus Willecke (Institute für Genetik, Bonn, Germany) [16]. For Cx45, HeLa cells were transfected with pCMV_mCx45_ΔIRES_Puro based on the pIRESpuro2 plasmid (Clontech). Cells were cultured in DMEM (GIBCO-BRL) supplemented with 10% FBS, 100 U/ml penicillin and 100 mg/ml streptomycin. Transfected cells were cultured in the presence of 1 μg/ml puromycin (Sigma).

Western blot analysis

Whole cell homogenates were prepared by lysing washed cell monolayers in SB20 (20% SDS, 0.15 mol/l Tris, pH 6.8). Protein estimation was carried out (Bio-Rad DC protein assay) and 15 μg total protein loaded per lane of a 10% polyacrylamide gel. The proteins were then transferred to PVDF membrane overnight and the resulting blot blocked with 4% milk powder in TBS/0.1% Tween20 for 1 h prior to sequential 1 h incubations with primary and alkaline phosphatase-conjugated secondary antibodies diluted in 2% milk powder in TBS/0.1% Tween20. Enzyme activity was detected using NBT/BCIP substrate (Perbio).

Antibodies

The anti-Cx45 (Q14E(GP42)), anti-Cx40 (S15C(R83)) and anti-Cx43 (S10C(CK84)) antibodies, affinity purified against the peptide immunogen to which they were raised, have previously been demonstrated to be specific for their respective connexins by western blot and immunofluorescence of transfected cells, and to label morphologically-defined gap junctions at the electron-microscopic level [10, 17]. The anti-Cx45 antibody (Q14E(GP42)) was used at 1 in 100 dilution, the anti-Cx40 antibody (S15C(R83)) at 1 in 500 dilution and the anti-Cx43 antibody (S10C(CK84)) at 1 in 50 dilution for immunofluorescence. A monoclonal antibody to Cx45 (Q14E-(mab19-11-5)) has also been developed (BioGenes GmbH, Berlin, Germany) using the same peptide sequence as that used to generate our original polyclonal guinea pig anti-Cx45 [10]. This monoclonal antibody labels a single band at 45 kDa (approximately) on western blots of Cx45-transfected cells but not non-transfected cells or cells transfected to express Cx43 or Cx40 (Fig. 1A). The band obtained on the Cx45-transfected cells is inhibited by incubation of the monoclonal antibody with the peptide immunogen. Immunofluorescent labelling of the transfected cells gives punctate labelling, consistent with gap junctions at points of cell-cell contact (Fig. 1B), which is also inhibited by the peptide (Fig. 1C). Furthermore, labelling of rat AV bundle by the new monoclonal antibody gives an identical labelling pattern to that observed with the previously characterised polyclonal antibody, Q14E-(GP42) [10] (Figs 1D and 1E). The monoclonal anti-Cx45 was used as a 1 in 40 dilution of culture supernatant for labelling the human tissue. The direct fluorophore-conjugated (FITC, CY3 and CY5) secondary antibodies were purchased from Chemicon (used at 1 in 250) except the anti-rabbit-FITC which was purchased from DAKO (used at 1 in 25).

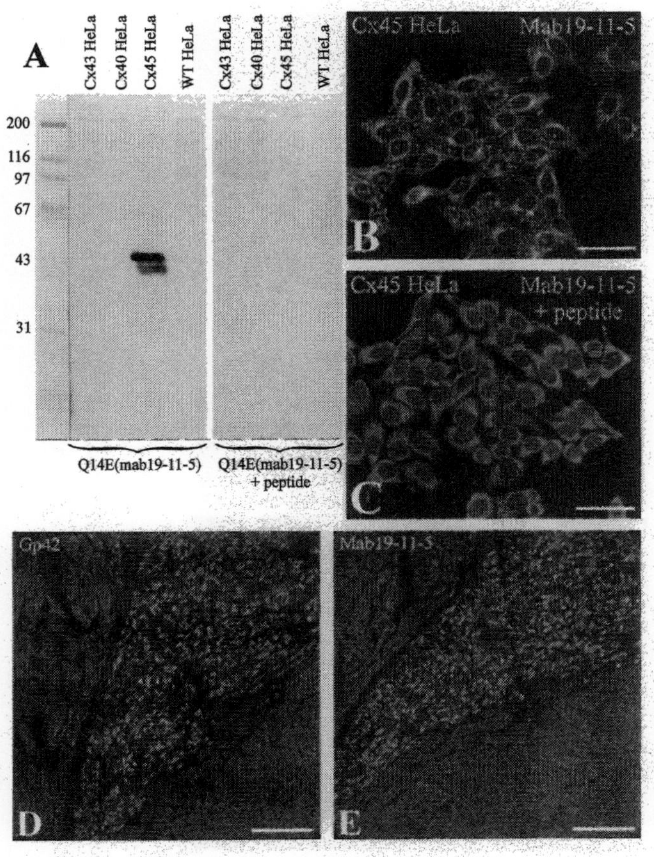

Fig. 1. Characterisation of a new monoclonal antibody to Cx45, Q14E-(mab19-11-5). (A) western blot of cells transfected with connexins 43, 40 and 45 and wild type cells probed with the new antibody. A reactive band is observed at 45 kDa only in the lane containing cells transfected with Cx45. This reaction is inhibited when the antibody is pre-incubated with the peptide immunogen. (B) The Cx45-transfected HeLa cell line immunolabelled with the new monoclonal antibody. Punctate label is observed between the cells. Again, this labelling is inhibited by pre-incubation with the peptide immunogen (C). Labelling of the rat AV bundle by the previously characterised anti-Cx45, GP42 (D), is reproduced by the new monoclonal antibody (E). These data confirm specificity of the new monoclonal antibody. Bar markers = 50 μm (B and C) and 100 μm (D and E).

Immunofluorescent labelling and confocal imaging

Frozen sections (12 μm) were cut from the hearts and mounted on poly-L-lysine-coated glass slides which were then stored at –80°C until use. The sections were fixed by immersing the slides in methanol at –20°C for 5 min and were then washed 3 times with PBS. Blocking was carried out for 1 h with 1% BSA in PBS before incubating with primary antibody (diluted in 1% BSA in PBS) for 2 h. After washing five times with PBS over 30 min, the sections were incubated with fluorophore-conjugated secondary antibodies (diluted in 1% BSA in PBS) for 1 h. The slides were washed a further five times with PBS over 30 min, mounted with Citifluor, and the coverslips sealed with clear nail varnish. Controls for the immunofluorescent labelling experiments were:(i) omission of the primary antibody, (ii) peptide inhibition in which antibodies were incubated with 50 μg/ml of the immunogen to which they were raised at room temperature for 30 min prior to application on tissue sections.

Immunolabelled sections were examined by confocal laser scanning microscopy using a Leica TCS SP system, equipped with argon and krypton lasers for the detection of FITC and CY3 fluorescence.

Results

Connexin45

At the earliest stage studied (E12.5), Cx45 was widespread throughout the mouse heart with more intense labelling occurring in the tissues of the outflow tract than in the surrounding myocardium (Fig. 2A). In the later stages of development (E15.5 and E17.5), Cx45 labelling declined in the working myocardium but remained prominent in the developing conduction tissues (Figs 2B and 2C). As in the mouse heart, labelling for Cx45 in the 9-week human foetal heart was widespread, again with greatest intensity in the developing conduction tissues (Fig. 2D).

Connexin40

In the E12.5 mouse heart, Cx40 labelling was most prominent in the trabeculated regions of the heart (Fig. 3A) with some label appearing in the more compact myocardium. By E15.5 (Fig. 3B) the Cx40 labelling was no longer present in the compact myocardium of the right ventricle and very little was detectable in that of the left ventricle. Cx40 label was most abundant in the trabeculated regions of the ventricles. By E17.5, Cx40 was mainly restricted to the ventricular conduction tissues (Fig. 3C). In the 9-week human foetal heart, Cx40 label was predominantly found in the trabeculated layer with low levels of signal detectable in the more compact regions of the ventricles.

Connexin43

Cx43 label was most apparent in the trabeculated tissues of the ventricles of the E12.5 mouse heart (Fig. 4A) with no detectable signal in the more compact ventricular tissues. This pattern of Cx43 labelling was still present in the E15.5 mouse heart (Fig. 4B) but by E17.5, Cx43 signal was present throughout the full thickness of the ventricular walls (Fig. 4C). In the 9-week human foetal heart, Cx43 label was most abundant in the trabeculated regions with less signal being apparent in the compact myocardium (Fig. 4D).

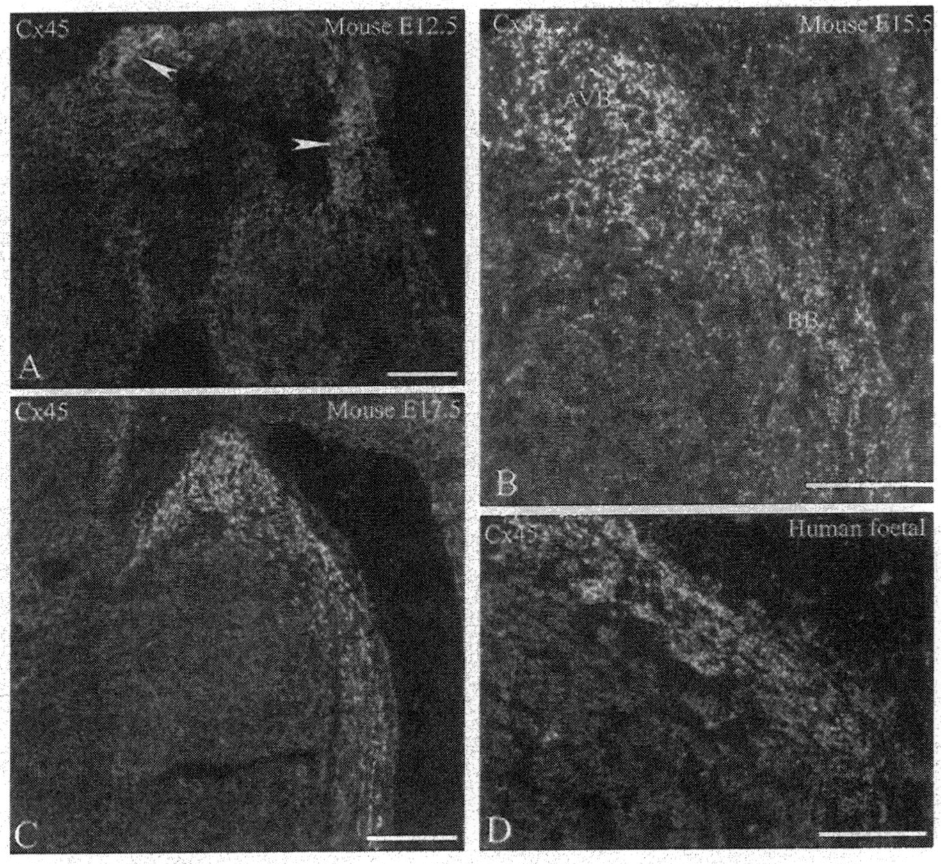

Fig. 2. Confocal micrographs showing the pattern of Cx45 in the developing mouse heart (A–C) and 9 week human foetal heart (D). In the E12.5 mouse heart (A), Cx45 is expressed throughout the heart but labelling is more intense in the tissues of the outflow tract (arrows). At E15.5 (B), Cx45 has declined in the majority of the heart but signal remains high in the developing conduction tissues such as the AV bundle (AVB) and bundle branches (BB). The Cx45 labelling in these structures becomes much more apparent in the E 17.5 mouse heart (C). Cx45 labelling of the 9-week human foetal heart (D) is again most prominent in the developing conduction tissues, as shown here for the bundle branch. Bar markers = 100 μm (A, C and D) and 50 μm (B).

Discussion

The results presented here show that distinct spatiotemporal patterns of connexin expression accompany mouse heart development and that with the data so far available [18] a similar progression of connexin expression patterns may be deduced during development in the human heart.

Connexin45 is the earliest detected connexin in the conduction tissues of the mouse heart, and has been shown to be present in the heart at E8.5 when the contractions are first initiated [14]. Although present throughout the heart in the early stages, there is a progressive down-regulation, with expression of Cx45 decreasing in the working ventricular myocardium but remaining high in the conduction tissues. In the 9-week human foetal heart, the pattern of Cx45 labelling appears to be similar to that found in the E15.5 mouse heart, with the signal being more intense in the conduction tissues than in the working myocardium. In the adult human heart, Cx45 signal is very low in working ventricular tissue [18]; thus, it is likely that the developmental pattern of Cx45 expression is similar in both human and mouse heart. Examination of other areas of the conduction tissues of the developing human heart will be required to determine if the pattern of Cx45 expression mirrors in every detail that in the mouse. Knocking-out Cx45 in the mouse results in death from heart failure at about E10, emphasising the importance of this connexin during development [19, 20].

At the earliest stages examined in the mouse heart, Cx40 label was most prominent in the trabeculated regions of the developing ventricles, with low levels of signal in the compact myocardium. As development progressed, the trabeculations became less apparent and Cx40 labelling became confined to the ventricular conduction tissues to give the expression pattern reported in adult hearts of many mammalian species [6]. Thus, it again appears that the pattern of Cx40 expression in the mouse and human are similar. In the normal adult heart there is no obvious expression of Cx40 in ventricular myocytes but in human end-stage ischaemic heart disease

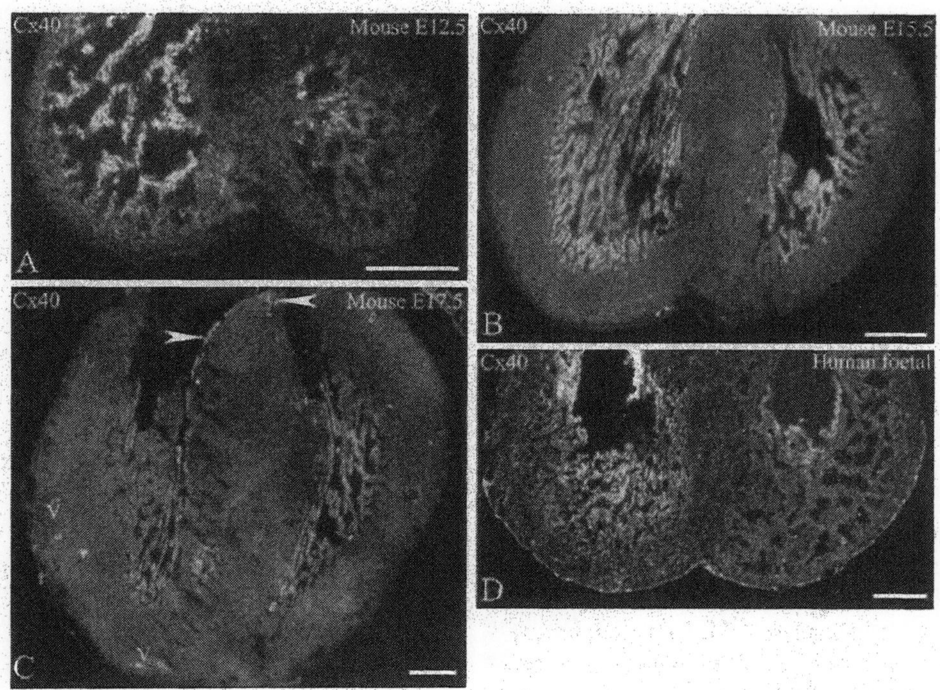

Fig. 3. Cx40 labelling in the developing mouse heart (A–C) and human foetal heart (D). At E12.5 (A), Cx40 is predominantly expressed in the trabeculated myocardium with some expression in the compact regions. This labelling of the compact myocardium decreases through development and is absent in the right ventricle by E15.5 (B) but remains in the left ventricle. By E17.5 (C), Cx40 labelling is restricted to the ventricular conduction tissues (arrows) and low levels of signal are present in the remaining trabeculated myocardium. Cx40 label also clearly seen in the endothelia of vessels (v) at this stage. The Cx40 labelling pattern in the 9-week human foetal heart (D) is similar to that observed in the E15.5 mouse heart. Bar markers = 250 μm.

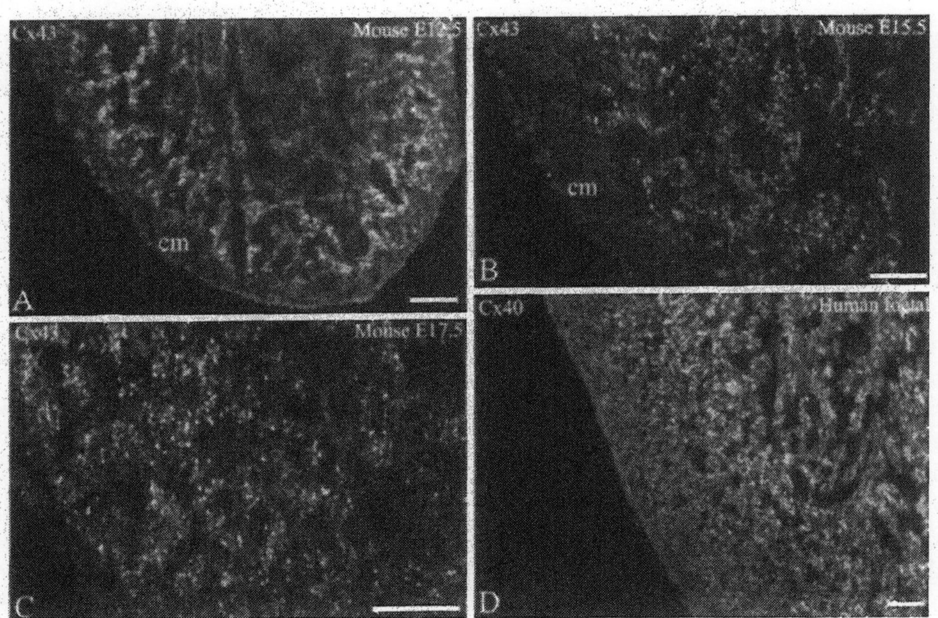

Fig. 4. Cx43 labelling in the developing mouse heart (A–C) and 9-week human foetal heart (D). At E12.5 (A), Cx43 is absent from the developing compact myocardium (cm) but is expressed in the trabeculated myocardium. The absence of Cx43 label in the compact tissue (cm) is more obvious at E15.5 (B). By E17.5 (C), Cx43 label is present throughout the ventricular tissue. In the 9-week human foetal heart (D), Cx43 label is much lower in the compact myocardium and absent from the very outer layers. Bar markers = 100 μm (A) and 50 μm (B– D).

expression of Cx40 increases in the sub-endocardial myocardium [21]. This increase could be likened to a partial reversion to a more foetal phenotype (where Cx40 is expressed in the compact myocardium of the ventricles) and could represent an adaptive response to the decrease in Cx43 which is characteristic of the diseased heart. The Cx40-knockout mouse has a much less severe phenotype than the Cx45 knockout, producing viable offspring which show impaired conduction properties through the bundle branches [22, 23]. From these results it could be inferred that while there is some redundancy in the system (i.e. Cx45 can maintain function in the Cx40 knockout), such redundancy does not apply equally (i.e. Cx45 knockout mice die at E10). Hence, the concept that different connexins have both shared and unique functions has grown, as elegantly illustrated by Plum *et al.* [24] who showed that whereas Cx32 and Cx40 were able to rescue function in the Cx43 knockout mice, the exact phenotypes did not match the wild type animals.

Cx43, though the major connexin in the ventricular myocardium of the adult mammalian heart, does not predominate during development. At E12.5, Cx43 is expressed in the trabeculated myocardium and not in the compact tissue. However, at this time there is expression of Cx45 and some Cx40 in the compact myocardium which will provide the necessary coupling between myocytes. By E17.5, Cx43 is uniformly expressed across the ventricular myocardium at which time Cx40 is undetectable and Cx45 is present in very low amounts. In the 9-week human foetal heart, Cx43 signal is absent from the outer layer of compact ventricular myocardium, probably equivalent to a stage between E15.5 and E17.5 in the mouse heart. Cx43 knockout mice die shortly after birth owing to obstruction of the pulmonary outflow tract [25] arising from defective migration of non-cardiac cells responsible for the correct formation of the outflow tract tissues [26, 27]. Therefore, most studies involving Cx43 knockout mice have been conducted on heterozygous ($Cx43^{+/-}$) mice which, in some instances, have been shown to have reduced conduction velocities [28, 29]. However, transgenic mice have now been bred with heart-specific knockout of Cx43 (Cx43 reduced by 86–95% at 4 weeks) and although the hearts were normal in terms of structure and function, the mice developed sudden cardiac death from spontaneous ventricular arrhythmias at about 2 months [30]. Such a result suggests that the low amounts of Cx45 found in the working myocardium may help maintain impulse propagation up to 2 months but not beyond, implying the existence of post-natal changes in connexin expression.

Overall, our findings suggest that major developmentally-regulated features of connexin expression patterns in the mouse broadly parallel those of the human heart. However, more subtle differences cannot be ruled out, especially in the specialised tissues such as the atrioventicular node where precise function may vary between species [31]. More detailed investigations of other developmental stages of the human heart are therefore required to confirm the full extent of the similarities and differences.

Acknowledgements

Work supported by the British Heart Foundation (grant No. PG 93136) and the European Commission (No. QLRT-1999-00516). Travel grants to SRC from The Biochemical Society and British Society for Cell Biology are also gratefully acknowledged.

References

1. Goodenough DA, Goliger JA, Paul DL: Connexins, connexons, and intercellular communication. Annu Rev Biochem 65: 475–502, 1996
2. Yamasaki H, Naus CCG: Role of connexin genes in growth control. Carcinogenesis 17: 1199–1213, 1996
3. Lo CW: The role of gap junction membrane channels in development. J Bioenerg Biomembr 28: 379–385, 1996
4. Gros DB, Jongsma HJ: Connexins in mammalian heart function. BioEssays 18: 719–730, 1996
5. Willecke K, Eiberger J, Degen J, Eckardt D, Romualdi A, Gueldenagel M, Deutsch U, Soehl G: Structural and functional diversity of connexin genes in the mouse and human genome. Biol Chem 5: 725–737, 2001
6. Severs NJ, Rothery S, Dupont E, Coppen SR, Yeh H-I, Ko Y-S, Matsushita T, Kaba R, Halliday D: Immunocytochemical analysis of connexin expression in the healthy and diseased cardiovascular system. Microsc Res Tech 52: 301–322, 2001
7. Veenstra RD: Size and selectivity of gap junction channels formed from different connexins. J Bioenerg Biomembr 28: 327–337, 1996
8. Coppen SR, Gourdie RG, Severs NJ: Connexin45 is the first connexin to be expressed in the central conduction system of the mouse heart. Exp Clin Cardiol 6: 17–23, 2001
9. Coppen SR, Severs NJ, Gourdie RG: Connexin45 (α6) expression delineates an extended conduction system in the embryonic and mature rodent heart. Dev Genet 24: 82–90, 1999
10. Coppen SR, Dupont E, Rothery S, Severs NJ: Connexin45 expression is preferentially associated with the ventricular conduction system in mouse and rat heart. Circ Res 82: 232–243, 1998
11. Delorme B, Dahl E, Jarry-Guichard T, Briand JP, Willecke K, Gros D, Theveniau-Ruissy M: Expression pattern of connexin gene products at the early developmental stages of the mouse cardiovascular system. Circ Res 81: 423–437, 1997
12. Van Kempen MJA, Vermeulen JLM, Moorman AFM, Gros D, Paul DL, Lamers WH: Developmental changes of connexin40 and connexin43 messenger RNA. Cardiovasc Res 32: 886–900, 1996
13. Gourdie RG, Litchenberg WH, Eisenberg LM: Gap junctions and heart development. In: W.C. De Mello (ed). Heart Cell Communication in Health and Disease. Kluwer, 1998, pp 19–44
14. Alcolea S, Theveniau-Ruissy M, Jarry-Guichard T, Marics I, Tzouanacou E, Chauvin JP, Briand JP, Moorman AF, Lamers WH, Gros DB: Downregulation of connexin 45 gene products during mouse heart development. Circ Res 84: 1365–1379, 1999
15. White TW, Paul DL: Genetic diseases and gene knockouts reveal diverse connexin functions. Annu Rev Physiol 61: 283–310, 1999
16. Elfgang C, Eckert R, Lichtenberg-Fraté H, Butterweck A, Traub O,

Klein RA, Hülser DF, Willecke K: Specific permeability and selective formation of gap junction channels in connexin-transfected HeLa cells. J Cell Biol 129: 805–817, 1995
17. Ko Y-S, Coppen SR, Dupont E, Rothery S, Severs NJ: Regional differentiation of desmin, connexin43 and connexin45 expression patterns in rat aortic smooth muscle. Arterioscler Thromb Vasc Biol 21: 355–364, 2001
18. Vozzi C, Dupont E, Coppen SR, Yeh H-I, Severs NJ: Chamber-related differences in connexin expression in the human heart. J Mol Cell Cardiol 31: 991–1003, 1999
19. Kumai M, Nishi K, Nakamura K, Takeda N, Suzuki M, Shibata Y: Loss of connexin45 causes a cushion defect in early cardiogenesis. Development 127: 3501–3512, 2000
20. Krüger O, Plum A, Kim J-S, Winterhager E, Maxeiner S, Hallas G, Kirchhoff S, Traub O, Lamers WH, Willecke K: Defective vascular development in connexin 45-deficient mice. Development 127: 4179–4193, 2000
21. Dupont E, Ko YS, Rothery S, Coppen SR, Baghai M, Haw M, Severs NJ: The gap-junctional protein, connexin40, is elevated in patients susceptible to post-operative atrial fibrillation. Circulation 103: 842–849, 2001
22. Kirchhoff S, Nelles E, Hagendorff A, Krüger O, Traub O, Willecke K: Reduced cardiac conduction velocity and predisposition to arrhythmias in connexin40-deficient mice. Curr Biol 8: 299–302, 1998
23. Simon AM, Goodenough DA, Paul DL: Mice lacking connexin40 have cardiac conduction abnormalities characteristic of atrioventricular block and bundle branch block. Curr Biol 8: 295–298, 1998
24. Plum A, Hallas G, Magin T, Dombrowski F, Hagendorff A, Schumacher B, Wolpert C, Kim J-S, Lamers WH, Evert M, Meda P, Traub O, Willecke K: Unique and shared functions of different connexins in mice. Curr Biol 10: 1083–1091, 2000
25. Reaume AG, De Sousa PA, Kulkarni S, Langille BL, Zhu D, Davies TC, Juenja SC, Kidder GM, Rossant J: Cardiac malformation in neonatal mice lacking connexin43. Science 267: 1831–1834, 1995
26. Huang GY, Wessels A, Smith BR, Linask KK, Ewart JL, Lo CW: Alteration in connexin 43 gap junction gene dosage impairs conotruncal heart development. Dev Biol 198: 32–44, 1998
27. Huang GY, Cooper ES, Waldo K, Kirby ML, Gilula NB, Lo CW: Gap junction-mediated cell-cell communication modulates mouse neural crest migration. J Cell Biol 143: 1725–1734, 1998
28. Guerrero PA, Schuessler RB, Davis LM, Beyer EC, Johnson CM, Yamada KA, Saffitz JE: Slow ventricular conduction in mice heterozygous for connexin43 null mutation. J Clin Invest 99: 1991–1998, 1997
29. Morley GE, Vaidya D, Samie FH, Lo C, Delmar M, Jalife J: Characterization of conduction in the ventricles of normal and heterozygous Cx43 knockout mice using optical mapping. J Cardiovasc Electrophysiol 10: 1361–1375, 1999
30. Gutstein DE, Morley GE, Tamaddon H, Vaidya D, Schneider MD, Chen J, Chien KR, Stuhlmann H, Fishman GI: Conduction slowing and sudden arrhythmic death in mice with cardiac-restricted inactivation of connexin43. Circ Res 88: 333–339, 2001
31. Coppen SR, Severs NJ: Diversity of connexin expression patterns in the atrioventricular node: Vestigial consequence or functional specialization. J Cardiovasc Electrophysiol 13: 625–626, 2002

Molecular and Cellular Biochemistry **242**: 129–134, 2003.

Ischemia-induced dephosphorylation of cardiomyocyte connexin-43 is reduced by okadaic acid and calyculin A but not fostriecin

Madhumathy Jeyaraman, Stéphane Tanguy, Robert R. Fandrich, Anton Lukas and Elissavet Kardami

Institute of Cardiovascular Sciences, St. Boniface General Hospital Research Centre, University of Manitoba, Winnipeg, Canada

Abstract

The gap junction protein connexin-43 (Cx43) exists mainly in the phosphorylated state in the normal heart, while ischemia induces dephosphorylation. Phosphatase(s) involved in cardiac Cx43 dephosphorylation have not as yet been identified. We examined the acute effects of ischemia on the dephosphorylation of the gap junction protein connexin-43 in isolated adult cardiomyocytes and isolated perfused hearts. In addition we tested the effectiveness of protein phosphatase 1 and 2A (PP1/2A) inhibitors in preventing Cx43 dephosphorylation. In both models, significant accumulation of the 41 kDa non-phosphorylated Cx43, accompanied by decreased relative levels of the 43–46 kDa phosphorylated Cx43, was observed at 30 min of ischemia. Okadaic acid decreased ischemia-induced Cx43 dephosphorylation; it also decreased the accumulation of non-phosphorylated Cx43 at the intercalated discs of myocytes in the whole heart. Calyculin A, but not fostriecin, also decreased ischemia-induced Cx43 dephosphorylation in isolated cardiomyocytes. It is concluded that isolated adult myocytes respond to ischemia in a manner similar to whole hearts and that ischemia-induced dephosphorylation of Cx43 is mediated, at least in part, by PP1-like phosphatase(s). (Mol Cell Biochem **242**: 129–134, 2003)

Key words: connexin43, ischemia, phosphorylation, protein phosphatases

Introduction

Gap junctions are intercellular channels mediating the passage of molecules less than 1000 Da between cells [1]. In the heart they ensure electrical coupling between myocytes, and thus coordinated activity of the muscle pump. In addition they are important for metabolic coupling between these cells [2]. Gap junctions are composed of connexins; connexin 43 is the major connexin of working cardiomyocytes [3]. Phosphorylation of connexins is believed to affect protein trafficking, assembly, turnover as well as conduction and metabolic coupling [4]. In the heart Cx43 exists mainly as a phosphoprotein [5], being predominantly phosphorylated on serine residues, although some tyrosine phosphorylation is also reported [6]. Various kinases have been implicated in the phosphorylation and thus regulation of cardiac connexin43, including protein kinase C (the -ε, and -α subtypes) [7, 8], as well as src kinase [9] and protein kinase A [10].

Ischemia is accompanied by loss of electrical coupling and development of arrhythmias, phenomena coinciding with Cx43 dephosphorylation at intercalated discs [11]. There is as yet no information as to the type(s) of phosphatases responsible for cardiac Cx43 dephosphorylation during ischemia.

There are several classes of serine/threonine phosphatases that might be involved in Cx43 dephosphorylation [12, 13], such as protein phosphatases type 1 (PP1), type 2A (PP2A), type 2B (PP2B) and type 2C (PP2C) [14]. Compounds such as okadaic acid or calyculin A are potent inhibitors of PP1 and PP2A type phosphatases [15], while fostriecin is a selective inhibitor of PP2A [16]. These inhibitors have been shown

Address for offprints: E. Kardami, Institute of Cardiovascular Sciences, St. Boniface General Hospital Research Centre R3008, 351 Tache Avenue, Winnipeg, Manitoba, Canada, R2H 2A6 (E-mail: ekardami@sbrc.umanitoba.ca)

to be cardioprotective under certain experimental conditions [15, 16]. We have thus examined the effect of PP1/2A inhibitors on cardiac Cx43 dephosphorylation in ischemia.

Materials and methods

Materials

Collagenase Type II was obtained from Worthington (Lakewood, NJ, USA). Phosphatase inhibitors, calyculin-A and fostriecin were purchased from CalBiochem (La Jolla, CA, USA), and okadaic acid was purchased from Sigma (St. Louis, MO, USA). Rabbit polyclonal antibodies, recognizing Cx43.res.367-382 (P.AB), have been described and characterized [7] and mouse monoclonal antibodies, #13-800, recognizing only the dephosphorylated Cx43 were characterized previously [17] and purchased from Zymed Labs (South San Francisco, CA, USA).

Animals

Hearts were obtained from Sprague Dawley rats (250 g) after decapitation. They were perfused immediately through the aorta at a constant pressure of 80 cm H_2O using oxygenated Krebs-Henseleit buffer (pH 7.4, 37.4°C) solution, as described [18]. Hearts were paced at 4 Hz (240 bpm) and function monitored via a water-filled balloon introduced in the left ventricle and connected to a pressure transducer. Hearts from 'control' or 'ischemic' groups were submitted to 20 min of stabilization under normoxic conditions. The 'ischemic' group was subjected to complete interruption of flow for 30 min under normothermic conditions. Animal use was in accordance to Canadian Council of Animal Care regulations.

Simulated ischemia of adult myocyte pellets

Calcium tolerant adult ventricular myocytes were isolated from Sprague Dawley (200–250 g) rats by collagenase perfusion as described previously [18]. Viability ranged between 80–90%.

The procedure followed for mineral overlay-induced simulated ischemia on myocyte pellets was as described [15]. The cell suspension from each preparation (one heart) was equally divided into 4 experimental groups (no ischemia/no inhibitor, no ischemia/+ inhibitor, ischemia/ no inhibitor, ischemia/ +inhibitor). Myocyte pellets were subjected to 30 min stabilization period, in which the cells were suspended in an oxygenated buffer in the absence of inhibitors. The cells were oxygenated continuously with 95% O_2 and 5% CO_2. After the stabilization period, myocytes were incubated for 10 min at 37°C in the absence or presence of phosphatase inhibitors. This was followed by mineral oil overlay to simulate ischemia for 15–30 min, as indicated [15].

Okadaic acid was dissolved in distilled water to a stock solution of 100 μM. Calyculin-A and fostriecin were dissolved in 100% ethanol to a stock solution of 100 μM.

Western blotting and immunofluorescence

These procedures were performed exactly as described previously [19]. Total lysates were obtained from minced ventricular tissue or myocyte pellets, homogenized and sonicated in SDS/PAGE sample buffer. Homogenates were centrifuged at 13,000 rpm for 20 min, to remove residual tissue. Lysates from myocytes or hearts were analyzed by Western blotting on either large format 9% or small format 10% polyacrylamide gels [20], at 20 μg protein/lane.

For immunofluorescence, sections (7 μM) were obtained by cryosectioning of frozen ischemic hearts (±okadaic acid). Sections were post-fixed for 15 min with freshly made 1% paraformaldheyde in cold phosphate buffered saline. For immunofluorescence, P.AB anti Cx43 and #13-800 were used at 1:2,000, 1:200 dilutions, respectively. For western blotting, they were used at 1:20,000, and 1:2,000, respectively. Sections were observed and photographed using a Nikon Diaphot microscope equipped with epifluorescence optics and appropriate filters. In western blots, antigen-antibody complexes were visualized by enhanced chemiluminescence (Pierce; Rockford, IL, USA).

Densitometry

Band densities of Cx43 in Western blots were determined by using the volume analysis tool of the software program, Molecular Analyst BioRad. The intensities of the phosphorylated and dephosphorylated bands were determined individually and combined.

Data analysis

Data are presented as means ± S.E.M. The Instat software program was used. Statistical analysis was performed using the student's *t*-test (paired). $P < 0.05$ was considered statistically significant.

Results

After equilibration with oxygenated buffer, perfused ex vivo hearts were subjected to global ischemia for 15 and 30 min.

Similarly, isolated adult myocytes were pelleted and subjected to simulated ischemia for 15 and 30 min. Total lysates from hearts and myocytes were analyzed by western blotting in large or small format acrylamide gels, and probed for Cx43.

As seen in Fig. 1, non-ischemic heart presented immunoreactive 43–45 kDa bands, characteristic of phosphorylated Cx43. Accumulation of the dephosphorylated 41 kDa Cx43 was clearly evident after 15 and 30 min of cardiac ischemia, in a time-dependent fashion. The more extensively phosphorylated 45 kDa Cx43 species became progressively converted to the less phosphorylated (43–44 kDa) or dephosphorylated Cx43 (Fig. 1A). Staining with #13-800 antibodies confirms the identification of the 41 kDa species as dephosphorylated Cx43.

Non-ischemic isolated myocytes exhibited primarily the phosphorylated Cx43 species, although some dephosphorylated Cx43 was also present (Fig. 1B (i)). Baseline levels of dephosphorylated Cx43 displayed some variability between different myocyte preparations (10–30% of total; data not shown). In all cases, however, ischemia increased relative levels of dephosphorylated-, and reduced those of phosphorylated- Cx43, in a time-dependent fashion (Fig. 1B). Multiple phosphorylated Cx43 bands were only evident when cardiac lysates were analysed in large format, 9% acrylamide gels. Cardiomyocyte phosphorylated Cx43 species, analyzed in small format, 10% acrylamide gels (Fig. 1B), tend to migrate as a broad band near the 45 kDa molecular weight marker.

Myocytes were pre-incubated with different types of PP1/2A inhibitors, such as okadaic acid, calyculin A or fostriecin, before being subjected to 30 min of simulated ischemia. Pilot experiments indicated that 1 μM of either okadaic acid or calyculin A were most effective in this system, without being toxic. Representative immunoblots are shown in Fig. 2, while cumulative data are shown in Fig. 3. The two antibody preparations gave identical results (Figs 2 and 3). Okadaic acid and calyculin A significantly ($p < 0.05$) reduced ischemia-induced dephosphorylation of cardiomyocyte Cx43. The two inhibitors reduced Cx43 dephosphorylation to a comparable extent, by 28% for calyculin A and by 32% for okadaic acid. Fostriecin, on the other hand, had no effect. Fostriecin was used at concentrations previously shown to be effective in cardiac tissue [16, 21]. Different batches of fostriecin gave similar results; increasing fostriecin dosage had no effect either (data not shown).

To examine whether our findings from isolated myocytes were valid in the whole heart context, isolated hearts were perfused with oxygenated Krebs buffer with or without 100 nM okadaic acid, and then subjected to 30 min of global ischemia. Subsequently, cardiac lysates were analysed for dephosphorylated and phosphorylated Cx43 by western blotting and densitometry. As shown in Fig. 4, the fraction of dephosphorylated Cx43 after 30 min of global ischemia was signifi-

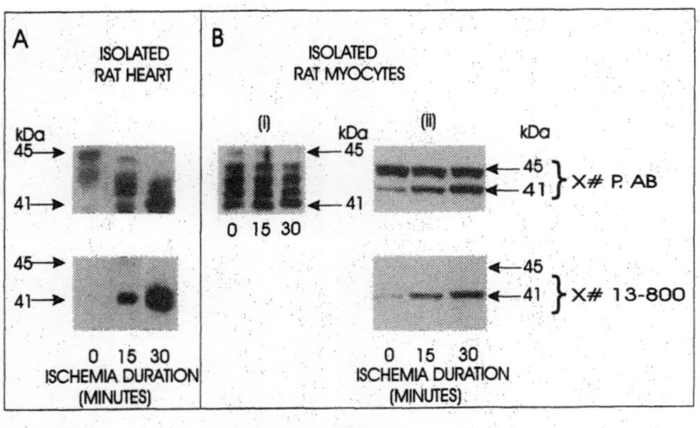

Fig. 1. Effects of global ischemia on the phosphorylation status of Cx43 in (A) the perfused intact heart and (B) isolated adult rat myocytes. Western blots were probed for total Cx43, using the polyclonal anti-Cx43 antibodies (#P.AB; upper blots) or for dephosphorylated Cx43 using the monoclonal #13-800 antibodies (lower blots), as indicated. Ischemia duration is indicated in minutes. Migration of molecular weight markers is indicated in kDa. Lysates were analyzed either on large format 9% acrylamide gels (A, B (i)), or small format 10% acrylamide gels (B (ii)).

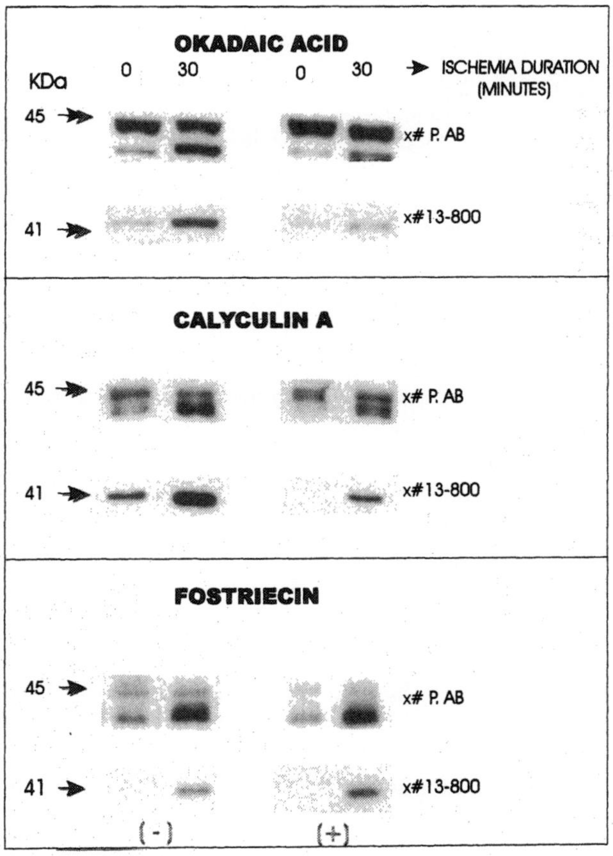

Fig. 2. Effects of PP1/2A inhibitors (okadaic acid, calyculin A, fostriecin) on ischemia-induced Cx43 dephosphorylation in isolated cardiac myocytes. Western blots were probed for total (#P.AB) or dephosphorylated (#13-800) Cx43, as indicated. (–) and (+) denote absence or presence of the inhibitor, respectively.

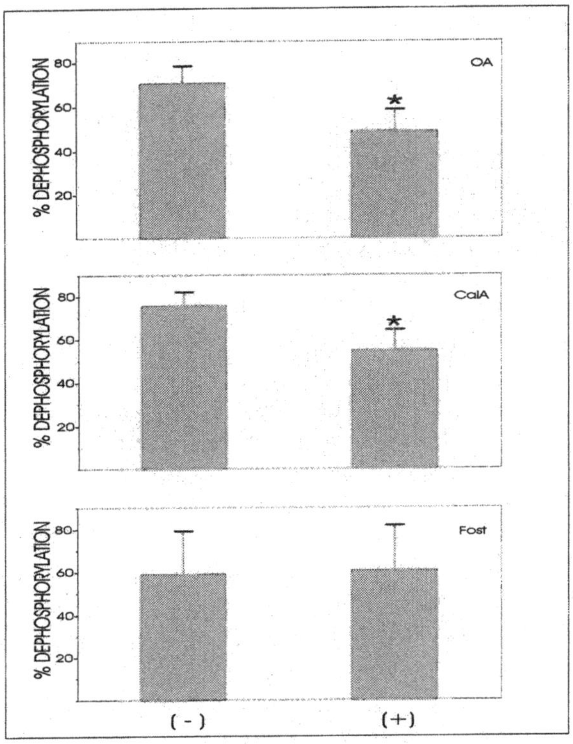

Fig. 3. Percent dephosphorylation of cardiomyocyte Cx43 during ischemia plotted as a function of absence (–) or presence (+) of phosphatase inhibitors, as indicated. OA, CalA, Fost denote use of okadaic acid, calyculin A or fostriecin, respectively. Asterisk indicates statistically significant differences from untreated samples ($p < 0.05$, unpaired student's *t*-test, n = 3–4, where n signifies independent experiment using a different myocyte preparation).

cantly reduced in the presence of okadaic acid; by 38% compared to dephosphorylation in the absence of the inhibitor.

Sections from hearts rendered ischemic in the presence or absence of okadaic acid pre-treatment (100 nM), were simultaneously probed for total and dephosphorylated Cx43 (Fig. 5). In the absence of okadaic acid, the intercalated discs of ischemic cardiomyocytes were immunostained by both antibodies, confirming the accumulation of dephosphorylated Cx43 at intercalated discs. In addition, anti-Cx43 staining of the intercalated discs was somewhat diffuse, suggesting loss of structural integrity. In hearts rendered ischemic in the presence of okadaic acid, the polyclonal antibody elicited the familiar punctate staining of intercalated discs; there was no evidence of loss of structural integrity. The intercalated disc staining by the monoclonal #13-800 antibody was substantially reduced (compared to that from untreated ischemic hearts) under these conditions (Fig. 5D).

Discussion

We have shown that isolated cardiomyocytes responded to simulated ischemia by Cx43 dephosphorylation in a manner

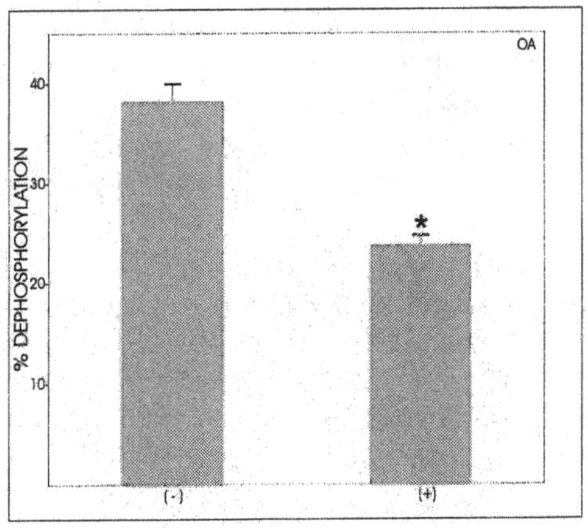

Fig. 4. Percent dephosphorylation of whole heart Cx43 after 30 min of global ischemia, plotted as a function of absence (–) or presence (+) of okadaic acid, as indicated. Asterisk indicates statistically significant differences from untreated samples ($p < 0.05$, unpaired student's *t*-test, n = 4, where n signifies individual hearts).

qualitatively similar to that of intact hearts and therefore that intact intercellular connections at the intercalated discs are not required for myocyte Cx43 dephosphorylation to occur. The PP1/2A inhibitors okadaic acid and caliculin A significantly reduced ischemia-induced Cx43 dephosphorylation in isolated myocytes, while the PP2A selective inhibitor fostriecin had no effect. Okadaic acid also prevented Cx43 dephosphorylation at the intercalated discs of globally ischemic whole hearts, again indicating similar behavior with isolated myocytes.

In the isolated myocyte system, okadaic acid and calyculin A were used at non- toxic concentrations that allowed maxi-

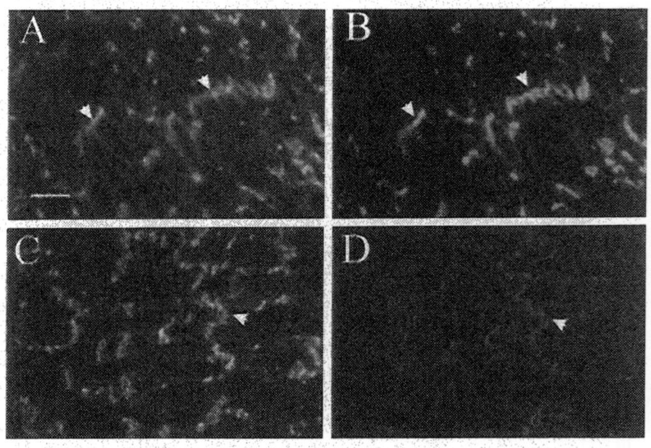

Fig. 5. Localization of Cx43 during global ischemia of the whole heart, as a function of phosphatase (okadaic acid, 100 nM) inhibition. Double immunofluorescence labeling for total Cx43 (A, C, x#P.AB) or dephosphorylated Cx43 (B, D, respectively, x #13-800). (A, B) Untreated ischemic hearts; (C, D) Ischemic hearts treated with okadaic acid. Arrows mark position of intercalated discs. Bar = 50 μM.

mal prevention of Cx43 dephosphorylation, as determined by pilot studies (data not shown). These concentrations, shown previously to protect myocytes against ischemic injury [15], would be expected to be effective against both PP1 as well as PP2A-type phosphatase(s). However, since fostriecin, a selective inhibitor of PP2A, had no effect on Cx43 dephosphorylation, it is concluded that the decrease of dephosphorylation by okadaic acid and calyculin A, observed in our system reflected primarily inhibition of PP1-like phosphatase(s).

Isolated perfused hearts responded to okadaic acid in a manner similar to isolated myocytes, by decreased Cx43 dephosphorylation during ischemia. Prevention of dephosphorylation was observed at the level of intercalated discs between cells, where functional gap junction channels are located. In the absence of the inhibitor, ischemic cardiomyocyte intercalated disc staining by the polyclonal anti-Cx43 antibody appeared somewhat frayed and disrupted, suggesting a disturbance of Cx43 plaques and cardiomyocyte structural integrity. Separation and disruption of intercalated discs during ischemia has been documented [22]. In comparison, anti-Cx43 staining of okadaic acid- treated ischemic hearts was indicative of improved structural integrity at the level of the intercalated disc. Prevention of Cx43 dephosphorylation in ischemia was therefore associated with better structural integrity of the cardiomyocyte, consistent with a protective effect. This would be in agreement with reports that okadaic acid (and calyculin A) prevents development of fragility of metabolically inhibited and/or ischemic cardiomyocytes [15]. The degree to which prevention of Cx43 dephosphorylation per se contributes to the okadaic acid induced preservation of structural integrity of the ischemic cardiomyocytes remains to be established.

The functional consequences of Cx43 dephosphorylation during ischemia are a matter of some controversy. It is thus not clear whether cardiac Cx43 dephosphorylation would result in increased or decreased intercellular coupling. Channels composed of dephosphorylated Cx43, in oocytes [23] or artificial membrane layers, have increased conductance properties [24]. A recent report indicated that gap junction permeable fluorescent dyes can still migrate between ischemic myocytes of the ex vivo heart; gap junction channels furthermore were shown to be capable of allowing the spread of contracture that occurs after 45 min of ischemia [25]. These conditions are accompanied by substantial dephosphorylation of Cx43 [11], and thus would imply that cardiac gap channels composed of dephosphorylated Cx43 are permeable to ions and small metabolites, and are thus are capable of conduction. This would also be in agreement with findings from ischemic astrocytes [26].

On the other hand, another group reported that Cx43 dephosphorylation seemed to coincide with loss of electrical coupling [11] and development of arrhythmias. Irreversible cardiomyocyte injury, brought about by calcium overload during ischemia would be expected to result in closure as well as loss of gap junctions [11]. ATP depletion (an event that would rapidly occur in ischemia, and thus be associated with Cx43 dephosphorylation) has been shown to decrease gap junctional conductance between myocyte pairs [27]. In these studies, PP1 inhibition (a manipulation that would prevent Cx43 dephosphorylation) prevented the run down in gap junctional communication, again providing a link between Cx43 dephosphorylation and decreased conductance [27].

The reason for these apparent discrepancies is not yet known. It should be noted that Cx43 is phosphorylated at multiple sites and can have different levels/states of phosphorylation. As shown in Fig. 1, early ischemia is associated with conversion of Cx43 to less phosphorylated species, but not complete dephosphorylation; complete dephosphorylation is observed at later time points (30-60 min ischemia; data not shown). The different phosphorylation states of Cx43 may have different properties; in addition, conductance and metabolic coupling may be regulated in a different manner [28].

Finally, okadaic acid and calyculin A did not fully prevent ischemia-induced Cx43 dephosphorylation. It is possible that other types of phosphatases, such as the calcium dependent calcineurin or PP2B are involved. This would be consistent with findings from other investigators reporting involvement of PP2B in the dephosphorylation of Cx43 in non-muscle cells [12, 13].

Acknowledgements

This work was funded (EK) by, the Canadian Institutes for Health Research. Stéphane Tanguy was supported by a post-doctoral fellowship by the Manitoba Health Research Council.

References

1. Goodenough DA, Goliger JA, Paul DL: Connexins, connexons, and intercellular communication. Annu Rev Biochem 65: 475–502, 1996
2. Beyer EC: Gap junctions. Int Rev Cytol 1–37, 1993
3. Saffitz JE, Green KG, Kraft WJ, Schechtman KB, Yamada KA: Effects of diminished expression of connexin43 on gap junction number and size in ventricular myocardium. Am J Physiol Heart Circ Physiol 278: H1662–H1670, 2000
4. Lampe PD, Lau AF: Regulation of gap junctions by phosphorylation of connexins. Arch Biochem Biophys 384: 205–215, 2000
5. Lau AF, Hatch-Pigott V, Crow DS: Evidence that heart connexin43 is a phosphoprotein. J Mol Cell Cardiol 23: 659–63, 1991
6. Toyofuku T, Yabuki M, Otsu K, Kuzuya T, Tada M, Hori M: Functional role of c-Src in gap junctions of the cardiomyopathic heart. Circ Res 85: 672–681, 1999
7. Doble BW, Ping P, and Kardami E: The epsilon subtype of protein kinase C is required for cardiomyocyte connexin-43 phosphorylation. Circ Res 86: 293–301, 2000

8. Bowling N, Huang X, Sandusky GE, Fouts RL, Mintze K, Esterman M, Allen PD, Maddi R, McCall E, Vlahos CJ: Protein kinase C-alpha and -epsilon modulate connexin-43 phosphorylation in human heart. J Mol Cell Cardiol 33: 789–798, 2001
9. Loo LW, Berestecky JM, Kanemitsu MY, Lau AF: pp60src-mediated phosphorylation of connexin 43, a gap junction protein. J Biol Chem 270: 12751–12761, 1995
10. Godwin AJ, Green LM, Walsh MP, McDonald JR, Walsh DA, Fletcher WH: *In situ* regulation of cell-cell communication by the cAMP-dependent protein kinase and protein kinase C. Mol Cell Biochem 127–128: 293–307, 1993
11. Beardslee MA, Lerner DL, Tadros PN, Laing JG, Beyer EC, Yamada KA, Kleber AG, Schuessler RB, Saffitz JE: Dephosphorylation and intracellular redistribution of ventricular connexin43 during electrical uncoupling induced by ischemia. Circ Res 87: 656–662, 2000
12. Cruciani V, Kaalhus O, Mikalsen SO: Phosphatases involved in modulation of gap junctional intercellular communication and dephosphorylation of connexin43 in hamster fibroblasts: 2B or not 2B? Exp Cell Res 252: 449–463, 1999
13. Li WE, Nagy JI: Connexin43 phosphorylation state and intercellular communication in cultured astrocytes following hypoxia and protein phosphatase inhibition. Eur J Neurosci 12: 2644–2650, 2000
14. Ingebritsen TS, Cohen P: Protein phosphatases: Properties and role in cellular regulation. Science 221: 331–338, 1983
15. Armstrong SC, Ganote CE: Effects of the protein phosphatase inhibitors okadaic acid and calyculin A on metabolically inhibited and ischaemic isolated myocytes. J Mol Cell Cardiol 24: 869–884, 1992
16. Weinbrenner C, Baines CP, Liu GS, Armstrong SC, Ganote CE, Walsh AE, Honkanen RE, Cohen MV, Downey JM: Fostriecin, an inhibitor of protein phosphatase 2A, limits myocardial infarct size even when administered after onset of ischemia. Circulation 98: 899–905, 1998
17. Nagy JI, Li WE, Roy C, Doble BW, Gilchrist JS, Kardami E, Hertzberg EL: Selective monoclonal antibody recognition and cellular localization of an unphosphorylated form of connexin43. Exp Cell Res 236: 1271-36, 1997
18. Padua RR, Merle PL, Doble BW, Yu CH, Zahradka P, Pierce GN, Panagia V, Kardami E: FGF-2-induced negative inotropism and cardioprotection are inhibited by chelerythrine: Involvement of sarcolemmal calcium-independent protein kinase C. J Mol Cell Cardiol 30: 2695–2709, 1998
19. Doble BW, Chen Y, Bosc DG, Litchfield DW, Kardami E: Fibroblast growth factor-2 decreases metabolic coupling and stimulates phosphorylation as well as masking of connexin43 epitopes in cardiac myocytes. Circ Res 79: 647–658, 1996
20. Laemmli UK: Cleavage of structural proteins during the assembly of the head of bacteriophage T4. Nature 227: 680–685, 1970
21. Armstrong SC, Kao R, Gao W, Shivell LC, Downey JM, Honkanen RE, Ganote CE: Comparison of *in vitro* preconditioning responses of isolated pig and rabbit cardiomyocytes: Effects of a protein phosphatase inhibitor, fostriecin. J Mol Cell Cardiol 29: 3009–3024, 1997
22. Ashraf M, Halverson C: Ultrastructural modifications of nexuses (gap junctions) during early myocardial ischemia. J Mol Cell Cardiol 10: 263–269, 1978
23. Moreno AP, Saez JC, Fishman GI, Spray DC: Human connexin43 gap junction channels. Regulation of unitary conductances by phosphorylation. Circ Res 74: 1050–1057, 1994
24. Kim DY, Kam Y, Koo SK, Joe CO: Gating connexin 43 channels reconstituted in lipid vesicles by mitogen-activated protein kinase phosphorylation. J Biol Chem 274: 5581–5587, 1999
25. Ruiz-Meana M, Garcia-Dorado D, Lane S, Pina P, Inserte J, Mirabet M, Soler-Soler J: Persistence of gap junction communication during myocardial ischemia. Am J Physiol Heart Circ Physiol 280: H2563–H2571, 2001
26. Lin JH, Weigel H, Cotrina ML, Liu S, Bueno E, Hansen AJ, Hansen TW, Goldman S, Nedergaard M: Gap-junction-mediated propagation and amplification of cell injury. Nat Neurosci 1: 494–500, 1998
27. Duthe F, Plaisance I, Sarrouilhe D, Herve JC: Endogenous protein phosphatase 1 runs down gap junctional communication of rat ventricular myocytes. Am J Physiol Cell Physiol 281: C1648–C1656, 2001
28. Kwak BR, van Veen TA, Analbers LJ, Jongsma HJ: TPA increases conductance but decreases permeability in neonatal rat cardiomyocyte gap junction channels. Exp Cell Res 220: 456–463, 1995

Molecular and Cellular Biochemistry **242**: 135–144, 2003.

Gap junction remodeling and altered connexin43 expression in the failing human heart

Sawa Kostin,[1] Markus Rieger,[1] Sebastian Dammer,[1] Stefan Hein,[2] Manfred Richter,[2] Wölf-Peter Klövekorn,[2] Erwin P. Bauer[2] and Jutta Schaper[1]

[1]*Department of Experimental Cardiology, Max-Planck-Institute, Bad Nauheim; and* [2]*Department of Cardiac Surgery, Kerckhoff-Clinic, Bad Nauheim, Germany*

Abstract

Gap junctions (GJ) are important determinants of cardiac conduction and the evidence has recently emerged that altered distribution of these junctions and changes in the expression of their constituent connexins (Cx) may lead to abnormal coupling between cardiomyocytes and likely contribute to arrhythmogenesis. However, it is largely unknown whether changes in the expression and distribution of the major cardiac GJ protein, Cx43, is a general feature of diverse chronic myocardial diseases or is confined to some particular pathophysiological settings. In the present study, we therefore set out to investigate qualitatively and quantitatively the distribution and expression of Cx43 in normal human myocardium and in patients with dilated (DCM), ischemic (ICM), and inflammatory cardiomyopathies (MYO). Left ventricular tissue samples were obtained at the time of cardiac transplantation and investigated with immunoconfocal and electron microscopy. As compared with the control group, Cx43 labeling in myocytes bordering regions of healed myocardial infarction (ICM), small areas of replacement fibrosis (DCM) and myocardial inflammation (MYO) was found to be highly disrupted instead of being confined to the intercalated discs. In all groups, myocardium distant from these regions showed an apparently normal Cx43 distribution at the intercalated discs. Quantitative immunoconfocal analysis of Cx43 in the latter myocytes revealed that the Cx43 area per myocyte area or per myocyte volume is significantly decreased by respectively 30 and 55% in DCM, 23 and 48% in ICM, and by 21 and 40% in MYO as compared with normal human myocardium. In conclusion, focal disorganization of GJ distribution and down-regulation of Cx43 are typical features of myocardial remodeling that may play an important role in the development of an arrhythmogenic substrate in human cardiomyopathies. (Mol Cell Biochem **242**: 135–144, 2003)

Key words: human heart, ischemic heart disease, dilated cardiomyopathy, myocarditis, gap junctions, arrhythmia

Introduction

Congestive heart failure is associated with an increased risk of life-threatening cardiac arrhythmias and sudden death [1]. Electrophysiological studies have established that these arrhythmias are mainly due to re-entrant ventricular circuits typically arising from regions of slow, heterogeneous conduction and unidirectional conduction block [2, 3]. In such cases, the active membrane properties may remain essentially normal [3, 4], implicating thus a reduced cell-to-cell coupling in the development of anatomic substrates of ventricular arrhythmias [5–7].

Central to the organizational pattern of cell-to-cell current transfer are the myocardial tissue architecture and specialised junctions responsible for electrical coupling, the gap junctions (GJs). The effect of myocardial architecture is determined by the size and shape of individual myocytes [8, 9] and their packing geometry in the ventricular myocardium [10]. The role of GJs in cardiac conduction is dependent on their constituent connexin isoforms (reviewed in [11–15]),

Address for offprints: S. Kostin, Max-Planck-Institute, Benekestrasse 2, D-61231 Bad Nauheim, Germany (E-mail: skostin@kerckhoff.mpg.de)

as well as the size, number and spatial distribution of these junctions [6, 10, 16–18]. The constituent proteins of GJs are connexins, of which connexin 43 (Cx43) is the major isoform in working ventricular myocytes.

In the last decade, evidence that the altered distribution and changes in the expression of Cx43 may lead to abnormal coupling between cardiomyocytes and to ventricular arrhythmogenesis in both experimental and human heart diseases has steadily accumulated [7, 19–29]. It is largely unknown, however, whether changes in the distribution and expression of Cx43 is a general feature of diverse chronic myocardial diseases or whether it is confined to particular pathophysiological settings. Although earlier and more recent studies showed reduced levels of Cx43 and changes in the distribution of GJs in failing hearts [22, 29], more studies are certainly needed, especially in human patients, to substantiate the consistency of these findings. In the present study, we therefore set out to investigate qualitatively and quantitatively the distribution and expression of Cx43 in normal human myocardium and in patients with idiopathic dilated, ischemic and inflammatory cardiomyopathies.

Materials and methods

Study groups

Human left ventricular (LV) samples were collected from the explanted hearts of patients undergoing orthotopic heart transplantation because of end-stage heart disease. All patients were severely symptomatic (NYHA grade IV) with poor LV systolic function. Patients either had normal coronary arteries with histologically proven active myocarditis (MYO, n = 6) according to the Dallas criteria, or without histological evidence of myocardial inflammation (idiopathic dilated cardiomyopathy, DCM, n = 7), or had severe coronary artery disease (ischemic cardiomyopathy, ICM, n = 7) with a history of previous myocardial infarction (7 of 7). Clinical data are summarized in Table 1. LV myocardium from four donor hearts that for technical reasons were not used for transplantation and intraoperative myocardial biopsies from 2 patients with atrial septal defect and 2 patients with a stenosed mitral valve but with normal LV function served as control tissues. The institutional Ethical Committee approved the study, and all patients gave written informed consent.

Table 1. Clinical data

Groups	n	Age	Sex (male/female)	EF (%)	(NYHA)
Control	8	51.5 ± 6.3	6/2	>60%	–
DCM	7	48.8 ± 7.7	7/0	14 ± 1.6	IV (7/7)
ICM	7	54.8 ± 2.4	7/0	21 ± 2.2	IV (7/7)
MYO	6	43.4 ± 5.8	4/2	23 ± 4.3	IV (6/6)

Tissue sampling

Tissue samples were either immediately frozen in liquid nitrogen for immunohistochemistry or immersed in 3% glutaraldehyde buffered with 0.1 mol/L Na cacodylate for electron microscopy as previously described [30, 31].

Immunolabeling and confocal microscopy

Five to eight samples were analyzed per each patient. Before immunolabeling, tissue characterization and orientation was recorded by hematoxylin–eosin staining. Frozen sections 12 μm thick were fixed for 10 min with 4% paraformaldehyde and then exposed for 10 min in 0.1% bovine serum albumin (Sigma), followed by incubation with the corresponding antibodies in single or double staining procedures. All antibodies used are listed in Table 2. Anti-mouse IgG-conjugated with FITC or TRITC (Dianova) and anti-rabbit IgG-conjugated with FITC (Dianova) or Cy3 (Chemicon) were used as detection systems. The nuclei were stained with 7-aminoactinomycin D (7-AAD, Molecular Probes). F-actin was fluorescently stained using phalloidin conjugated with TRITC (Sigma).

The samples were examined with a confocal scanning laser microscope Leica TCSNT, equipped with argon/krypton and helium/neon lasers. Series of confocal optical sections were taken through the depth of the tissue samples at 0.5–1-μm intervals. In order to improve image quality and to obtain a high signal/noise ratio each image from the series was signal-averaged. After data acquisition, the images were transferred to a Silicon Graphics Indy workstation (Silicon

Table 2. Primary antibodies

Antibody	Type	Clone	Company
Connexin 43	Mono-	1E9 and 4E6.2	Biotrend
Connexin 43	Poly-	–	Zymed
N-cadherin	Mono-	GC-4	Sigma
N-cadherin	Poly-	–	Sigma
Desmoplakin	Mono-	DP 1&2–2.15	Boehringer
Desmoplakin	Poly-	–	NatuTec
Myomesin	Mono-	B4	Gift from Dr. H. Eppenberger
α-actinin	Mono-	EA-53	Sigma
CD-3	Mono-	T3-4B5	Dako
Collagen I	Mono-	COL-1	Sigma
Collagen III	Poly-	–	Sigma
Collagen VI	Poly-	–	Rockland

Graphics) for image restoration and reconstruction using Imaris®, the multichannel image processing software (Bitplane, Zürich, Switzerland). The principles of this method have been previously described [16, 32].

Techniques for Cx43 quantification

Areas selected for quantitative analysis of Cx43 consisted of well-preserved compact bundles of LV mid-myocardial myocytes cut in a plane parallel to the long axis of the cells. To reduce variability between preparations, all sections were immunolabeled simultaneously using identical aliquots of diluted primary and secondary antibodies and other reagents. Before quantification, all tissue sections were shortly inspected to ensure that the images collected demonstrate the full range of the fluorescence intensity from 0–255 intensity levels. The standardized parameters of imaging, zoom, pinholes, objective, and laser power were kept constant for recording of data in all measurements. From each optical field (size 200 × 200 μm), 10 confocal slides were obtained at a 1-μm interval. Quantification of Cx43 was performed blindly having on the screen only one channel showing F-actin labeling. The values of myocyte area and myocyte volume were immediately determined using the Leica TCNT software and were based on clearly delineated cell borders by F-actin labeling. From each cell, one histogram of Cx43 fluorescence intensity of a single confocal section (the section number 5 from the total 10 sections forming a stack of confocal images) was obtained and was converted into Macintosh Excel data for calculation of Cx43 signal. The area of positive Cx43 labeling was defined as the number of pixels with Cx43 signal intensity exceeding the threshold of 50 on the 0–255 gray intensity scale as described previously [28, 33–35]. Thereafter the total number of positive pixels was converted into square micrometers. The quantity of Cx43 per myocyte area was expressed as the percent of myocyte surface area occupied by Cx43 positive label. The quantity of Cx43 per myocyte volume (GJ surface density) was calculated from measurements of the total GJ area per myocyte surface divided by the values of the myocyte volume and expressed as $\mu m^2/\mu m^3$. Per each patient, 41.8 ± 5.8 LV myocytes (range 34–53 cells) with clearly defined cellular borders were investigated for Cx43 expression.

Transmission electron microscopy

After overnight fixation in 3% glutaraldehyde, small LV tissue samples were embedded in Epon following routine procedures. Ultrathin sections were stained with uranyl acetate and lead citrate, and viewed and photographed in a Philips CM 10 electron microscope.

Statistical analysis

Results are reported as means ± S.D. Differences between groups were analyzed using ANOVA, followed by secondary analysis by use of Bonferroni's corrections to Student's *t*-test. Differences between groups were considered significant at $p < 0.05$.

Results

GJ remodeling in the failing human heart

In normal human myocardium, Cx43 and N-cadherin labeling were predominantly confined to the intercalated discs (Figs 1a and 1b) in a pattern as widely reported previously. In patients with ICM, altered patterns of intercellular junction distribution were conspicuously observed in myocytes at the border of healed infarcts. The immunolabeled GJs for Cx43 (Figs 1c and 1e) were found to be randomly distributed over the lateral surfaces of the cells. Abnormal localization of Cx43 in these regions were accompanied by similar disrupted patterns of the distribution of immunolabeled fascia adherens for N-cadherin (Fig. 1d) and immunolabeled desmosomes for desmoplakin (Fig. 1f). The extent of the border zone myocytes exhibiting spatial derangements of both GJs and adhesive junctions increased in parallel with the size of healed infarcts.

Another peculiar feature of some infarct zones was the presence of viable myocytes connected via GJs and extending across healed infarcts (Fig. 2). A Cx43 signal was commonly detected in such myocytes (Fig. 2a) despite the absence of recognizable intercalated discs (Figs 2b and 2c). Further details of the appearance of surviving myocytes within healed infarcts were obtained using electron microscopy. Figures 2d through 2f show that these myocytes are extremely distorted and attenuated, but still intact. It is worthy to note that these myocytes were found to be interconnected via ultrastructurally intact GJs, confirming thus immunoconfocal findings.

In patients with DCM abnormal patterns of Cx43 distribution were detected in myocytes bordering small microscopic zones of replacement fibrosis (Figs 3a and 3b) and in bizarrely shaped and disorganized myocytes (Figs 3c and 3d). In patients with MYO severe disruptions of GJs were observed in areas of myocardial inflammation (Figs 3e and 3f).

In areas of diseased human hearts free from any histologically detectable structural damage, the distribution of Cx43 labeling qualitatively resembled that of normal LV myocardium. However, as compared with normal myocardium, the LV myocytes in the failing hearts demonstrated a statistically significant increase in cell area by 57 ± 18% in patients with DCM, by 92 ± 48% in ICM, and by 54 ± 47% in patients with MYO.

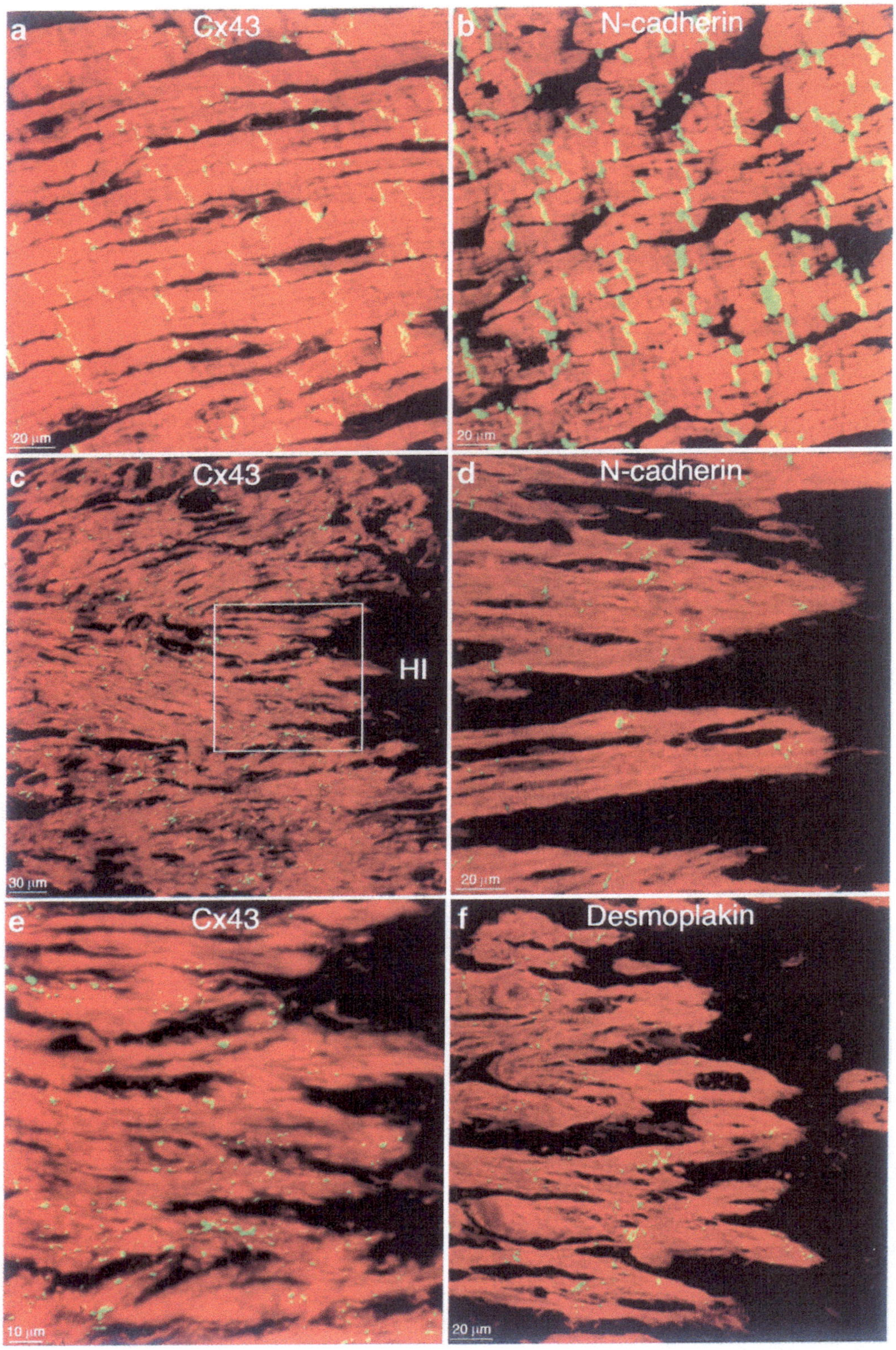

Fig. 1. Immunolocalization of GJs and adhesive junctions in normal human myocardium (panels a and b) and in myocytes bordering healed myocardial infarcts (panels c–f). Panels a and b are confocal images of longitudinally sectioned myocytes showing that in normal human myocardium the GJ protein – Cx43 (a, green) and the major fascia adherens protein – N-cadherin (b, green) are localized as transverse lines representing intact intercalated discs. Cx43 labeling (c, green) in myocytes bordering healed myocardial infarcts (HI) is extensively dispersed over the lateral myocyte borders. Panel e is an enlargement of the boxed area shown in panel c. Panels d and f show that the border zone myocytes display a disruption of adhesive interconnections visualized by N-cadherin (fascia adherens junctions) and desmoplakin (desmosomal plaque protein) labeling (green). Myofibrils are stained red with TRITC-conjugated phalloidin (panels a–d) or with an anti-α-actinin antibody (panels e and f).

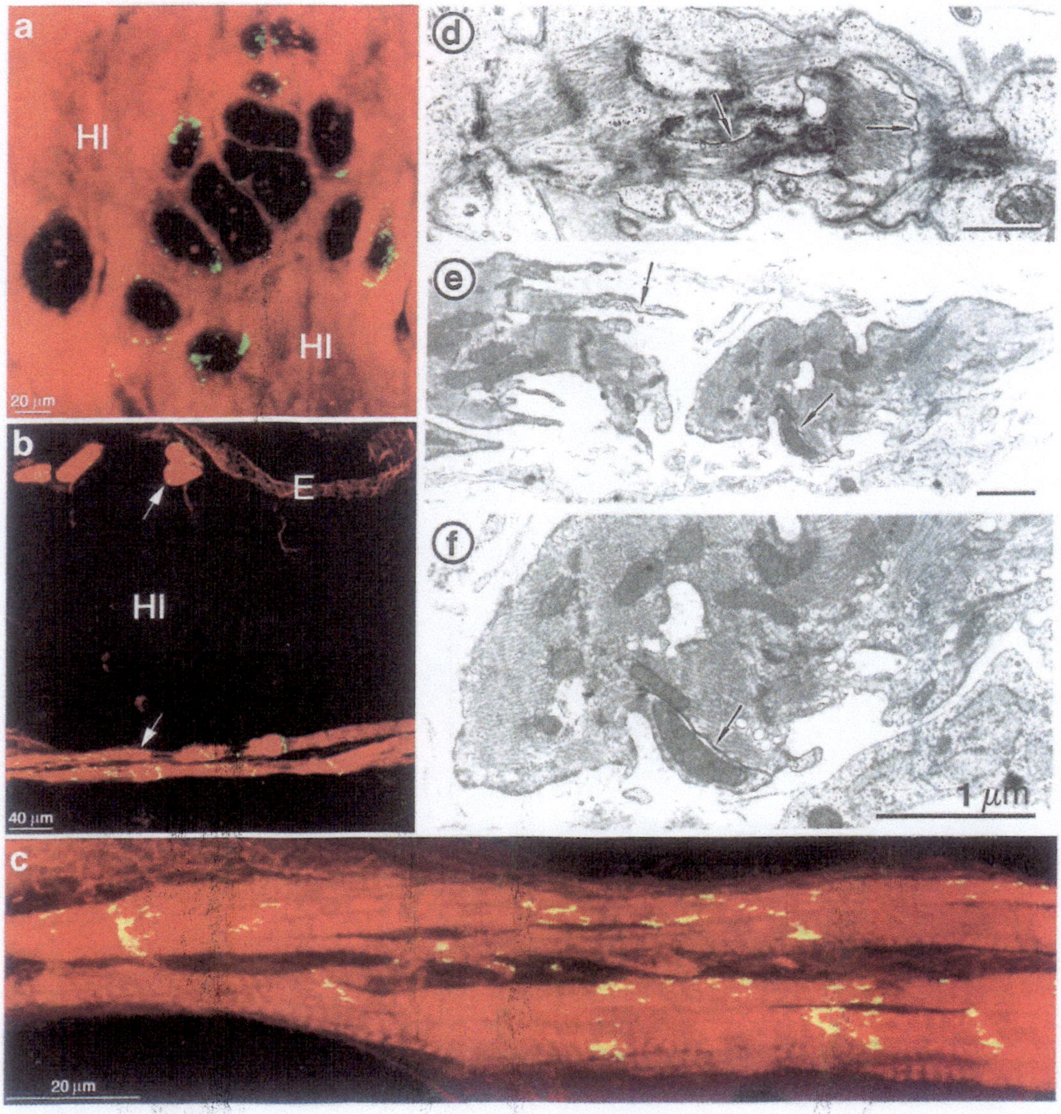

Fig. 2. Immunoconfocal and ultrastructural features of surviving myocytes within healed myocardial infarcts (HI) in patients with ICM. Shown in panel a is a transversally sectioned scar tissue displaying isolated myocytes positively labeled for Cx43 (green). Note that these myocytes are totally engulfed in a scar tissue visualized with collagen VI staining (red). Panel b shows an subendocardial region (E – endocardium) of a healed myocardial infarct containing viable myocytes (arrows) labeled for F-actin (red) and Cx43 (green). Panel c is an enlargement of the myocytes shown with the arrow on the left bottom in panel b. Note that the myocytes forming a penetrating bridge into the infarct zone display a disordered pattern of GJs. Note also that the width of this bridge comprises only two layers of myocytes. Shown in panels d and e are representative electron microscopic images taken at different steps of a single chain of survived myocytes located deeply within an infarct scar. Note that these myocytes are interconnected via ultrastructurally intact GJs (arrows). Panel f is an enlargement of the myocyte shown on the right side in panel e. Bars in panels d–f = 1 µm.

Quantitative analysis of GJ in normal and diseased human myocardium

Results of the quantitative analysis of the LV myocyte area occupied by a positive Cx43 signal and the Cx43-GJ surface density per individual patient are given in Table 3. In adult normal human LV myocardium, the Cx43 area occupies 1.68 ± 0.10% of the myocyte area resulting in a GJ surface density of 0.0055 ± 0.00063 $\mu m^2/\mu m^3$ myocyte volume. These values are highly consistent with previously reported data in normal LV myocytes of canine or human hearts [22, 36]. Figure 4 compares mean values of Cx43 per cell area (Fig. 4A) or per cell volume (Fig. 4B) in diseased human LV myocytes relative to those obtained in LV control tissue. Diseased myocardium showed a statistically significant reduction of the Cx43 area per myocyte surface area or per myocyte volume respectively by 30.3 ± 8.3% and 55.5 ± 18.7% in patients with DCM, by 23.2 ± 6.1% and 48.3 ± 13.6% in ICM,

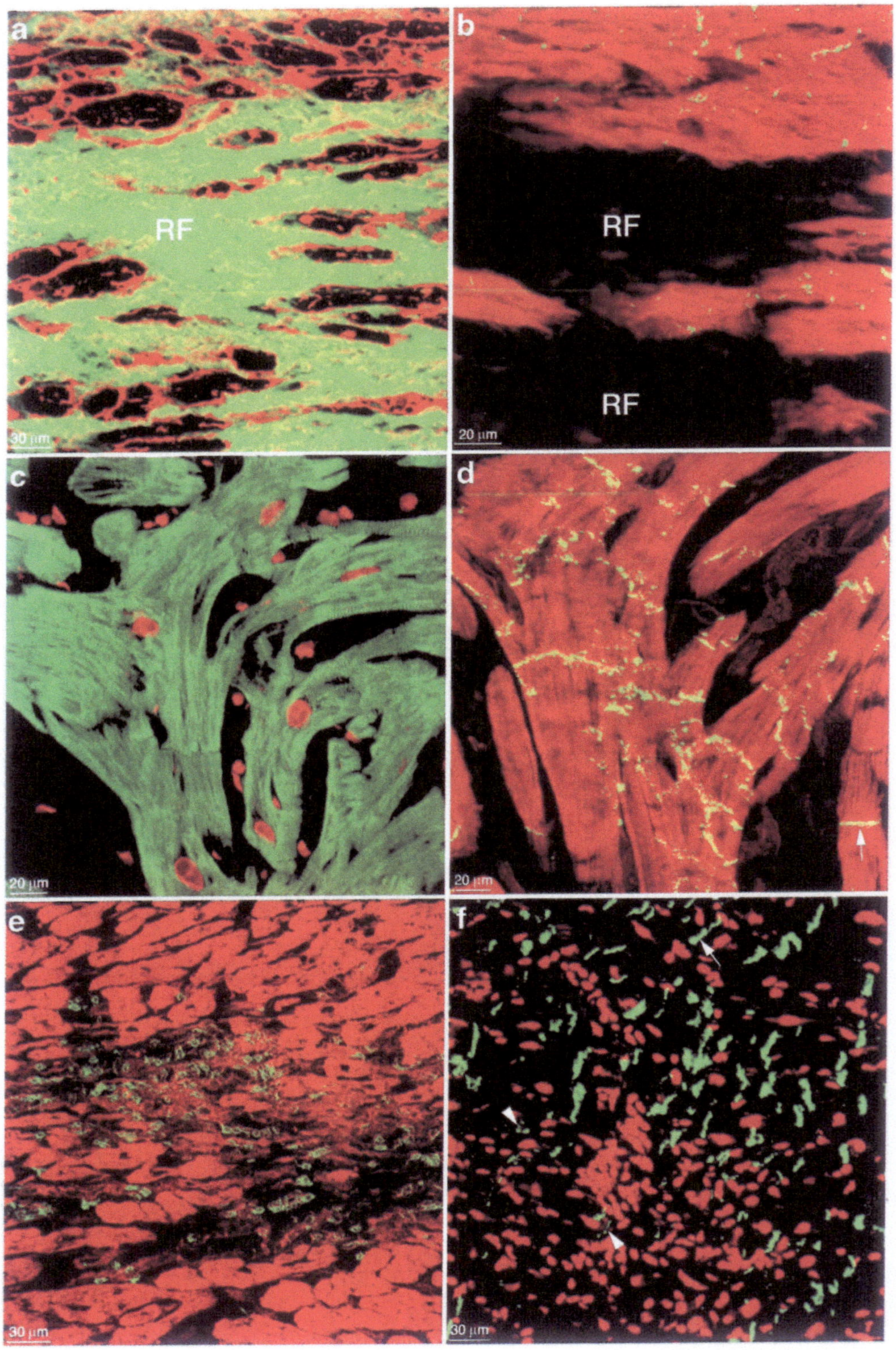

Fig. 3. Patterns of focal disorganization of Cx43 label in patients with DCM and MYO. Panel a is a confocal micrograph showing a region of replacement fibrosis (RF) double labeled for collagen III (red) and collagen I (green) in a patient with DCM. Panel b is a serial section of the myocardial tissue shown in panel a demonstrating a disrupted pattern of Cx43 label (green) in the myocytes bordering replacement fibrosis. Panel c shows a region of myofiber disarray displaying bizarrely shaped myocytes labeled for myomesin (green) in a patient with DCM. Panel d is a serial section of the myocardium shown in panel c demonstrating an abnormal pattern of Cx43 label (green) in spatially disorganized myocytes. Panel e shows a focal interstitial accumulation of lymphocytes labeled for CD-3 (green) in a patient with MYO. Panel f is a serial section of the myocardium shown in panel e demonstrating a disrupted pattern of Cx43 (green) distribution (arrowheads). Shown with arrows in panels d and f are preserved patterns of GJ distribution at the intercalated discs. In panels b, d and f myofibrils are stained red with TRITC-phalloidin. In panels c and f nuclei are stained red with 7-AAD.

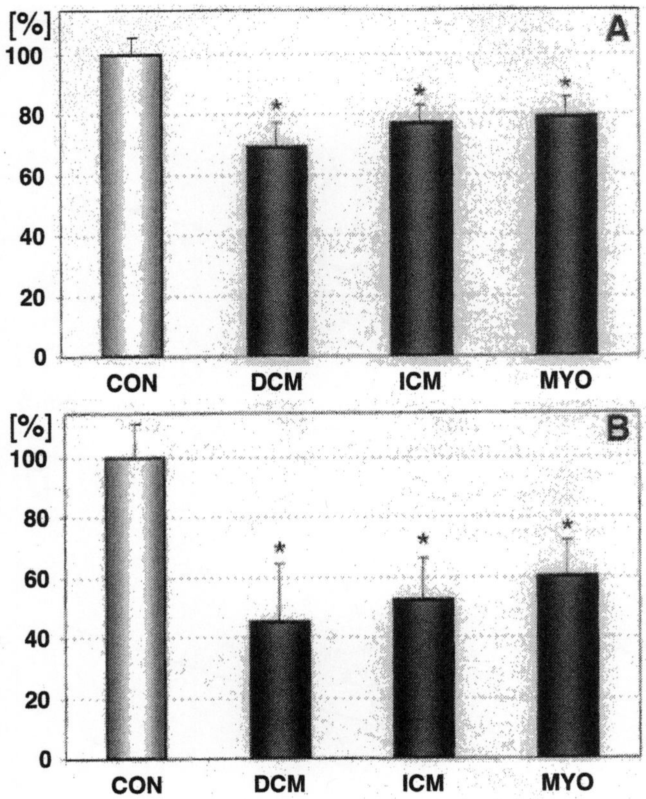

Fig. 4. Mean relative values of Cx43 area per myocyte surface area (A) or per myocyte volume (B) in the control group (CON) and in patients with DCM, ICM and MYO. Values of each patient have been normalized and expressed as percent to the values obtained in the control group. *p < 0.05 as compared with control.

and by 20.6 ± 6.4% and 39.8 ± 12.4% in patients with MYO as compared with the control tissue. Taken together, the data shown in Table 3 and Fig. 4 indicate that in the vast majority of investigated patients, the amount of Cx43 per myocyte in the failing hearts is significantly less than in control hearts.

Figure 5 shows relationships between total GJ area per cell and myocyte surface area. In control human myocardium, total GJ area per cell showed a linear increase with myocyte area (r = 0.80). In diseased myocardium, this relationship was much weaker: r = 0.17, in DCM; r = 0.22, in ICM; and r = 0.24 in patients with MYO. These data indicate that hypertrophic response of myocytes distant from focal histological lesions is maladaptive in terms that an increase in cell size is not associated with an adequate increase in Cx43 protein.

Table 3. Results of quantitative morphometric analysis of GJs

Groups	Patient no.	% Myocyte area occupied by GJs	GJ surface density ($\mu m^2/\mu m^3$)
CON	1	1.6504	0.00559
	2	1.5124	0.00434
	3	1.8580	0.00643
	4	1.6212	0.00576
	5	1.7782	0.00616
	6	1.7174	0.00559
	7	1.6821	0.00527
	8	1.6224	0.00491
	Mean ± S.D.	1.68 ± 0.10	0.0055 ± 0.00063
DCM	1	1.4654	0.00432
	2	1.2685	0.00378
	3	1.0823	0.00198
	4	1.1152	0.00149
	5	1.0595	0.00194
	6	1.0487	0.00199
	7	1.1543	0.00166
	Mean ± S.D.	1.17 ± 0.14*	0.0024 ± 0.0010*
ICM	1	1.4214	0.00289
	2	1.2719	0.00335
	3	1.3217	0.00234
	4	1.2902	0.00281
	5	1.2512	0.00301
	6	1.0888	0.00148
	7	1.3987	0.00408
	Mean ± S.D.	1.29 ± 0.10*	0.0028 ± 0.007*
MYO	1	1.2690	0.00304
	2	1.1974	0.00214
	3	1.2653	0.00328
	4	1.3772	0.00365
	5	1.5289	0.00444
	6	1.3714	0.00331
	Mean ± S.D.	1.33 ± 0.11*	0.0033 ± 0.007*

*p < 0.05 as compared with control.

Discussion

Previous studies in ischemic human hearts showed two major abnormalities in GJs: disruption of the normal distribution of GJs between myocytes at infarct scar border zones, and a significant reduction of the quantity of Cx43 in LV myocytes distant from infarct scars [20, 22]. These findings were confirmed in the present study. A number of experimental and clinical studies have reported that myocytes in the vicinity of healed infarcts undergo severe alterations in morphology, associated with a remodeling of their intercellular junctions ([7, 19, 26], reviewed in [8, 12–15, 37, 38]) that likely contribute to slow conduction, conduction block, and complex fractioned electrograms typical of these regions [3, 7]. As shown in the present study, both GJs and adhesive junctions are affected by this process. Mapping studies in experimental animals and human patients have directly linked this pattern of GJ remodeling to location and development of re-entrant ventricular arrhythmias [3, 7, 39].

A noteworthy histopathological feature of the healed infarct that may be relevant to explaining fatal arrhythmias and

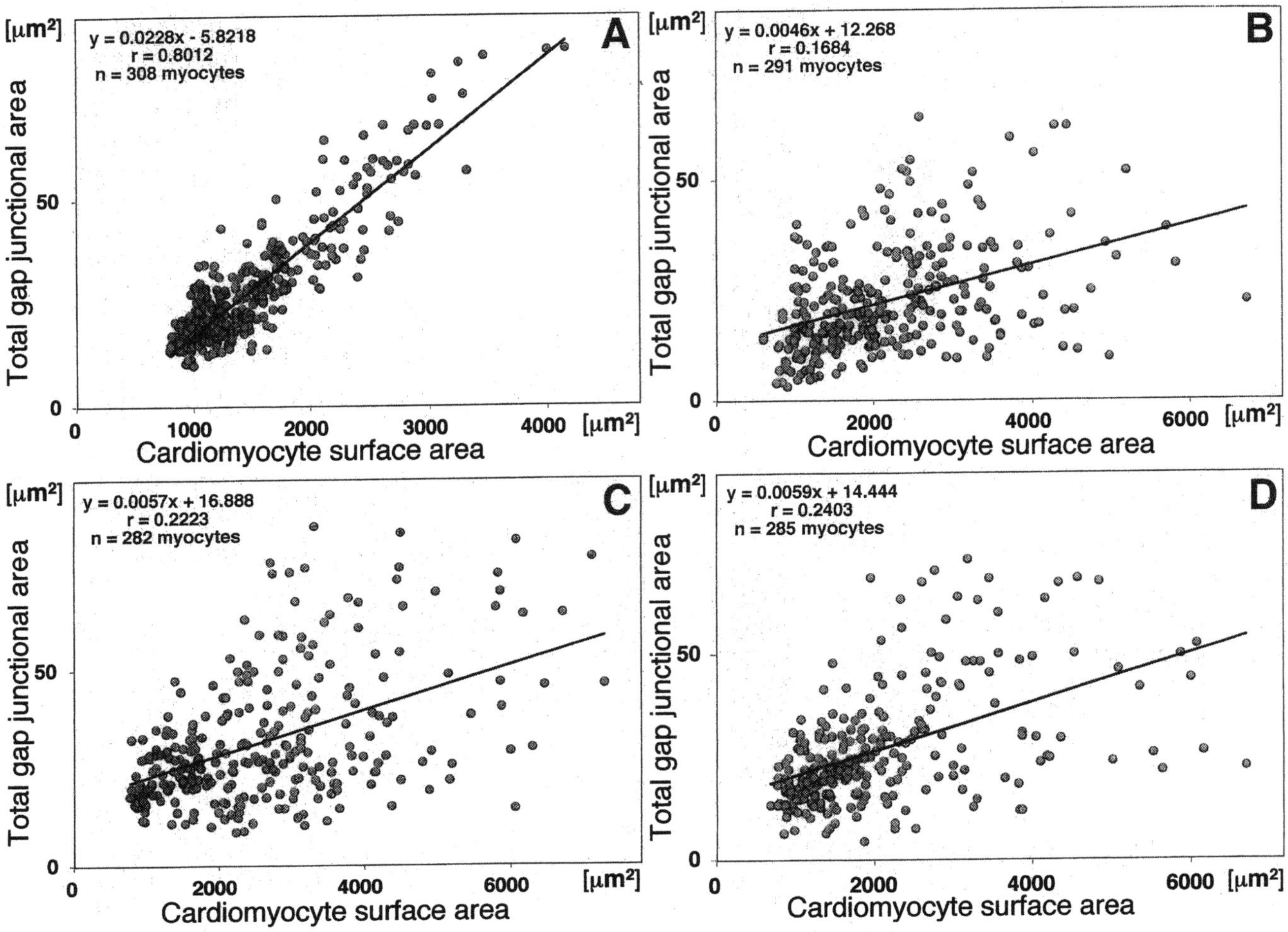

Fig. 5. Relationship between total GJ area and cardiomyocyte surface area in control tissues (A) and in patients with DCM (B), ICM (C), and MYO (D).

sudden death in patients with ischemic heart disease is the presence of myocyte strands penetrating and transversing infarct scars. The occurrence of such 'peninsulas' or bridges across infarcted regions has been described in the late 70s [40]. In 1991, Smith *et al.* [20] using immunolabeling for Cx43 and confocal microscopy have elegantly demonstrated that the myocytes penetrating scar tissue connect healthy myocardium across the infarct zone, and most importantly, that the myocytes forming these bridges are electrically coupled via GJs. The present immunoconfocal study confirmed these observations. In particular, similiar with the findings of Smith *et al.*, we show that the myocytes forming these bridges display abnormal patterns of Cx43 labeling specified by laterally disposed and dispersed GJs. In addition, using transmission electron microscopy we have demonstrated that these myocytes are viable and are interconnected with structurally intact GJs even in the absence of identifiable intercalated discs. The functional consequence of the presence of such bridges comprising abnormally interconnected but viable myocytes would be predicted to provide an accessory pathway of slow propagation rendering the impulse to re-enter post-refractory healthy myocardial tissue and initiate an extra beat and tachycardia.

Apart from abnormalities in GJ distribution observed at the border zone of macroscopically visible infarct scars, the present study demonstrates that disruptions of the normal pattern of GJ distribution also occurs in myocytes bordering smaller microscopic areas of myocardial damage, namely replacement fibrosis and myocardial inflammation. Given the evidence that spatial non-uniformities of the propagation, local conduction block and re-entry can originate even in small myocardial regions [41, 42], our findings of focally disrupted patterns of GJs in the failing hearts would predict an impairment of the impulse conduction at the microscopic level thereby creating stable sources for micro-re-entry.

In addition to the regions of altered GJ distribution associated with foci of myocardial damage, we found severe disorganization of GJ distribution in myocardial areas featuring myofiber disarray. Although myofiber disarray is more specific for hypertrophic cardiomyopathy, this cellular disorganization can also be found in other cardiac diseases, including

DCM [43, 44]. In line with the previous observations of severe alterations of GJ distribution in regions of myofiber disarray in patients with hypertrophic cardiomyopathy [24], that are prone to arrhythmias and sudden death [45], the present findings of similar severe disarrangements of GJ distribution in bizarrely branched myocytes in our patients with DCM can thus be envisaged as another potential morphological substrate of ventricular arrhythmias [46].

While focal disruptions of GJ distribution consistently found in our patients with end-stage heart failure may undoubtedly predispose to the increased frequency of arrhythmia and sudden death in this clinical setting, the present findings of reduced levels of Cx43 in myocytes distant from any histological lesions provide an additional explanation for the arrhythmogenic behaviour of the failing heart. Our conclusion of reduced levels of Cx43 in the failing hearts relies on immunoconfocal measurements of Cx43 expression per individual myocyte which provided meaningful comparisons between normal and diseased hearts. Thus, the values of GJ surface density in longitudinal sections of normal human LV myocardium (0.0055 $\mu m^2/\mu m^3$ myocyte volume), as determined in the present study, are in good agreement with previously reported values obtained in transverse sections of normal human LV using immunofluorescent labeling and confocal microscopy (0.0051 $\mu m^2/\mu m^3$ myocyte volume) [22], or with those obtained in longitudinal sections of normal canine myocardium using an immunohistochemical technique and light microscopy (0.0052 $\mu m^2/\mu m^3$ myocyte volume) [36]. Moreover, a 40–55% reduction of Cx43 per myocyte volume observed in our patients with heart failure of different origins falls within the range of values previously reported from quantitative immunoconfocal analysis of intraoperative biopsies from ischemic (47% reduction), hypertrophied ventricles (40% reduction) [22, 47], and hibernating myocardium (33% reduction of Cx43 per intercalated disc) [25], or from quantitative immunoblotting analysis of Cx43 in explanted hearts of patients with ICM and DCM (50% reduction of the Cx43 protein) [29]. Significantly decreased levels of Cx43 have also been reported in animal models of decompensated heart failure induced by chronic pressure overload [23, 48].

Our novel finding of decreased levels of Cx43 per myocyte in patients with inflammatory heart disease are in good agreement with the observations of reduced GJ conductance and immunohistochemically detectable Cx43 in a culture model of infected myocytes with *Trypanosoma cruzi* – a causative agent of inflammatory cardiomyopathy (Chagas disease) [21]. Furthermore, it has recently been reported that myocardial inflammation in cardiac allograft rejection is also associated with a marked down-regulation of Cx43 [49].

Taken together, the present study demonstrates that heart failure is associated with reduced expression of Cx43, the major intercellular coupling protein between ventricular myocytes, which may contribute to arrhythmias and sudden death in this clinical setting. That reduced levels of Cx43 can result in slowed conduction and ventricular arrhythmia has been demonstrated in studies carried out in mice heterozygous for a null mutation of the Cx43 gene with an ≈50% reduction in Cx43 content [28, 50], that is comparable to the degree of reduction of Cx43 observed in our patients. Furthermore, a significant slowing in ventricular conduction and the occurrence of sudden arrhythmic death has recently been reported in another model of mice with cardiac-restricted inactivation of the Cx43 gene [51].

In conclusion, focal disorganization of GJ distribution and down-regulation of Cx43 are typical features of myocardial remodeling that may play an important role in the development of an arrhythmogenic substrate in the human failing hearts.

References

1. Levy D, Anderson KM, Savage DD, Balkus SA, Kannel WB, Castelli WP: Risk of ventricular arrhythmias in cardiac failure and hypertrophy: The Framingham study. Am J Cardiol 60: 560–565, 1987
2. Janse MJ, Wit AL: Electrophysiological mechanisms of ventricular arrhythmias resulting from myocardial ischemia and infarction. Phisiol Rev 69: 1049–1169, 1989
3. Gardner PI, Ursell PC, Fenoglio JJ, Wit AL: Electrophysiologic and anatomic basis for fractionated electrograms recorded from healed myocardial infarcts. Circulation 72: 596–611, 1985
4. Boyden PA, Gardner PI, Wit AL: Action potentials of cardiac muscle in healing infarcts: Response to norepinephrine and caffeine. J Mol Cell Cardiol 20: 525–537, 1988
5. Saffitz JE, Corr PB, Sobel BE: Arrhythmogenesis and ventricular dysfunction after myocardial infarction. Is anomalous cellular coupling the elusive link? Circulation 87: 1742–1745, 1993
6. Severs NJ: Cardiac muscle cell interaction: From microanatomy to the molecular make-up of the gap junction. Histol Histopathol 10: 481–501, 1995
7. Peters NS, Coromilas J, Severs NJ, Wit AL: Disturbed connexin43 gap junction distribution correlates with the location of reentrant circuits in the epicardial border zone of healing canine infarcts that cause ventricular tachycardia. Circulation 95: 988–996, 1997
8. Peters NS, Wit AL: Myocardial architecture and ventricular arrhythmogenesis. Circulation 97: 1746–1754, 1998
9. Spach MS, Heidlage F, Dolber PC, Barr RC: Electrophysiological effects of remodeling cardiac gap junctions and cell size. Experimental and model studies of normal cardiac growth. Circ Res 86: 302–311, 2000
10. Saffitz JE, Kanter HL, Green KG, Tolley TK, Beyer EC: Tissue-specific determinants of anisotropic conduction velocity in canine atrial and ventricular myocardium. Circ Res 74: 1065–1070, 1994
11. Gros DB, Jongsma HJ: Connexins in mammalian heart function. Bioessays 18: 719–730, 1996
12. Jongsma HJ, Wilders R: Gap junctions in cardiovascular disease. Circ Res 86: 1193–1197, 2000
13. van Veen TAB, van Rijen HVM, Opthof T: Cardiac gap junction channels: Modulation of expression and channel properties. Cardiovasc Res 51: 217–229, 2001

14. Kanno S, Saffitz JE: The role of myocardial gap junctions in electrical conduction and arrythmogenesis. Cardiovasc Pathol 10: 169–177, 2001
15. Severs NJ, Rothery S, Dupont E, Coppen SR, Yeh H-I, Ko Y-S, Matsushita T, Kaba R, Halliday D: Immunohistochemical analysis of connexin expression in the healthy and diseased cardiovascular system. Microsc Res Tech 52: 301–322, 2001
16. Kostin S, Schaper J: Tissue-specific patterns of gap junctions in adult rat atrial and ventricular cardiomyocytes *in vivo* and *in vitro*. Circ Res 88: 933–939, 2001
17. Saffitz JE, Davis LM, Darrow BJ, Kanter HL, Laing JG, Beyer EC: The molecular basis of anisotropy: Role of gap junctions. J Cardiovasc Electrophysiol 6: 498–510, 1995
18. Severs NJ, Dupont E, Kaprielian RR, Yeh HI, Rothery S: Gap junctions and connexins in the cardiovascular system. In: M.H. Yacoub, A. Carpentier, J. Pepper, J.N. Fabiani (eds). Annual of Cardiac Surgery 1996, 9th edition. Current Science, London, 1996, pp 31–44
19. Luke RA, Saffitz JE: Remodeling of ventricular conduction pathways in healed canine infarct border zones. J Clin Invest 87: 1594–1602, 1990
20. Smith JH, Green CR, Peters NS, Rothery S, Severs NJ: Altered patterns of gap junction distribution in ischemic heart disease: An immunohistochemical study of human myocardium using laser scanning confocal microscopy. Am J Pathol 139: 801–821, 1991
21. Campos de Carvalho AC, Tanowitz HB, Wittner M, Dermietzel R, Roy C, Hertzberg EL, Spray DC: Gap junction distribution is altered between cardiac myocytes infected with *Trypanosoma cruzi*. Circ Res 70: 733–742, 1992
22. Peters NS, Green CR, Poole-Wilson PA, Severs NJ: Reduced content of connexin 43 gap junctions in ventricular myocardium from hypertrophied and ischemic human hearts. Circulation 88: 864–875, 1993
23. Bastide B, Neyses L, Ganten D, Paul M, Willecke K, Traub O: Gap junction protein connexin40 is preferentially expressed in vascular endothelium and conductive bundles of rat myocardium and is increased under hypertensive conditions. Circ Res 73: 1138–1149, 1993
24. Sepp R, Severs N, Gourdie R: Altered patterns of cardiac junction distribution in hypertrophic cardiomyopathy. Heart 76: 412–417, 1996
25. Kaprielian RR, Gunning M, Dupont E, Sheppard MN, Rothery SM, Underwood R, Pennel DJ, Fox Kim, Pepper J, Poole-Wilson PA, Severs NJ: Downregulation of immunodetectable connexin43 and decreased gap junction size in the human left ventricle. Circulation 97: 651–660, 1998
26. Matsushita T, Oyamada M, Fujimoto K, Yasuda Y, Masuda S, Wada Y, Oka T, Takamatsu T: Remodeling of cell–cell and cell–extracellular matrix interactions at the border zone of rat myocardial infarcts. Circ Res 85: 1046–1055, 1999
27. Uzzaman M, Honjo H, Takagishi Y, Emdad L, Magee AI, Severs NJ, Kodama I: Remodeling of gap junctional coupling of rats with monocrotaline-induced pulmonary hypertension. Circ Res 86: 871–878, 2000
28. Saffitz JE, Green KG, Kraft WJ, Schechtman KB, Yamada KA: Effects of diminished expression of connexin 43 on gap junction number and size in ventricular myocardium. Am J Physiol 278: H1662–H1670, 2000
29. Dupont E, Matsushita T, Kaba RA, Vozzi C, Coppen SR, Khan N, Kaprelian R, Yacoub MH, Severs NJ: Altered connexin expression in human congestive heart failure. J Mol Cell Cardiol 33: 359–371, 2001
30. Kostin S, Scholz D, Shimada T, Maeno Y, Mollnau H, Hein S, Schaper J: The internal and external protein scaffold of the T-tubular system in cardiomyocytes. Cell Tissue Res 294: 449–460, 1998
31. Kostin S: The structural correlate of reduced cardiac function in failing human hearts. In: N. Takeda, M. Nagano, N.S. Dhalla (eds). The Hypertrophied Heart. Kluwer Academic Publishers, Boston, MA, 2000, pp 423–439
32. Kostin S, Hein S, Bauer EP, Schaper J: Spatio-temporal development and distribution of the intercellular junctions in adult rat cardiomyocytes in culture. Circ Res 85: 154–167, 1999
33. Gourdie RG, Green CR, Severs NJ: Gap junction distribution in mammalian myocardium revealed by an anti-peptide antibody and laser scanning confocal microscopy. J Cell Sci 99: 41–45, 1991
34. Green CR, Peters NS, Gourdie RG, Rothery S, Severs NJ: Validation of immunohistochemical quantification in confocal scanning laser microscopy: A comparative assessment of gap junction size with confocal and ultrastructural techniques. J Histochem Cytochem 41: 1339–1349, 1993
35. Severs NJ, Gourdie RG, Harfst E, Peters NS, Green CR: Intercellular junctions and the application of microscopical techniques: The cardiac gap junction as a case model. J Microsc 169: 299–328, 1993
36. Luke RA, Beyer EC, Hoyt RH, Saffitz JE: Quantitative analysis of intercellular connections by immunohistochemistry of the cardiac gap junction protein connexin43. Circ Res 65: 1450–1457, 1989
37. Severs NJ: Gap junction alteration in the failing human heart. Eur Heart J 15: 53–57, 1994
38. Saffitz JE, Schuessler RB, Yamada KA: Mechanisms of remodeling of gap junction distributions and the development of anatomic substrates of arrhythmias. Cardiovasc Res 42: 309–317, 1999
39. DeBakker MJT, Van Capelle FJL, Janse MJ, Tasserson S, Vermeulen JT, de Jonge N, Lahpor JR: Slow conduction in the infarcted human heart. 'Zigzag' course of activation. Circulation 88: 915–926, 1993
40. Factor SM, Sonnenblick EH, Kirk ES: The histologic border zone of acute myocardial infarction – islands or peninsulas? Am J Pathol 92: 111–124, 1978
41. Spach MS, Josephson ME: The role of nonuniform anisotropy in small circuits. J Cardiovasc Electrophysiol 5: 182–209, 1994
42. Spach MS, Boineau JP: Microfibrosis produces electrical load variations due to loss of side-to-side connections: A major mechanism of structural heart disease. Pacing Clin Electrophysiol 20: 397–413, 1997
43. Maron BJ, Anan TJ, Roberts WC: Quantitative analysis of the distribution of cardiac muscle cell disorganization in the left ventricular wall of patients with hypertrophic cardiomyopathy. Circulation 63: 882–894, 1981
44. Kostin S: Morphological and morphometric characteristics of hypertrophic cardiomyopathy. Ark Patol 51: 47–53, 1989
45. McKenna WJ, Camm AJ: Sudden death in hypertrophic cardiomyopathy. Assessment of patients at high risk. Circulation 80: 1489–1492, 1989
46. Kucera JP, Rudy Y: Mechanistic insights into very slow conduction in branching cardiac tissue. Circ Res 89: 799–806, 2001
47. Peters NS: New insights into myocardial arrhythmogenesis: Distribution of gap-junctional coupling in normal, ischaemic and hypertrophied human hearts. Clin Sci 90: 447–452, 1996
48. Wang X, Gerdes AM: Chronic pressure overload cardiac hypertrophy and failure in quinea pigs: III. Intercalated disc remodeling. J Mol Cell Cardiol 31: 333–343, 1999
49. Lerner DL, Chapman Q, Green KG, Saffitz JE: Reversible down-regulation of connexin43 expression in acute cardiac allograft rejection. J Heart Lung Transplant 20: 93–97, 2001
50. Guerrero PA, Schuessler RB, Davis LM, Beyer EC, Johnson CM, Yamada KA, Saffitz JE: Slow ventricular conduction in mice heterogeneous for a connexin43 null mutation. J Clin Invest 99: 1991–1998, 1997
51. Gutstein DE, Morley GE, Tamaddon H, Vaidya D, Schneider MD, Chen J, Chien KR, Stuhlmann H, Fishman GI: Conduction slowing and sudden arrhythmic death in mice with cardiac-restricted inactivation of connexin43. Circ Res 88: 333–339, 2001

Molecular and Cellular Biochemistry **242:** 145–152, 2003.

Alterations in protein kinase C isoenzyme expression and autophosphorylation during the progression of pressure overload-induced left ventricular hypertrophy

Allison L. Bayer,[1] Maria C. Heidkamp,[2] Nehu Patel,[2] Michael Porter,[2] Steve Engman[2] and Allen M. Samarel[1,2]

[1]*The Cardiovascular Institute and the Department of Physiology;* [2]*Department of Medicine, Loyola University Chicago Stritch School of Medicine, Maywood, IL, USA*

Abstract

Cardiomyocytes express several isoenzymes of protein kinase C (PKC), which as a group have been implicated in the induction of left ventricular hypertrophy (LVH) and its transition to heart failure. Individual PKC isoenzymes also require transphosphorylation and autophosphorylation for enzymatic activity. To determine whether PKC isoenzyme expression and autophosphorylation are altered during LVH progression *in vivo*, suprarenal abdominal aortic coarctation was performed in Sprague-Dawley rats. Quantitative Western blotting was performed on LV tissue 1, 8 and 24 weeks after aortic banding, using antibodies specific for total PKCα, PKCδ and PKCε, and their C-terminal autophosphorylation sites. Aortic banding produced sustained hypertension and gradually developing LVH that progressed to diastolic heart failure over time. PKCε levels and autophosphorylation were not significantly different from sham-operated controls during any stage of LVH progression. PKCα expression levels were also unaffected during the induction of LVH, but increased 3.2 ± 0.8 fold during the transition to heart failure. In addition, there was a high degree of correlation between PKCα levels and the degree of LVH in 24 week banded animals. However, autophosphorylated PKCα was not increased at any time point. In contrast, PKCδ autophosphorylation was increased prior to the development of LVH, and also during the transition to heart failure. The increased PKCδ autophosphorylation in 1 week banded rats was not accompanied by an increase in total PKCδ, whereas total PKCδ levels were markedly increased (6.0 ± 1.7 fold) in 24 week banded animals. Furthermore, both phosphorylated and total PKCδ levels were highly correlated with the degree of LVH in 24 week banded rats. In summary, we provide indirect evidence to indicate that PKCδ may be involved in the induction of pressure overload LVH, whereas both PKCδ and PKCα may be involved in the transition to heart failure. (Mol Cell Biochem **242**: 145–152, 2003)

Key words: signal transduction, heart failure, autophosphorylation, cardiomyocytes

Introduction

During the induction of cardiac hypertrophy and its transition to heart failure, cardiomyocytes rely on a cascade of protein kinases to transmit mechanical and neurohormonal signals from the cell surface to the nucleus and other organelles [1]. At present, the most compelling evidence for a role for specific intracellular signaling kinases in regulating cardiomyocyte hypertrophy involves studies of the protein kinase C (PKC) family of serine-threonine protein kinases. Our knowledge of the structure and function of PKCs has increased dramatically since publication of the first experiments investigating load- and neurohormonally-induced PKC activation in hypertrophic signal transduction [2–7]. There are

Address for offprints: A.M. Samarel, The Cardiovascular Institute, Building 110, Room 5222, Loyola University Medical Center, 2160 South First Avenue, Maywood, IL 60153, USA (E-mail: asamare@lumc.edu)

at least 12 different PKC isoenzymes divided into three subfamilies based on their lipid and Ca^{2+} requirements. The 'classic' subfamily of PKCs includes α, β1, β2 and γ, which require diacylglycerol (DAG), phosphatidyl serine, and Ca^{2+} for activation. The 'novel' subfamily includes δ, ε, η, μ, and θ, and differ from cPKCs in that they do not require Ca^{2+} for activation. Less is known about the "atypical" aPKCs, except that they include ζ, ι, and λ and are not activated by either Ca^{2+} or DAG. Of these, the main isoenzymes found in rat cardiomyocytes are α, δ, ε, and ζ [8, 9]. Each isoenzyme may have distinct subcellular locations, biological substrates, and differential responses to hypertrophic stimuli.

Translocation of cPKCs and nPKCs from the soluble to particulate fraction of cell or tissue homogenates is a standard technique that has been used to assess PKC activation [1]. This translocation is due to association of active PKCs with isoenzyme-specific binding proteins present within distinct subcellular compartments. However, not all of the potential PKC binding proteins are localized to typical membrane components, calling into question the validity of the translocation assay to detect activation of all PKC isoenzymes [10]. Adding to this complexity is the fact that individual PKC isoenzymes require transphosphorylation and autophosphorylation for enzymatic activity (for review, see [11]). Three serine phosphorylation sites within the classical and novel PKCs have been identified. One site within the activation loop is transphosphorylated by 3-phosphoinositide-dependent protein kinase-1 (PDK1), and this phosphorylation facilitates subsequent PKC autophosphorylation at two additional sites within the C-terminus. Phosphorylation of all three sites is required for activation of PKCs by second messengers [12]. Recently developed antibodies that detect individual PKC isoenzymes only when phosphorylated on specific residues should therefore prove useful as another tool to detect PKC activation in complex cell or tissue homogenates.

Few studies have addressed how PKC isoenzyme expression and/or phosphorylation change during the induction of cardiac hypertrophy and its progression to heart failure. Gu and Sanford [13] demonstrated that PKC activity and concentration increased during the induction of pressure overload-induced left ventricular hypertrophy (LVH) in a rat model of ascending aortic banding. The increased PKC isoenzyme concentration in LV tissue homogenates was mainly limited to PKCβ_{1-2}, and PKCε. Similarly, PKCα, PKCε, and PKCγ protein abundance were increased in a continuous fashion from compensated LVH to heart failure in tissue extracts of guinea pig LV myocardium following descending thoracic aortic banding [14]. In contrast, Bowling *et al.* [15] showed that the expression levels of the Ca^{2+}-dependent PKC isoenzymes PKCβ_{1-2} and PKCα were markedly increased in failing human LV cardiomyocytes, whereas PKCε levels remained unchanged. Using differential centrifugation and Western blotting, these studies also attempted to analyze which PKC isoenzymes might be activated during the induction of LVH and its transition to heart failure. However no study to date has evaluated PKC autophosphorylation during LVH progression.

In the present report, we examined PKCα, PKCδ, and PKCε isoenzyme expression and serine autophosphorylation in an *in vivo* rat model of pressure overload-induced LVH. This model of suprarenal aortic coarctation has been extensively characterized in our laboratory [16–19] and consists of three stages. During the first stage, banded animals develop concentric LVH slowly over an 8 week period, followed by an established phase of compensatory LVH. Once established, LVH progresses to heart failure (16–24 weeks), with interstitial and perivascular fibrosis, slowed isovolumic relaxation, and increased LV end-diastolic pressure. Characteristic alterations in LV collagen content, and β-myosin heavy chain, ANF, and SERCA2 mRNA and protein levels are evident during disease progression [16, 17, 19]. Thus, this model is useful for examining *in vivo* PKC isoenzyme expression and autophosphorylation in relation to the induction of LVH and its transition to heart failure. Our results indicate that there are distinct differences in the expression and autophosphorylation levels of the three PKC isoenzymes examined, and suggest that each PKC isoenzyme may have different roles in LVH progression.

Materials and methods

Reagents

Antibodies specific for PKCε, PKCδ, and PKCα were obtained from Transduction Laboratories (Lexington, KY, USA). Anti-phospho-PKCα (pPKCα) antibody, specific for PKCα phosphorylated at Serine 657, was obtained from Upstate Biotechnology (Lake Placid, NY, USA). Anti-phospho-PKCδ (pPKCδ), specific for PKCδ phosphorylated at Serine 643, was obtained from Cell Signaling Technology (Beverly, MA, USA). Anti-phospho-PKCε (pPKCε) antibody, specific for PKCε phosphorylated at Serine 719, was also obtained from Upstate. All three serine residues represent C-terminal autophosphorylation sites for their respective PKCs. All other reagents were from Sigma Chemical (St. Louis, MO, USA), or Baxter S/P (McGaw Park, IL, USA).

Experimental animals

Animals used in these experiments were handled in accordance with the National Institute of Health Guide for the Care and Use of Laboratory Animals (DHHS Publication No.

(NIH) 85-23, Revised 1985). Male Sprague-Dawley rats weighing 160–170 g (Harlan Industries, Indianapolis, IN, USA) were anesthetized with an intramuscular injection of ketamine (90 mg/kg) and xylazine (10 mg/kg). Constriction of the suprarenal abdominal aorta was produced using a tantalum hemoclip (Edward Weck and Co.). The applicator was modified to allow passage of a 25-gauge needle through the hemoclip when completely closed. Sham-operated animals underwent dissection of the abdominal aorta without application of the suprarenal band. Postoperatively, all animals received food and water ad libitum. Prior to sacrifice, arterial blood pressure was recorded by carotid artery catheterization. There was a minimum of 4 animals in each group.

Western blotting

LV tissue was homogenized in lysis buffer (50 mM HEPES pH 7.4, 150 mM NaCl, 10% glycerol, 1.5 mM $MgCl_2$, 1.0 mM EGTA, 1.0 mM Na_3VO_4, 10 mM Na pyrophoshate, 100 mM NaF, 1% Triton X-100, 1% sodium deoxycholate, 0.1% SDS, 10 μg/ml leupeptin and aprotinin, and 1 mM Pefabloc). Equal amounts of extracted protein (200 μg) were separated on 10% SDS-polyacrylamide gels with 5% stacking gels. Proteins were transferred to PVDF membrane using the recommended transfer buffer. Western blots were probed with antibodies specific for PKCα, PKCδ, or PKCε, or the autophosphorylated forms of each isoenzyme (i.e. pPKCα, pPKCδ, or pPKCε). Primary antibody binding was detected with horseradish-peroxidase-conjugated goat anti-mouse or anti-rabbit secondary antibody and visualized by enhanced chemiluminescence (Amersham, Arlington Heights, IL, USA). Band intensity was quantified using laser densitometry.

Data analysis

Results were expressed as means ± S.E.M. Normality was assessed using the Kolmogorov-Smirnov test. Data were compared using unpaired *t*-test, rank sum test, or linear regression analysis, where appropriate. Differences among means were considered significant at $p < 0.05$. Data were analyzed using SigmaStat Statistical Software (Ver. 1.0, Jandel Scientific, San Rafael, CA, USA).

Results

Suprarenal aortic coarctation produced sustained hypertension and LVH

As documented in our previous studies [16–19], suprarenal aortic coarctation produced sustained hypertension and LVH which progressed to diastolic heart failure. Heart weight to body weight (HWBW) ratios were similar in 1-week sham and banded animals (3.9 ± 0.1 vs. 4.0 ± 0.1 g/kg for 1-week sham vs. 1-week banded rats) during the induction phase of LVH. During the established phase of LVH, HWBW ratios were significantly increased (3.1 ± 0.1 vs. 3.9 ± 0.4 g/kg in 8-week sham vs. 8-week banded rats; $p < 0.05$). However, the degree of LVH did not further increase during the transition to heart failure (2.9 ± 0.04 vs. 3.8 ± 0.1 g/kg in 24-week sham vs. and banded animals; $p < 0.05$).

PKCε expression and phosphorylation were unaffected during LVH progression

We first examined PKCε expression levels during each stage of LVH progression. As seen in Figs 1A and 1B, PKCε levels were not significantly increased in LV tissue extracts of banded rats. We also did not detect a disproportionate increase in the level of pPKCε during LVH progression (Fig 1C and 1D). Furthermore, there was no correlation between the degree of hypertrophy (as assessed by HWBH ratio) and either pPKCε or total PKCε concentration at any time point during the development of LVH and its transition to heart failure (data not shown).

PKCε expression but not phosphorylation was increased during the transition to heart failure

As was the case with PKCε, neither the level of expression, nor the phosphorylation state of PKCα was increased during the induction and compensatory phases of LVH progression (Figs 2A and 2B). However, in agreement with previous studies involving guinea pigs [14] and humans [15] with heart failure, PKCα expression was markedly increased (3.2 ± 0.8 fold) during the transition to heart failure. Further analysis of this group of animals indicated that PKCα expression was highly correlated with the degree of LVH. Animals in the transition phase with the greatest degree of LVH had the greatest increase in PKCα expression (Fig. 2C). Despite the increase in PKCα levels, however, there was no significant increase in pPKCα tissue levels during any phase of LVH progression (Figs 3A and 3B).

Alterations in PKCδ expression and phosphorylation during LVH progression

As seen in Figs 4A and 4B, total PKCδ levels were not significantly increased in LV tissue homogenates of 1-week or 8-week banded rats. However, as was the case for PKCα, total

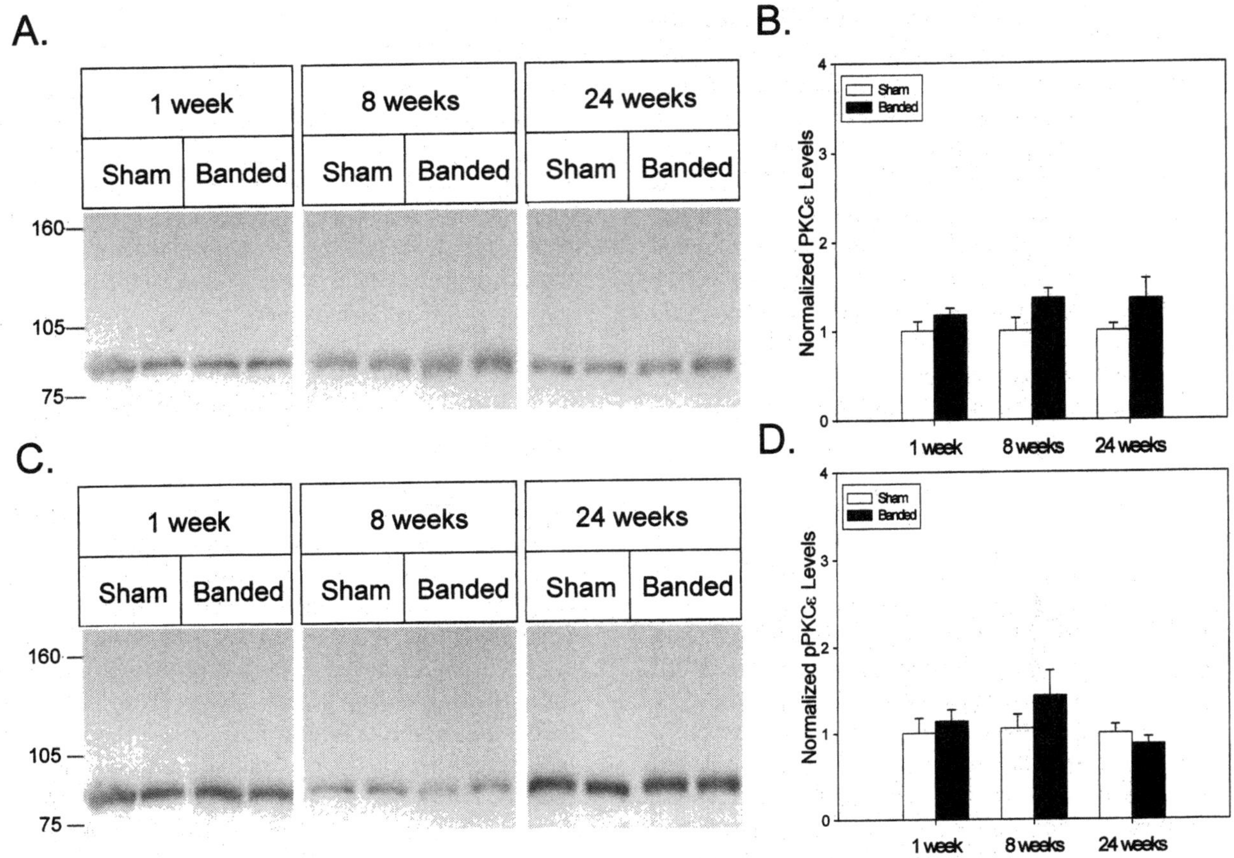

Fig. 1. PKCε expression and phosphorylation during LVH progression. LV tissue extracts from 1, 8, or 24 week-sham or banded animals were separated by SDS-PAGE followed by Western blotting with anti-PKCε mAb. (A) Representative Western blots for PKCε (200 μg tissue protein) are depicted. The position of molecular weight markers is indicated to the left of the Western blot. (B) The quantitative analysis of Western blotting experiments from 1, 8, and 24 week animals is depicted. PKCε levels were normalized to the levels observed in time-matched, sham-operated controls. Data are means ± S.E.M. for 4–13 animals in each group. (C) The same LV tissue extracts from 1, 8, or 24 week-sham or banded animals were probed with a pAb specific for PKCε phosphorylated at Serine 719 (pPKCε). Representative Western blots for pPKCε (200 μg tissue protein) are depicted. The position of molecular weight markers is indicated to the left of the Western blots. (D) The quantitative analysis of Western blotting experiments from 1, 8, and 24 week animals is depicted. pPKCε levels were normalized to the levels observed in time-matched, sham-operated controls. Data are means ± S.E.M. for 4–13 animals in each group.

PKCδ levels were markedly increased (6.0 ± 1.7 fold) during the transition to heart failure. Similarly, animals in the transition phase with the greatest degree of LVH had the greatest increase in PKCδ expression (Fig. 4C). Despite the lack of a significant increase in total PKCδ levels within the first week following abdominal aortic coarctation (Fig. 4B), pPKCδ were found to be significantly increased by 2.8 ± 0.8 fold as compared to sham-operated control animals (Figs 5A and 5B). This increase in pPKCδ preceded the onset of significant LVH. Once established, pPKCδ levels returned to control levels, only to increase again (along with the increase in total PKCδ levels) during the transition to heart failure. As was the case with total PKCδ expression, animals in the transition phase with the greatest degree of LVH had the greatest increase in pPKCδ (Fig. 5C).

Discussion

LV pressure overload causes a marked increase in systolic wall stress, which is a potent stimulus for the induction of cardiomyocyte hypertrophy [20]. The ensuing increase in contractile protein mass provides a means to increase LV systolic pressure in order to compensate for the increased afterload, thus maintaining cardiac output at near normal levels for several months. Despite this compensatory growth, a variety of alterations in cardiomyocyte structure and function develops over time that ultimately leads to the failure of the LV to maintain normal cardiac output without significantly elevated filling pressures (i.e. diastolic heart failure). Clearly, investigators interested in the induction of LVH and its transition to heart failure have focused on identifying the

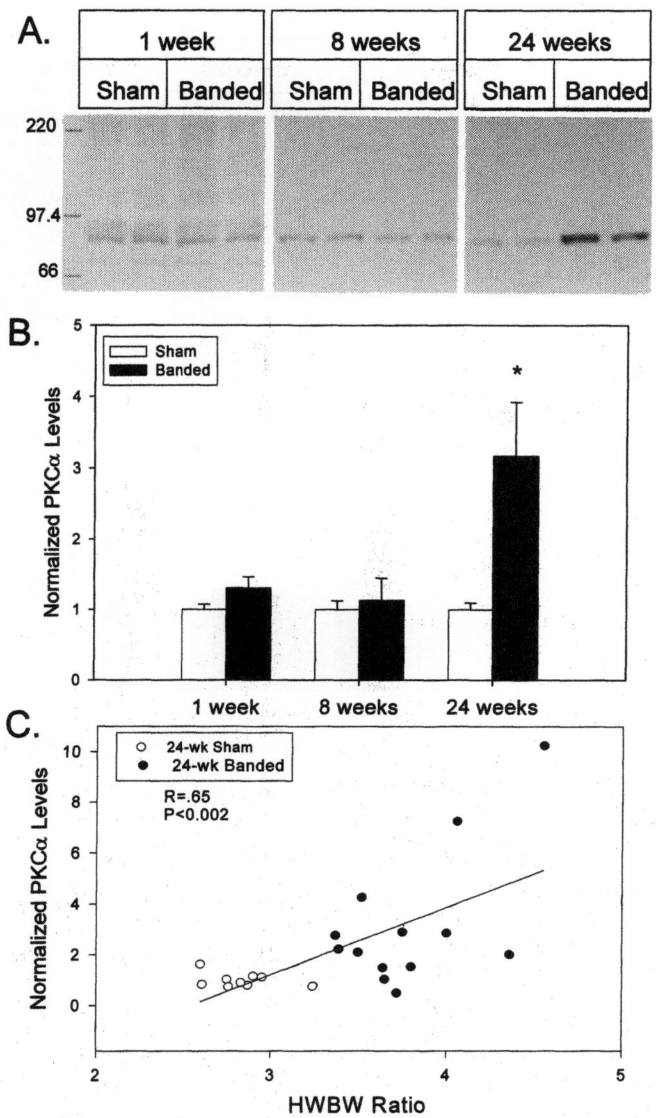

Fig. 2. PKCα expression during LVH progression. (A) LV tissue extracts from 1, 8, or 24 week-sham or banded animals were separated by SDS-PAGE followed by Western blotting with anti-PKCα mAb. Representative Western blots for PKCα (200 μg tissue protein) are depicted. The position of molecular weight markers is indicated to the left of the Western blot. (B) The quantitative analysis of Western blotting experiments from 1, 8, and 24 week animals is depicted. PKCα levels were normalized to the levels observed in time-matched, sham-operated controls. Data are means ± S.E.M. for 4–13 animals in each group. *$p < 0.05$ vs. sham-operated controls. (C) Normalized PKCα expression levels are plotted vs. HWBW ratios in 24 week-sham (open circles) and banded (closed circles) animals. Data were analyzed by linear regression for 9–13 animals in each group.

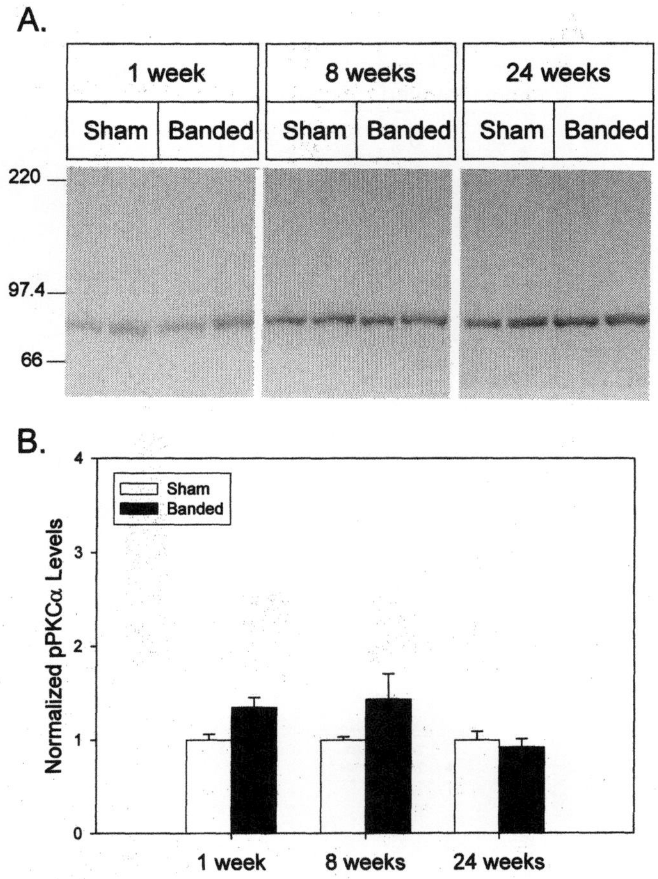

Fig. 3. pPKCα levels during LVH progression. LV tissue extracts from 1, 8, or 24 week-sham or banded animals were probed with a pAb specific for PKCα phosphorylated at serine 657 (pPKCα). (A) Representative Western blots for PKCα (200 μg tissue protein) are depicted. The position of molecular weight markers is indicated to the left of the Western blots. (B) The quantitative analysis of Western blotting experiments from 1, 8, and 24 week animals is depicted. pPKCα levels were normalized to the levels observed in time-matched, sham-operated controls. Data are means ± S.E.M. for 4–13 animals in each group.

signaling pathways that are responsible for converting mechanical stimuli (i.e. increased wall stress) into biochemical signals resulting in adaptive cardiomyocyte hypertrophy. However, it is conceivable that additional signaling pathways may herald the transition from compensated hypertrophy to contractile dysfunction. In this report, we provide indirect evidence to indicate that PKCδ may be involved in the induction of pressure overload LVH, whereas both PKCδ and the PKCα may be involved in the transition to heart failure.

Surprisingly few studies have interrogated the role of PKCδ in cardiomyocyte hypertrophic signaling. Recently, Chen *et al.* [21] described a transgenic mouse with sustained activation of PKCδ due to cardiac-specific overexpression of a pseudoRACK (Receptor for Activated C Kinase) peptide activator of PKCδ (ψδRACK mice). These animals developed LVH with normal LV systolic function, which was similar to another line of mice (ψεRACK mice) with cardiac-specific overexpression of a pseudoRACK peptide activator of PKCε [22]. Thus, sustained activation of either novel PKC was sufficient to induce cardiomyocyte hypertrophy. We previously showed that relatively modest activation of PKCδ in the absence of substantial activation of PKCε was

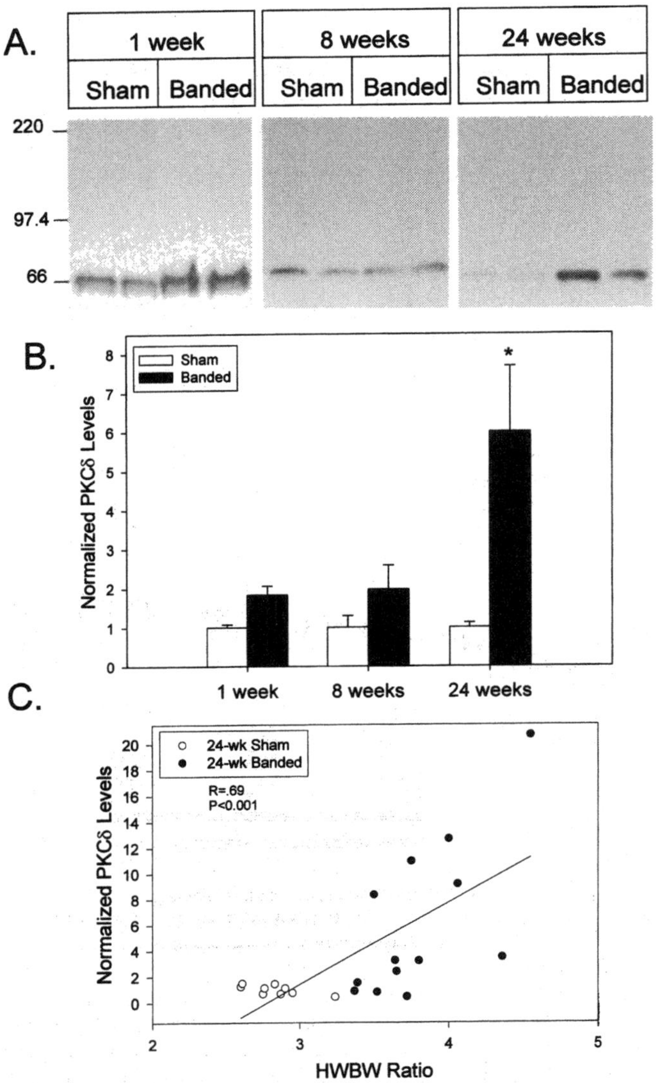

Fig. 4. PKCδ expression during LVH progression. LV tissue extracts from 1, 8, or 24 week-sham or banded animals were separated by SDS-PAGE followed by Western blotting with anti-PKCδ mAb. (A) Representative Western blots for PKCδ (200 μg tissue protein) are depicted. The position of molecular weight markers is indicated to the left of the Western blot. (B) The quantitative analysis of Western blotting experiments from 1, 8, and 24 week animals is depicted. PKCδ levels were normalized to the levels observed in time-matched, sham-operated controls. Data are means ± S.E.M. for 4–13 animals in each group. *$p < 0.05$ vs. sham-operated controls. (C) Normalized PKCδ expression levels are plotted vs. HWBW ratios in 24 week-sham (open circles) and banded (closed circles) animals. Data were analyzed by linear regression for 9–13 animals in each group.

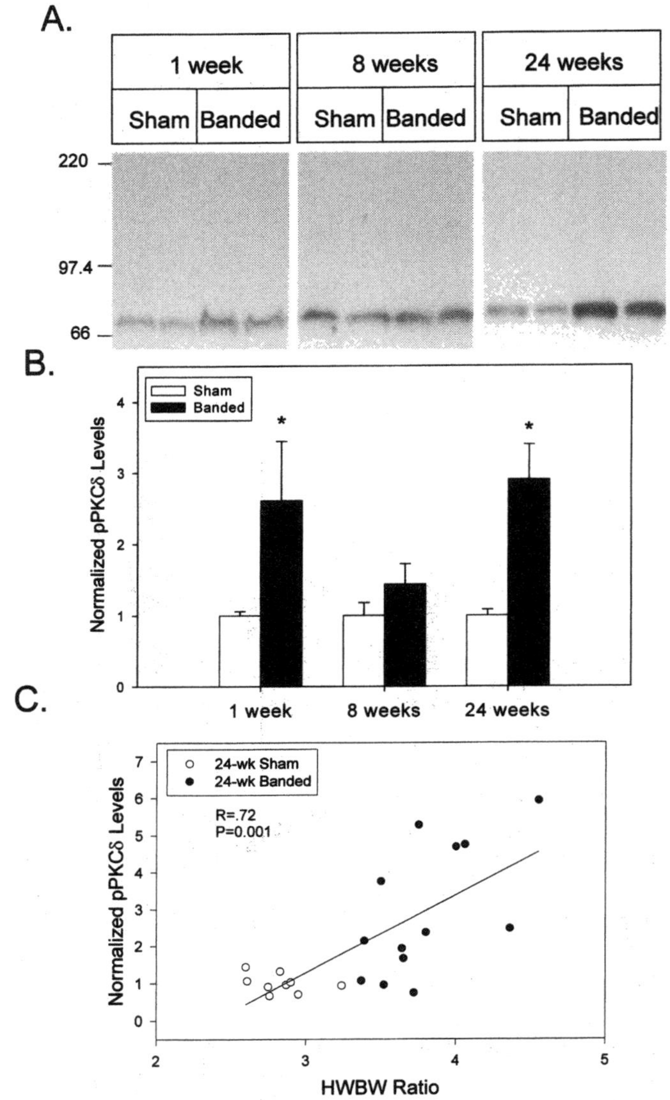

Fig. 5. pPKCδ levels during LVH progression. LV tissue extracts from 1, 8, or 24 week-sham or banded animals were probed with a pAb specific for PKCδ phosphorylated at serine 643 (pPKCδ). (A) Representative Western blots for pPKCδ (200 μg tissue protein) are depicted. The position of molecular weight markers is indicated to the left of the Western blots. (B) The quantitative analysis of Western blotting experiments from 1, 8, and 24 week animals is depicted. pPKCδ levels were normalized to the levels observed in time-matched, sham-operated controls. Data are means ± S.E.M. for 4–13 animals in each group. $p < 0.05$ vs. sham-operated controls. (C) Normalized pPKCδ expression levels are plotted vs. HWBW ratios in 24 week-sham (open circles) and banded (closed circles) animals. Data were analyzed by linear regression for 9–13 animals in each group.

sufficient to induce hypertrophy in cultured cardiomyocytes [23]. Similarly, overexpression of a constitutively active mutant of PKCε was sufficient to induce some features of pressure overload-induced LVH in neonatal cardiomyocyte cultures [24, 25], and in transgenic mice [26]. However, we also showed that PKCε activation may not be necessary for this response when PKCδ is also activated [24]. As indicated above, we did not detect a substantial increase in PKCε expression or autophosphorylation during any stage of LVH progression, whereas PKCδ expression and phosphorylation were markedly affected. Thus, we provide additional evidence to suggest that there is substantial crosstalk in the PKCε- and PKCδ-dependent pathways that are involved in hypertrophic signal transduction.

Whereas the novel PKCs have long been considered to be important in hypertrophic signaling, the Ca^{2+}-dependent PKC isoenzymes have been implicated in the development of heart failure. Most attention has been focused on the PKCβ isoenzymes, although normal adult rat cardiomyocytes express very little, if any PKCβ [8, 27, 28]. However, $PKC\beta_1$ and $PKC\beta_2$ concentrations may increase somewhat in diseased tissue [15]. Wakasaki *et al.* [29] have shown that cardiac-specific overexpression of constitutively active $PKC\beta_2$ induced dilated cardiomyopathy and overt heart failure in transgenic mice, whereas both PKCα and $PKC\beta_{1-2}$ levels were increased in failing hearts of patients with ischemic or dilated cardiomyopathies [15]. Again, considerably less is known about the role of PKCα in the induction of hypertrophy and its transition to heart failure. PKCα does not appear to be activated by most stimuli which induce hypertrophy in cultured cardiomyocytes [23, 30, 31], but it should be emphasized that these activation studies were conducted using differential sedimentation and Western blotting to measure PKC activation. Total PKCα levels were markedly increased, and PKCα (but not PKCε) was enriched in the membrane fraction of failing human hearts [15]. A similar pattern of translocation was observed in mice with LVH and heart failure induced by overexpression of L-type Ca^{2+} channels [32]. It would be particularly interesting to examine whether PKCα autophosphorylation was also enhanced under these conditions. In our study, we identified a 3.2-fold increase in PKCα expression, but no increase in PKCα autophosphorylation during the transition to heart failure. These results suggest that PKCα was not substantially 'primed' and signaling at this time point in LVH progression. However, it is conceivable that PKCα autophosphorylation would also increase as the animals developed worsening heart failure over time. Furthermore, it is unclear whether the increased PKCα (or PKCδ) abundance originated from cardiomyocytes, or from proliferating cardiac fibroblasts [16]. Indeed, a detailed analysis of PKC isoenzyme expression, translocation, and phosphorylation (at the activation loop as well as the C-terminal autophosphorylation sites of each enzyme) in pure populations of cardiomyocytes isolated during each stage of LVH progression would be required to further elucidate the significance of these changes.

Acknowledgements

This work was supported in part by NIH grants HL34328 and HL63711, and a gift to the Cardiovascular Institute from the Dr. Ralph and Marian Falk Trust for Medical Research. Drs. Bayer and Heidkamp were recipients of NIH National Research Service Awards F32 HL10313 and F32 HL68476, respectively, and Mr. Engman was a recipient of an AHA-Midwest Predoctoral Fellowship during the time these studies were performed.

References

1. Bogoyevitch MA, Sugden PH: The role of protein kinases in adaptational growth of the heart. Int J Biochem Cell Biol 28: 1–12, 1996
2. Dunnmon PM, Iwaki K, Henderson SA, Sen A, Chien KR: Phorbol esters induce immediate-early genes and activate cardiac gene transcription in neonatal rat myocardial cells. J Mol Cell Cardiol 22: 901–910, 1990
3. Samarel AM, Engelmann GL: Contractile activity modulates myosin heavy chain-β expression in neonatal rat heart cells. Am J Physiol 261: H1067–H1077, 1991
4. Komuro I, Katoh Y, Kaida T, Shibazaki Y, Kurabayashi M, Hoh E, Takaku F, Yazaki Y: Mechanical loading stimulates cell hypertrophy and specific gene expression in cultured rat cardiac myocytes. Possible role of protein kinase C activation. J Biol Chem 266: 1265–1268, 1991
5. Kariya K, Karns LR, Simpson PC: Expression of a constitutively activated mutant of the β-isozyme of protein kinase C in cardiac myocytes stimulates the promoter of the β-myosin heavy chain isogene. J Biol Chem 266: 10023–10026, 1991
6. Allo SN, Carl LL, Morgan HE: Acceleration of growth of cultured cardiomyocytes and translocation of protein kinase C. Am J Physiol 263: C319–C325, 1992
7. Sadoshima J, Takahashi T, Jahn L, Izumo S: Roles of mechano-sensitive ion channels, cytoskeleton, and contractile activity in stretch-induced immediate-early gene expression and hypertrophy of cardiac myocytes. Proc Natl Acad Sci USA 89: 9905–9909, 1992
8. Rybin VO, Steinberg SF: Protein kinase C isoform expression and regulation in the developing rat heart. Circ Res 74: 299–309, 1994
9. Clerk A, Bogoyevitch MA, Fuller SJ, Lazou A, Parker PJ, Sugden PH: Expression of protein kinase C isoforms during cardiac ventricular development. Am J Physiol 269: H1087–H1097, 1995
10. Ron D, Kazanietz MG: New insights into the regulation of protein kinase C and novel phorbol ester receptors. FASEB J 13: 1658–1676, 1999
11. Parekh DB, Ziegler W, Parker PJ: Multiple pathways control protein kinase C phosphorylation. EMBO J 19: 496–503, 2000
12. Dempsey EC, Newton AC, Mochly-Rosen D, Fields AP, Reyland ME, Insel PA, Messing RO: Protein kinase C isozymes and the regulation of diverse cell responses. Am J Physiol Lung Cell Mol Physiol 279: L429–L438, 2000
13. Gu X, Bishop SP: Increased protein kinase C and isozyme redistribution in pressure-overload cardiac hypertrophy in the rat. Circ Res 75: 926–931, 1994
14. Jalili T, Takeishi Y, Song G, Ball NA, Howles G, Walsh RA: PKC translocation without changes in Gαq and PLC-β protein abundance in cardiac hypertrophy and failure. Am J Physiol 277: H2298–H2304, 1999
15. Bowling N, Walsh RA, Song G, Estridge T, Sandusky GE, Fouts RL, Mintze K, Pickard T, Roden R, Bristow MR, Sabbah HN, Mizrahi JL, Gromo G, King GL, Vlahos CJ: Increased protein kinase C activity and expression of Ca^{2+}-sensitive isoforms in the failing human heart. Circulation 99: 384–391, 1999
16. Eleftheriades EG, Durand JB, Ferguson AG, Engelmann GL, Jones SB, Samarel AM: Regulation of procollagen metabolism in the pressure-overloaded rat heart. J Clin Invest 91: 1113–1122, 1993
17. Qi M, Shannon TR, Euler DE, Bers DM, Samarel AM: Downregulation of sarcoplasmic reticulum Ca^{2+}-ATPase during progression of left ventricular hypertrophy. Am J Physiol 272: H2416–H2424, 1997

18. Delbridge LM, Satoh H, Yuan W, Bassani JW, Qi M, Ginsburg KS, Samarel AM, Bers DM: Cardiac myocyte volume, Ca^{2+} fluxes, and sarcoplasmic reticulum loading in pressure-overload hypertrophy. Am J Physiol 272: H2425–H2435, 1997
19. McCall E, Ginsburg KS, Bassani RA, Shannon TR, Qi M, Samarel AM, Bers DM: Ca flux, contractility, and excitation-contraction coupling in hypertrophic rat ventricular myocytes. Am J Physiol 274: H1348–H1360, 1998
20. Imamura T, McDermott PJ, Kent RL, Nagatsu M, Cooper GT, Carabello BA: Acute changes in myosin heavy chain synthesis rate in pressure versus volume overload. Circ Res 75: 418–425, 1994
21. Chen L, Hahn H, Wu G, Chen CH, Liron T, Schechtman D, Cavallaro G, Banci L, Guo Y, Bolli R, Dorn GW, Mochly-Rosen D: Opposing cardioprotective actions and parallel hypertrophic effects of δPKC and εPKC. Proc Natl Acad Sci USA 98: 11114–11119, 2001
22. Mochly-Rosen D, Wu G, Hahn H, Osinska H, Liron T, Lorenz JN, Yatani A, Robbins J, Dorn GW: Cardiotrophic effects of protein kinase Cε: Analysis by *in vivo* modulation of PKCε translocation. Circ Res 86: 1173–1179, 2000
23. Strait JB, Samarel AM: Isoenzyme-specific protein kinase C and c-Jun N-terminal kinase activation by electrically stimulated contraction of neonatal rat ventricular myocytes. J Mol Cell Cardiol 32: 1553–1566, 2000
24. Strait JB, Martin JL, Bayer A, Mestril R, Eble DM, Samarel AM: Role of protein kinase C-ε in hypertrophy of cultured neonatal rat ventricular myocytes. Am J Physiol Heart Circ Physiol 280: H756–H766, 2001
25. Heidkamp MC, Bayer AL, Martin JL, Samarel AM: Differential activation of mitogen-activated protein kinase cascades and apoptosis by protein kinase C ε and δ in neonatal rat ventricular myocytes. Circ Res 89: 882–890, 2001
26. Takeishi Y, Ping P, Bolli R, Kirkpatrick DL, Hoit BD, Walsh RA: Transgenic overexpression of constitutively active protein kinase Cε causes concentric cardiac hypertrophy. Circ Res 86: 1218–1223, 2000
27. Bogoyevitch MA, Parker PJ, Sugden PH: Characterization of protein kinase C isotype expression in adult rat heart. Protein kinase C-ε is a major isotype present, and it is activated by phorbol esters, epinephrine, and endothelin. Circ Res 72: 757–767, 1993
28. Clerk A, Bogoyevitch MA, Parker PJ, Sugden PH: Down-regulation of protein kinase C isotypes during postnatal development of rat heart. Biochem Soc Trans 21: 385S, 1993
29. Wakasaki H, Koya D, Schoen FJ, Jirousek MR, Ways DK, Hoit BD, Walsh RA, King GL: Targeted overexpression of protein kinase $C\beta_2$ isoform in myocardium causes cardiomyopathy. Proc Natl Acad Sci USA 94: 9320–9325, 1997
30. Puceat M, Hilal-Dandan R, Strulovici B, Brunton LL, Brown JH: Differential regulation of protein kinase C isoforms in isolated neonatal and adult rat cardiomyocytes. J Biol Chem 269: 16938–16944, 1994
31. Clerk A, Bogoyevitch MA, Anderson MB, Sugden PH: Differential activation of protein kinase C isoforms by endothelin-1 and phenylephrine and subsequent stimulation of p42 and p44 mitogen- activated protein kinases in ventricular myocytes cultured from neonatal rat hearts. J Biol Chem 269: 32848–32857, 1994
32. Muth JN, Bodi I, Lewis W, Varadi G, Schwartz A: A Ca^{2+}-dependent transgenic model of cardiac hypertrophy: A role for protein kinase Cα. Circulation 103: 140–147, 2001

Molecular and Cellular Biochemistry **242:** 153–161, 2003.

CaM kinase IIδC phosphorylation of 14-3-3β in vascular smooth muscle cells: Activation of class II HDAC repression

Joel J. Ellis, Thomas G. Valencia, Hong Zeng, L. Don Roberts, Rebecca A. Deaton and Stephen R. Grant

Laboratory of Cardiac and Vascular Molecular Genetics, Cardiovascular Research Institute and Department of Integrative Physiology, University of North Texas Health Science Center at Fort Worth, Fort Worth, TX, USA

Abstract

The myocyte enhancer factor-2 (MEF2) family of transcription factors regulates transcription of muscle-dependent genes in cardiac, skeletal and smooth muscle. They are activated by calcium/calmodulin (CaM)-dependent protein kinases I and IV and silenced by CaM KIIδC. MEF2 is held in an inactive form by the class II histone deacetylases (HDAC) until phosphorylated by either CaM kinase I or IV. Upon phosphorylation, HDAC is transported out of the nucleus via a 14-3-3 dependent mechanism freeing MEF2 to drive transcription. The 14-3-3 chaperone protein exists as a homodimer. In the region of homodimerization, there are two canonical CaM kinase II phosphorylation sites (ser60 and ser65). *In vitro* phosphorylation assay results indicate that 14-3-3β is indeed a substrate for CaM kinase II. We hypothesize that CaM kinase IIδC phosphorylation of 14-3-3β will disrupt homodimer formation resulting in the return of HDAC to the nucleus and their reassociation with MEF2. To test this, we mutated serines 60 and 65 of 14-3-3β to aspartates to mimic the phosphorylated state. In MEF2 enhancer-reporter assays in smooth muscle cells, expression of the 14-3-3β double mutant attenuated MEF2-enhancer activity driven by CaM kinase I or IV. The intracellular fate of HDAC4 was followed by transfection of smooth muscle cells with an HDAC4-Green Fluorescent Protein fusion hybrid. The 14-3-3β double mutant prevented HDAC4 cytoplasmic localization in the presence of active CaM kinase I or IV. These data suggest that the mechanism of CaM kinase IIδC silencing of MEF-2-dependent genes is by phosphorylation of 14-3-3β, which allows HDAC to return to the nucleus to reform a complex with MEF2, thereby silencing MADS box-dependent gene induction in smooth muscle. (Mol Cell Biochem **242**: 153–161, 2003)

Key words: CaM kinase II, 14-3-3, HDAC, transcriptional silencing, vascular smooth muscle, MEF2

Introduction

The myocyte enhancer factor-2 (MEF2) family of transcription factors is known to be involved in the transcriptional regulation of skeletal, cardiac and vascular smooth muscle-specific gene expression. The MADS box domain, CTA(A/T)$_4$TAG, is found in the proximal promoter regions of the majority of muscle-specific genes. In vascular smooth muscle, MEF2 has been demonstrated to regulate expression of Smooth Muscle Myosin Heavy Chain (SM-MHC) [1] and is required for differentiation of smooth muscle cells *in vivo* [2].

MEF2 transcription factors have been demonstrated to be regulated by Class II Histone Deacetylases (HDACs) [3–5]. HDAC4, HDAC5 and HDAC7 bind to MEF2 transcription factors within the nucleus thereby repressing MEF2-dependent gene transcription. In addition, activation of the CaM kinase kinase cascade (specifically CaM KI and CaM KIV), which is known to activate MEF2-dependent transcription, results in the destabilization of the transcriptionally inactive MEF2/HDAC complex [6]. Release of MEF2 from this complex, through a CaM kinase dependent phosphorylation, activates MEF2 as a transcription factor, thereby allowing

Address for offprints: S.R. Grant, Laboratory of Cardiac and Vascular Molecular Genetics, Cardiovascular Research Institute and Department of Integrative Physiology, University of North Texas Health Science Center at Fort Worth, 3500 Camp Bowie Boulevard, Fort Worth, TX 76107, USA

MEF2-dependent transcription to occur. Unbound HDAC is sequestered by the chaperone protein 14-3-3, which actively exports the HDAC out of the nucleus and into the cytoplasm [7, 8]. The cytoplasmic sequestering of class II HDACs by 14-3-3 attenuates class II repression of MEF2-driven muscle-specific gene expression.

Our laboratory has recently demonstrated that another member of the CaM kinase family, CaM kinase II, silences the activation of MEF2-dependent transcription mediated by CaM KI or CaM KIV. Interestingly, there are two CaM KII consensus phosphorylation motifs (Arg-X-X-Ser/Thr) within the amino acid sequence of 14-3-3β (serines 60 and 65). These consensus phosphorylation motifs lie within the dimerization region of 14-3-3β monomers immediately adjacent to glutamate residues in the partnering dimer. In order to function as an efficient nucleocytoplasmic chaperone, 14-3-3β must exist as a dimer whereby one monomer serves to bind a protein to be transported and the other binds an anchored ligand such as CRM1, which would dictate compartmentalization [8–12]. We have hypothesized that the phosphorylation of these serine residues by CaM KIIδC in the cytoplasm will result in (i) the destabilization of the 14-3-3β dimer, (ii) a block in cytoplasmic sequestration of class II HDACs by 14-3-3β, and (iii) re-entry of class II HDACs into the nucleus. Re-entry of the HDACs will promote sequestration of MEF2 transcription factors, and thus silence MEF2-dependent transcription. In the present study, we examined the existence of the CaM kinase-MEF2 signaling pathway in a vascular smooth muscle cell line, PAC-1. We provide evidence that CaM KI and IV activate expression of a prototypical smooth muscle specific gene, SM-MHC. In addition, we demonstrate that CaM KII can silence the activation of this promoter. In addition, we provide data demonstrating that CaM KII directly phosphorylates 14-3-3β. Finally, we investigated the affects of a conservative amino acid substitution of serine 60 and 65 (two putative CaM KII phosphorylation motifs of 14-3-3β) for aspartate residues. We anticipated that this substitution would mimic the negative charge of phosphorylation of these residues by CaM KII. Indeed, the 14-3-3β double mutant blocks CaM KI, CaM KIV and PE-mediated activation of MEF2-dependent transcription. In addition, we demonstrate that the 14-3-3β double mutant alters the subcellular localization of HDAC4 allowing it to re-enter the nucleus where it would silence MEF2-dependent transcription.

Materials and methods

Cell culture and reagents

Rat pulmonary arterial smooth muscle (PAC-1) cells [13] were cultured in monolayers using medium 199 supplemented with 10% fetal bovine serum (FBS). Fetal bovine serum, LipofectAMINE™ transfection solution, and medium 199 were purchased from Life Technologies, Inc. (Rockville, MD, USA); Medium 199 consists of Earle's salts, L-glutamine, and 2.2 grams of sodium bicarbonate per liter; Luciferase substrate was purchased from Promega (Madison, WI, USA); Phenylephrine was purchased from Sigma (St. Louis, MO, USA); γ[P^{32}] ATP was obtained from Perkin Elmer Life Sciences (Boston, MA, USA).

In vitro *CaM kinase II phosphorylation of the chaperone protein 14-3-3β*

Confluent monolayers of PAC-1 cells were cultured in a 100 mm culture dish and incubated at 37°C in a humidified atmosphere of 5% CO_2 until 70% confluency. The cells were then transfected with CMV driven 14-3-3β plasmid DNA using the LipofectAMINE™ commercial protocol. The cells were lysed 24 h post transfection and 14-3-3β was immunoprecipitated using a c-myc antibody (Upstate Biotechnology). Purified CaM KII, 14-3-3β, γ[P^{32}] ATP, Ca^{2+} and calmodulin were co-incubated in a 5 mL tube at 37°C for 30 min in the presence and absence of EGTA, a Ca^{2+} chelating agent. The *in vitro* kinase reactions were run on a 4–12% gradient NuPage SDS polyacrylamide gel (Invitrogen) along with a molecular weight standard. The results were visualized by autoradiography.

PAC-1 transfection assays

PAC-1 cells were seeded in 12-well plates at a density of 2.8 × 10^5 cells per plate and cultured until 60–70% confluent. The cells were then transfected for 16 h with varying treatments of DNA using the LipofectAMINE™ commercial protocol. The cells were washed with serum free 199 media and refed with 0.2% FBS in 199 media. After 24 h of incubation at 37°C, the cells were washed with 1 × PBS (Life Technologies, Inc.) and lysed with 1 × lysis buffer (Promega). The lysed cells were scraped from the well bottom and placed into a luminometer cuvette. Luciferase substrate was added and relative luciferase activity was analyzed using a Turner TD20e luminometer.

HDAC4 subcellular localization

PAC-1 cells were cultured as previously described. Cells were transfected (using LipofectAMINE™) with GFP alone, GFP-HDAC4 alone or GFP-HDAC4 in the absence or presence of active CaM KI, CaM KIV, CaM KII, 14-3-3β wild type, 14-3-3β mutant or combinations of these enzymes. Twenty-four hours post transfection, cells were washed twice with 1X PBS and treated with 3.5% formaldehyde for 30 min at room temperature. Cells were again washed twice with 1X PBS and incubated with 300 nM DAPI stain for 10 min. Cells were rinsed with 1X PBS 3 times and mounted onto glass slides using Prolong antifade reagent (Molecular Probes). Fluorescence was visualized using a Nikon Microphot FXA microscope.

DNA constructs

The plasmids containing GFP-HDAC4 and c-myc tagged 14-3-3β were gifts from Stuart L. Schreiber (Harvard University, MA, USA) and were described previously [11]. The constitutively active constructs for CaM KI (pSRa/CaM KI (1–294)) and CaM KIV (pSG5/CaM KIV (1–313)) were provided by Thomas Soderling (Vollum Institute, Oregon Health Science Center, OR, USA) and were described previously [14, 15]. CaM KII constructs were provided by Howard Schulman (Stanford School of Medicine, CA, USA) and were described previously [16]. A 1.2 Kb SM-MHC promoter-reporter construct was provided by Gary Owens (University of Virginia School of Medicine, VA, USA) and was described previously [17]. In order to quantify MEF2 mediated gene transcription, a luciferase-based MEF2 enhancer-reporter plasmid construct, containing three MADS box repeats upstream from a minimal promoter and structural luciferase gene, was provided by Eric Olson (University of Texas Southwestern Medical Center, TX, USA) and was previously described [18]. 14-3-3β was conservatively point mutated at serine 60 and serine 65 to aspartate using a commercial mutagenesis kit (QuickChange™, Stratagene).

Results

CaM KI and CaM KIV activate the 1.2 Kb SM-MHC promoter-reporter in PAC-1 cells

Recent studies by us [19, 20] and others [6, 7] have shown that activation of the CaM kinase kinase signaling cascade in myocytes leads to the induction of MEF2-dependent transcription of muscle-specific genes. Activation of MEF2 was shown to be the direct result of phosphorylation of the transcriptionally inactive HDAC/MEF2 complex resulting in the sequestration of HDAC in the cytoplasm by the chaperone protein, 14-3-3 [7, 11]. In other studies from our laboratory using both constitutive and inducible expression vectors harboring an active form of the cytoplasmic enzyme CaM KII, results have demonstrated that promoter-reporter activity for MEF2-dependent transcription can be completely silenced in rat neonate primary cardiomyocyte cultures. While this cascade has been characterized in cardiac and skeletal muscle, it has yet to be established in vascular smooth muscle.

In this study, we set out to investigate if MEF2-dependent gene expression in the smooth muscle cell line, PAC-1, could be controlled in a positive manner by CaM kinase kinase signaling and silenced by CaM KIIδC activity. SM-MHC promoter-reporter activity was measured in transiently transfected PAC-1 cell cultures. Results of this experiment are shown in Fig. 1. In the absence of either active CaM Kinase I or CaM Kinase IV, SM-MHC reporter activity was quite low (empty vector control). When active CaM KI or active CaM KIV was over-expressed, reporter activity was increased 8-fold (CaM KI) and 4-fold (CaM KIV) over baseline values. Exogenous expression of active CaM KIIδC silenced reporter activity well below control values. When active CaM KI or CaM KIV was co-expressed with active CaM KIIδC, reporter activity was decreased 4-fold below CaM KI induced values and decreased approximately 2.5-fold below CaM KIV values. These results verify that the expression of active forms of either CaM KI or CaM KIV in smooth muscle cell cultures strongly up-regulated SM-MHC promoter-reporter activity. The data support the argument that CaM KI and IV will induce transcription of this predominant smooth muscle contractile protein. Importantly, these results also demonstrate that the expression of active CaM KIIδC silenced CaM KI- or CaM KIV-mediated induction of SM-MHC promoter-reporter activity in PAC-1 cultures.

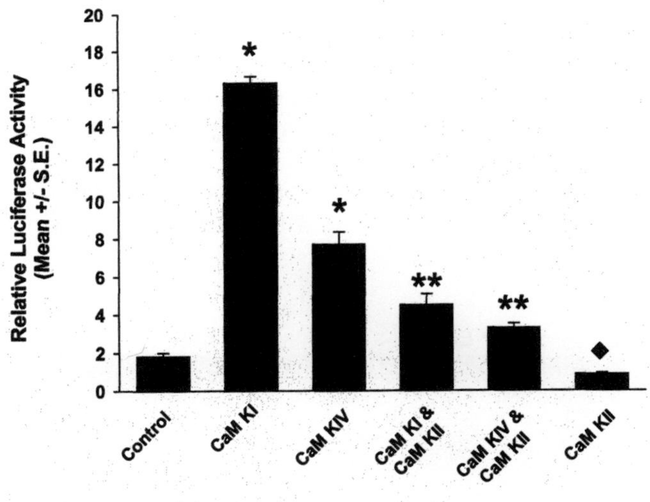

Fig. 1. Activation of the SM-MHC promoter by CaM KI and IV in PAC-1 cells. Cells were transiently transfected with active CaM KI, CaM KIV, and CaM KII and assayed as previously described under 'Materials and methods'. The results are reported as relative luciferase activity of the 1.2 Kb SM-MHC promoter-reporter. (Means ± S.E. of 6 separate treatment groups). *Indicates a significant increase from control ($p < 0.0001$). **Indicates a significant decrease from CaM KI or IV ($p < 0.0001$) respectively. ♦ indicates a significant decrease from control ($p < 0.01$).

CaM kinase II phosphorylation of 14-3-3β

It has been well documented that MEF2-dependent transcription is tightly regulated through the phosphorylation of the MEF2/HDAC complex by CaM kinase kinase signaling. Phosphorylation disrupts the complex, resulting in the export of HDAC mediated by chaperone protein, 14-3-3. However, it is unknown how HDACs are released from 14-3-3 in the cytoplasm to re-establish class II HDAC repression of MEF2-dependent transcription.

Interestingly, there are three CaM KII phosphorylation consensus motifs (Arg-x-x-Thr/Ser) in the primary amino acid sequence of the nuclear chaperone protein, 14-3-3β. Two of these phosphorylation motifs lie within the dimerization region of 14-4-3β and when phosphorylated, should disrupt its dimerization. In order to determine whether 14-3-3β is a substrate for CaM KII, it was immunoprecipitated from PAC-1 cells and subjected to an *in vitro* phosphorylation assay using purified wild-type CaM KII under conditions optimal for CaM KII activity. The chaperone protein was analyzed for incorporation of radiolabeled phosphate. The results of the phosphorylation of 14-3-3β by active CaM KII is shown in Fig. 2. A distinct radioactive band of phosphorylated 14-3-3β protein was observed at approximately 30 kDa. In the presence of EGTA (a Ca^{2+} chelating agent and thus an inhibitor of CaM KII activity) the incorporation of phosphate on 14-3-3β protein was dramatically decreased, demonstrating that the phosphorylation of 14-3-3β was indeed dependent upon a calcium-dependent kinase. These results provide the first evidence that endogenous 14-3-3β may be a natural substrate for CaM KII activity in the smooth muscle cell.

14-3-3β double mutant silences CaM kinase I and IV induction of MEF2- mediated transcription in PAC-1 cells

CaM kinase I and IV have been shown to activate MEF2-dependent transcription and CaM KII silences this activity in myocytes [6]. Activation of MEF2 as a transcription factor requires 14-3-3 export and sequestration of class II HDACs in the cytoplasm. Since 14-3-3β is a substrate for CaM KII (as shown in Fig. 2) and these proteins reside in the same subcellular compartment, we were interested to determine if CaM KII phosphorylation of 14-3-3β would result in re-establishing class II HDAC repression. Each of the two serine residues (serine 60 and 65) in the putative CaM KII phosphorylation motifs within the dimerization region of 14-3-3β were changed to aspartate through conservative amino acid substitution by site-directed mutagenesis. To assess activity of these 14-3-3β mutants, a MEF2-luciferase based enhancer-reporter (containing three MEF2 consensus binding sites upstream of a minimal promoter) was employed. PAC-1 cell cultures were co-transfected with active CaM KI or CaM KIV in the absence or presence of the 14-3-3β double mutant and relative luciferase activity was measured. Transient transfection of PAC-1 cells with active CaM kinase I showed induction of MEF2 enhancer-reporter activity well above baseline values (Fig. 3A). However, when PAC-1 cells were co-transfected with active CaM KI and the 14-3-3β double mutant, MEF2 enhancer-reporter activity was reduced to baseline values (Fig. 3A). Similarly, transient transfection of PAC-1 cells with active CaM KIV also showed an induction of MEF2 reporter (Fig. 3B). Again, when the 14-3-3β double mutant was co-transfected with active CaM KIV, MEF2 reporter activity was returned to baseline values (Fig. 3B). These *in situ* experiments suggest the 14-3-3β double mutant protein may no longer sequester HDAC in the cytoplasm, resulting in the silencing of MEF2-dependent gene transcription. These data also argue that phosphorylation of 14-3-3β by CaM KII destabilizes the 14-3-3/HDAC complex and may provide a mechanism for re-establishing class II HDAC repression.

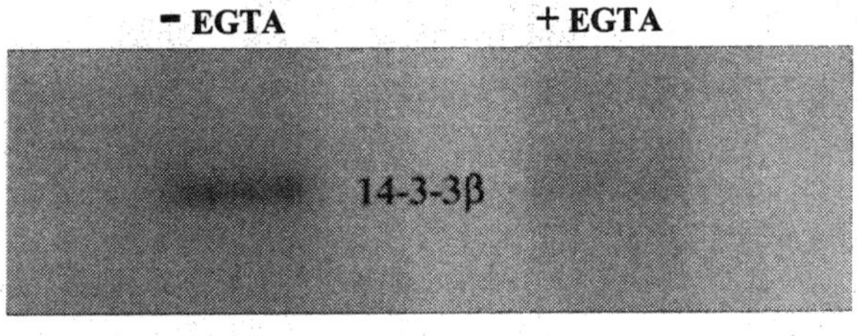

Fig. 2. Phosphorylation of 14-3-3β by CaM kinase II. 14-3-3β was phosphorylated *in-vitro* by co-incubation with wild-type CaM KII in the presence of [γ^{32}P]-ATP, calcium and calmodulin. A chelating agent, EGTA, was used to nullify the activation of CaM kinase II by Ca^2. The phosphorylated bands detected were verified by a molecular weight standard to be 14-3-3 at 30 kDa. Radioactivity of the phosphorylated bands was detected by autoradiography.

14-3-3β blocks phenylephrine stimulation in PAC-1 cells

Phenylephrine (PE) is an α_1-agonist that activates CaM kinase kinase signaling, leading to the induction of MEF2-mediated transcription [21, 22]. To verify that the 14-3-3β double mutant may re-establish class II HDAC repression, we have assessed the activity of these mutants on PE-stimulated MEF2-enhancer activity. PAC-1 cells were co-transfected with either the 14-3-3β single mutant (serine 60 only) or double mutant along with the MEF2-luciferase enhancer-reporter. Following transfection, cells were stimulated with 10 μM PE for 24 h and assayed for luciferase activity. The results show PE induction of reporter activity was reduced to baseline values by co-transfection of 10 ng/well of 14-3-3β double mutant. Furthermore, co-transfection of 20 ng/well of 14-3-3β double mutant reduced PE-stimulated reporter activity well below baseline values (Fig. 4). In contrast, cells transfected with 14-3-3β single mutant did not alter PE in-

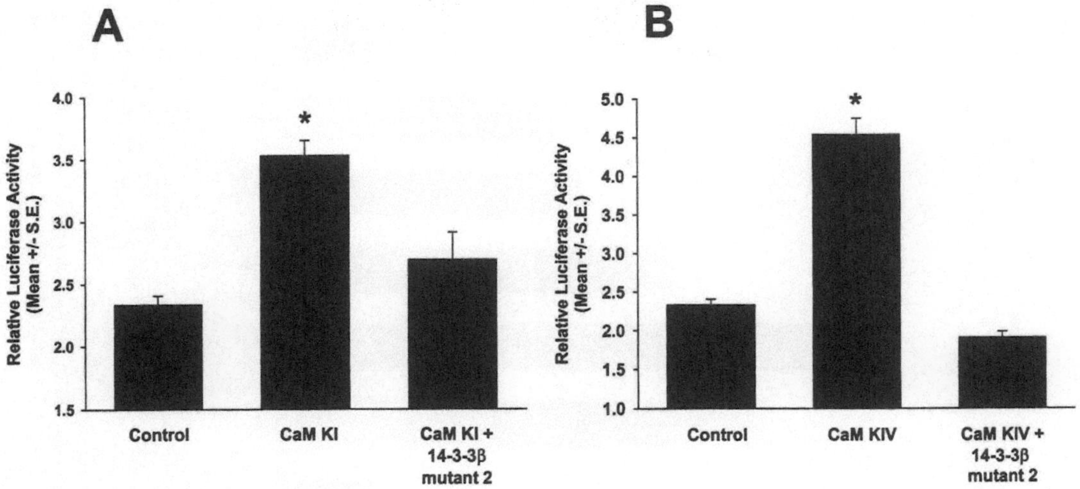

Fig. 3. Silencing of CaM KI and IV induced MEF-2 mediated luciferase activity by 14-3-3β double mutant. Luciferase activity was determined as described under 'Materials and methods'. (A) CaM KI was transiently transfected either alone or co-transfected with 14-3-3β double mutant. (B) CaM KIV was transfected alone or with 14-3-3β double mutant. The results are presented as MEF-2 enhancer-reporter activity (means ± S.E. of 3 separate treatment groups). *Indicates a significant increase from control ($p < 0.01$) in both A and B.

duced reporter activity (Fig. 4). These results are consistent with those from our previous reporter assays demonstrating that the 14-3-3β double mutant was able to silence MEF2-mediated gene transcription. Taken together, the results argue that mutation of 14-3-3β on the CaM KII consensus phosphorylation motifs may disrupt its ability to retain class II HDACs in the cytoplasm, thereby re-establishing HDAC repression of MEF2-dependent transcription.

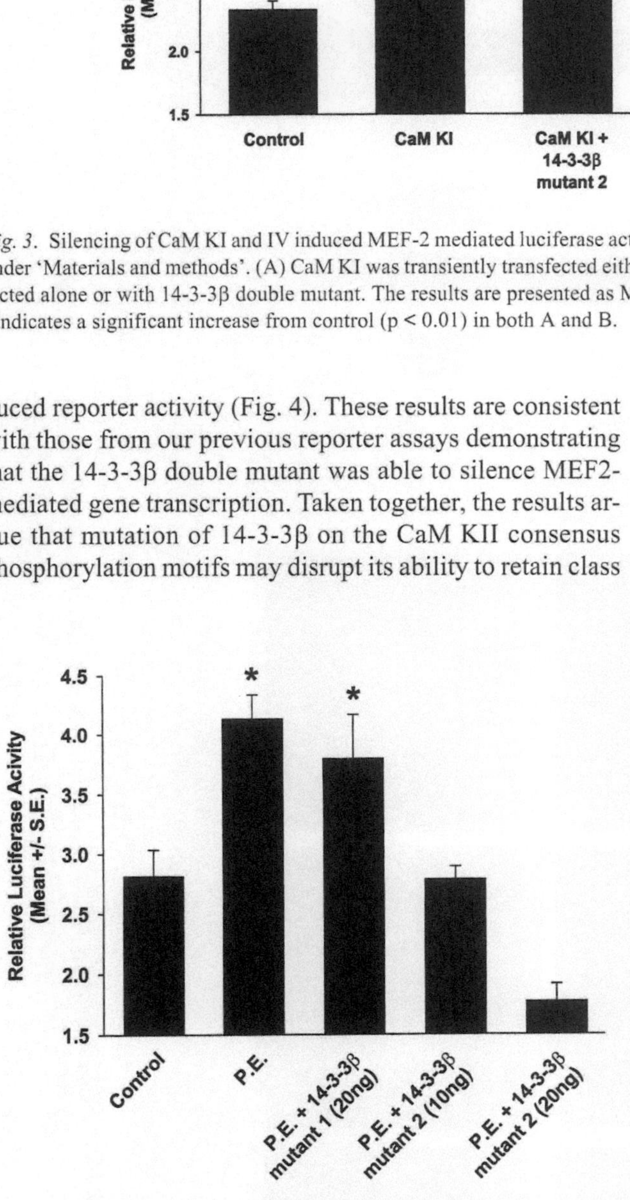

Fig. 4. 14-3-3β double mutant blocks the stimulation of PAC-1 cells by phenylephrine. Cells were cultured as previously described and transfected with either empty vector, 14-3-3β single mutant or 14-3-3β double mutant. After 16 h of transfection the PAC-1 cells were stimulated with phenylephrine at a 10 μM for 24 h in the re-feeding 199 media supplemented with 0.2% FBS. The results are presented as MEF-2 enhancer-reporter activity (means ± S.E. of 3 separate treatment groups). *Indicates a significant increase from control ($p < 0.01$).

14-3-3β double mutant is unable to sequester HDAC4 in the cytoplasm of PAC-1 cells

The results from Figs 3 and 4 argue that CaM KII phosphorylation of 14-3-3β may re-establish class II repression. We next evaluated whether over-expression of the 14-3-3β double mutant would prevent class II HDAC sequestration in the cytoplasm by endogenous 14-3-3 and promote their relocation to the nucleus. To assess this, we followed the subcellular localization a HDAC4/GFP chimeric fusion protein (a generous gift of S. Schreiber, Cambridge, MA, USA). HDAC4/GFP was transiently co-transfected with active CaM kinase and/or 14-3-3β wild type or double mutant, into PAC-1 cells and its subcellular distribution was established. Under control conditions where only HDAC4/ GFP was transfected, HDAC4 was localized predominantly within the nucleus (Fig. 5A, panel A). When wild-type 14-3-3β was co-transfected with HDAC4/GFP, HDAC4 shifted to a predominantly cytoplasmic localization (Fig. 5A, panel B). When HDAC4/GFP was co-transfected with the 14-3-3β double mutant, HDAC4 accumulated in the nucleus (Fig. 5A, panel C). Co-transfection of active CaM KI or CaM KIV resulted in significant cytoplasmic localization of HDAC4 (Fig. 5A, panels D and E, respectively). However, when active CaM KI or IV was co-transfected with 14-3-3β double mutant, there was a significant shift toward nuclear localization of HDAC4 (Fig. 5A, panels G and H, respectively) compared to CaM KI or CaM KIV alone. Interestingly, increased nuclear localization

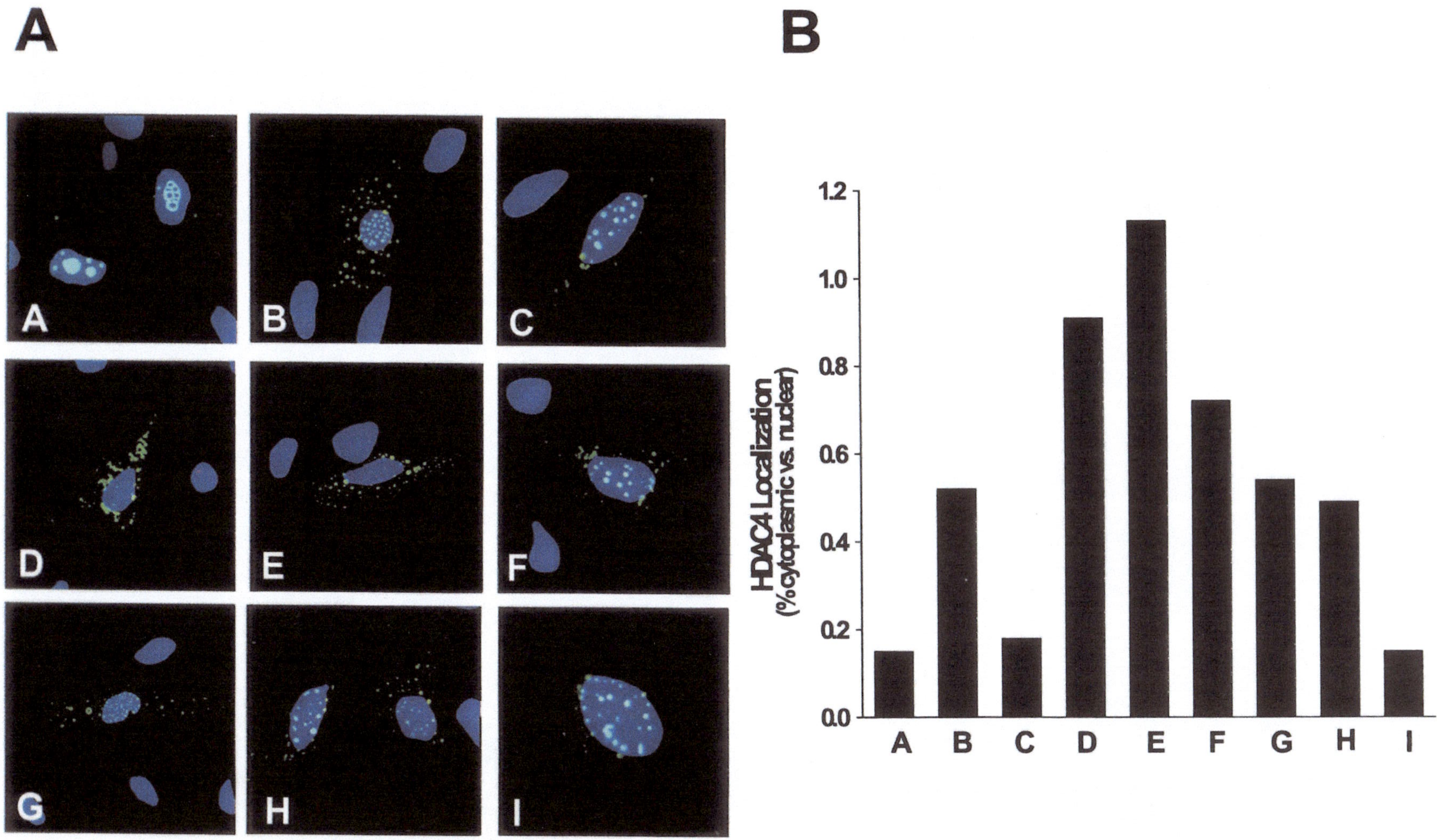

Fig. 5. (A) CaM KII δ_c and 14-3-3β double mutant promote nuclear localization of HDAC4/ GFP. Localization of HDAC4 was shown by fluorescence of GFP, within PAC-1 cells, and the nuclei were visualized using DAPI stain. Cells were transiently transfected with HDAC4/GFP, active CaM kinases, 14-3-3β wild-type, 14-3-3β double mutant, or various combinations of these enzymes. Panel A – pSG5 (empty vector), panel B – 14-3-3β wild-type, panel C – 14-3-3β double mutant, panel D – CaM KI, panel E – CaM KIV, panel F – 14-3-3β wild-type + CaM KIIδ_c, panel G – 14-3-3β double mutant + CaM KI, panel H – 14-3-3β double mutant + CaM KIV, panel I – 14-3-3β double mutant + 14-3-3β wild-type. Each panel in Fig. 5A is representative of raw data presented in Fig. 5B. The data shown in Fig. 5B are the ratio of the percentage of total cells with cytoplasmic HDAC4/GFP to the percentage of cells with nuclear HDAC4/GFP for all GFP expressing cells. N ≥ 150 cells were counted for each treatment group.

of HDAC4 was seen when 14-3-3β wild type and CaM KII were co-expressed (Fig. 5A, panel F compared to panel B). When the 14-3-3β double mutant was co-expressed with 14-3-3β wild-type, the mutant prevented wild-type sequestration of HDAC4 in the cytoplasm, and re-established significant nuclear localization (Fig. 5A, panel I).

Quantification of HDAC4/GFP subcellular localization, observed in Fig. 5A, were compiled from 4 independent experiments and are represented in Fig. 5B. Total fluorescent cells were counted for each treatment and quantified as the ratio of the percent cytoplasmic versus nuclear HDAC4/GFP. Clearly, results from Fig. 5 establish that wild-type 14-3-3β sequesters HDAC4 in the cytoplasm and that active CaM KII or 14-3-3β double mutant disrupts this endogenous interaction. As expected, CaM KI and IV promote sequestration of HDAC4 in the cytoplasm. This process requires 14-3-3β dimerization. When 14-3-3β double mutant was co-expressed in the presence of active CaM KI or IV, cytoplasmic sequestering was diminished and nuclear entry was re-established. Surprisingly, 14-3-3β wild-type sequestering of HDAC4 into the cytoplasm was completely blocked by co-expression of 14-3-3β double mutant. This data is consistent with our model of CaM KII mediated activation of class II HDAC repression.

Discussion

The 14-3-3 family of proteins is ubiquitous and plays a key regulatory role in signal transduction pathways, cell cycle regulation and apoptosis [23, 24]. 14-3-3β has been shown to be intimately involved in regulation of the activity of the MEF2 family of transcription factors through CaM kinase signaling [6, 7]. In this work, we have explored the role of 14-3-3β in MEF2 signaling in PAC-1 smooth muscle cells with respect to cytoplasmic CaM kinase IIδC activity. The sequestration of class II HDACs by 14-3-3β upon CaM kinase I and IV signaling is paramount to the ability of MEF2 to function as a viable transcription factor. CaM kinase IIδC, when co-expressed with either CaM kinase I or IV, completely silenced MEF2-driven transcriptional events (data not shown). Inspection of the primary amino acid sequence of 14-3-3β revealed two CaM kinase II phosphorylation sites (Ser60 and Ser65) in the homodimerization domain. Here, we use a mutant form of 14-3-3β whereby Ser60 and Ser65 were conservatively substituted with aspartate, which presumably rendered 14-3-3β monomers unable to homodimerize and would therefore be unable to function effectively as a chaperone protein. Previous work has focused on rendering 14-3-3 ligands unable to bind 14-3-3 [25, 26]. Here, we render 14-3-3 unable to dimerize and function as a class II HDAC chaperone, an event that we show accounts for CaM kinase IIδC's ability to silence MEF2-driven transcriptional events.

Expression of SM-MHC was previously shown to be dependent on MEF2 activation [1]. In the present study, transfection of PAC-1 cell cultures with active CaM kinase I or IV resulted in the up-regulation of the SM-MHC promoter-reporter (Fig. 1). Similarly, transfection of PAC-1 cells with active CaM kinase I and IV up-regulated the MEF2 enhancer-reporter (Figs 3A and 3B). When CaM KIIδC was co-transfected with either CaM kinase I or IV, SM-MHC promoter-reporter activity was completely silenced (Fig. 1). These data serve to validate the ability of the CaM kinases to function in the same manner in smooth muscle myocytes as was previously observed in other myocyte cell types with respect to MEF2-dependent transcription.

The intriguing issue of the ability of cytoplasmic CaM kinase IIδC to silence the induction of muscle cell specific genes led us to search for a target protein that could explain this activity. Recent work revealed that MEF2 is complexed with class II HDACs in the nucleus and is transcriptionally inactive in cardiac and skeletal myocyte [7]. Upon CaM kinase I and IV activation, the HDAC component of the complex is phosphorylated and shuttled to the cytoplasm via 14-3-3, resulting in the activation of MEF2 [3]. When we examined the amino acid sequence of 14-3-3β, we discovered two tandem CaM kinase II phosphorylation sites (Ser60 and Ser65) in the region critical for 14-3-3β homodimerization. We hypothesized that phosphorylation of 14-3-3β at these serine residues would result in electrostatic repulsion too great to allow dimer formation due to adjacent glutamic acid residues on the partnering monomer. We first set out to determine whether or not 14-3-3β was a substrate for CaM kinase II. Here we show that indeed, 14-3-3β can be phosphorylated by CaM kinase II *in vitro* (Fig. 2).

In order to determine the effect of CaM kinase II-phosphorylated 14-3-3β on MEF2 signaling *in situ*, we conservatively substituted Ser60 and Ser65 with aspartate residues. This substitution is commonly used to mimic the steric and charge characteristics of a phosphorylated serine residue [27–31]. When we co-transfected the aspartate substituted 14-3-3β double mutant with either CaM KI or IV, MEF2 enhancer-reporter activity was silenced (Figs 3A and 3B). Phenylephrine (PE) is an α_1-adrenergic agonist that is known to activate CaM kinase I and IV [21, 22]. We show that PE is able to stimulate MEF2 enhancer activity in PAC-1 cells and more importantly, the double mutant 14-3-3β can completely silence PE induction of the MEF2 enhancer-reporter (Fig. 4).

With this, we hypothesized that the MEF2 enhancer-reporter was silenced due to the inability of the 14-3-3β double mutant to sequester class II HDACs in the cytoplasm even in the presence of active CaM KI or IV. To test this, we used HDAC4/GFP (Green Fluorescent Protein) fusion hybrids to follow the cellular fate of the class II HDACs during CaM kinase signaling (Fig. 5). In the absence of signal, HDAC4 is predominantly nuclear. Addition of CaM KI or IV causes

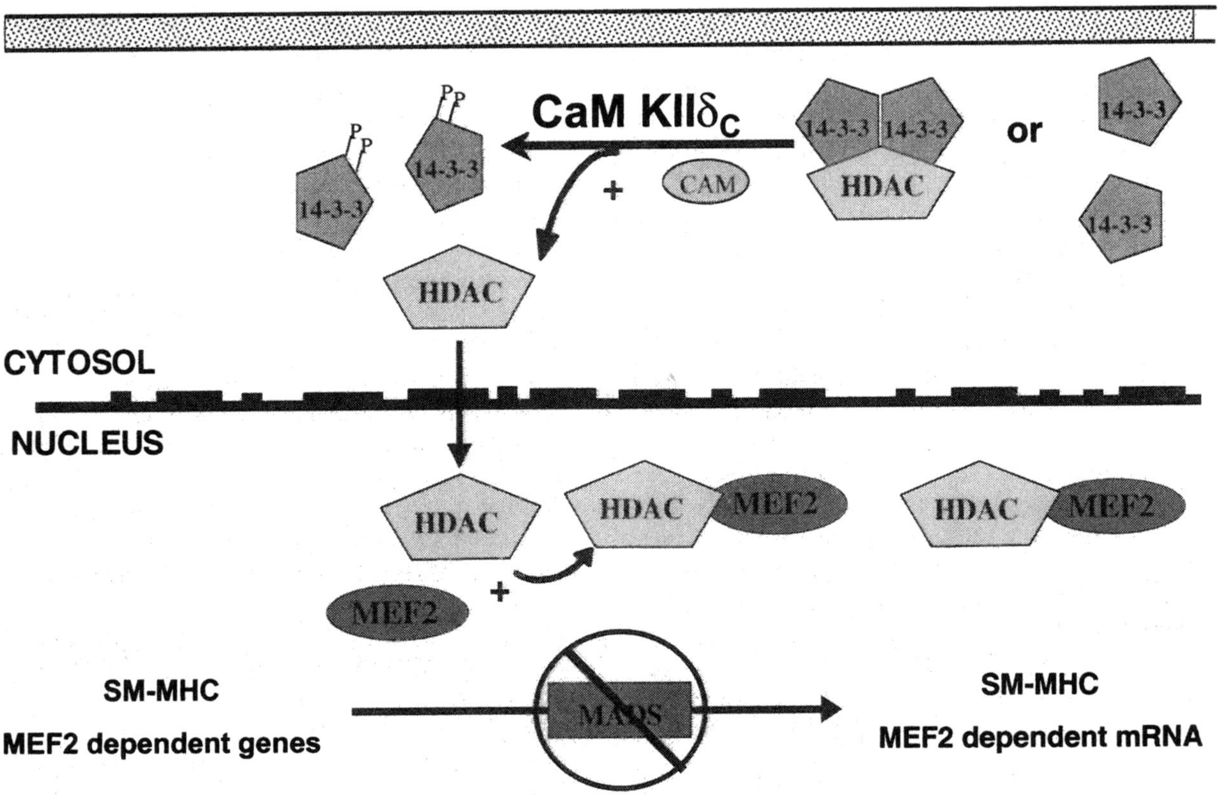

Fig. 6. Proposed model of CaM kinase II transcriptional silencing of MEF2-dependent transcription through the destabilization of the 14-3-3/HDAC complex in the cytoplasm.

cytoplasmic localization of HDAC4 as was observed previously in other muscle cell types. However, the 14-3-3β double mutant diminishes cytoplasmic sequestration in the presence of active CaM KI or IV and attempts to re-establish HDAC4 in the nucleus. HDAC4 is predominantly cytoplasmic when co-expressed with wild-type 14-3-3β alone, however, when co-expressed with both 14-3-3β wild-type and active CaM KIIδC, HDAC4 nuclear localization was increased. When both 14-3-3β wild-type and double mutant were co-expressed, we were surprised to find that the 14-3-3β double mutant completely prevented 14-3-3β wild-type sequestration of HDAC4 in the cytoplasm and shifted its localization to the nucleus.

The data presented here allowed us to construct a model that we believe accounts for the ability of CaM kinase IIδC to silence SM-MHC reporter activity and MEF2 enhancer activity in PAC-1 cells (Fig. 6). As previously reported, MEF2 is bound by class II HDACs in a transcriptionally inactive complex. Upon phosphorylation by active CaM KI or IV, the HDAC is shuttled to the cytoplasm by 14-3-3. Our model provides the reverse mechanism in which CaM kinase IIδC phosphorylates 14-3-3β monomers, prevents dimerization, re-localizes class II HDACs to the nucleus and re-activates class II HDAC repression of MEF2-dependent transcription. To the best of our knowledge, this is the first evidence delineating the mechanism for myocyte activation of class II HDAC repression.

Acknowledgements

We would like to thank Dr. Thomas Soderling (Vollum Institute, Portland, OR, USA) for CaM KI and IV constructs, Dr. Howard Schulman (Stanford School of Medicine, Stanford, CA, USA) for CaM KII constructs, Dr. Stuart Schreiber and Dr. Christina Grozinger (Harvard, Cambridge, MA, USA) for HDAC/GFPs and 14-3-3 constructs, Dr. Eric Olson (UT Southwestern Medical Center, Dallas, TX, USA) for the MEF2-luciferase enhancer-reporter, and Dr. Gary Owens (University of Virginia School of Medicine, Charlottesville, VA, USA) for the SM-MHC1.2 promoter-reporter construct. This work was supported by grants from the American Heart Association Texas Affiliate 0150766, The National Institutes of Health, NHLB (RO-1 HL67152) to S.R.G. In addition, this work was supported by a sponsored research agreement from Myogen Inc., (Westminster, CO, USA) to S.R.G.

References

1. Katoh Y, Molkentin JD, Dave V, Olson EN, Periasamy M: MEF2B is a component of a smooth muscle-specific complex that binds an A/T-rich element important for smooth muscle myosin heavy chain gene expression. J Biol Chem 273: 1511–1518, 1998
2. Lin Q, Lu J, Yanagisawa H, Webb R, Lyons GE, Richardson JA, Olson EN: Requirement of the MADS-box transcription factor MEF2C for vascular development. Development 125: 4565–4574, 1998
3. Lu J, McKinsey TA, Nicol RL, Olson EN: Signal-dependent activation of the MEF2 transcription factor by dissociation from histone deacetylases. Proc Natl Acad Sci USA 97: 4070–4075, 2000
4. Lu J, McKinsey TA, Zhang CL, Olson EN: Regulation of skeletal myogenesis by association of the MEF2 transcription factor with class II histone deacetylases. Mol Cell 6: 233–244, 2000
5. Kao HY, Verdel A, Tsai CC, Simon C Juguilon H, Khochbin S: Mechanism for nucleocytoplasmic shuttling of histone deacetylase 7. J Biol Chem 276: 47496–47507, 2001
6. Passier R, Zeng H, Frey N, Naya FJ, Nicol RL, McKinsey TA, Overbeek P, Richardson JA, Grant SR, Olson EN: CaM kinase signaling induces cardiac hypertrophy and activates the MEF2 transcription factor *in vivo*. J Clin Invest 105: 1339–1342, 2000
7. McKinsey TA, Zhang CL, Olson EN: Activation of the myocyte enhancer factor-2 transcription factor by calcium/calmodulin-dependent protein kinase-stimulated binding of 14-3- 3 to histone deacetylase 5. Proc Natl Acad Sci USA 97: 14400–14405, 2000
8. Wang AH, Kruhlak MJ, Wu J, Bertos NR, Vezmar M, Posner BI, Bazett-Jones DP, Yang XJ: Regulation of histone deacetylase 4 by binding of 14-3-3 proteins. Mol Cell Biol 20: 6904–6912, 2000
9. Vincenz C, Dixit VM: 14-3-3 proteins associate with A20 in an isoform-specific manner and function both as chaperone and adapter molecules. J Biol Chem 271: 20029–20034, 1996
10. Rittinger K, Budman J, Xu J, Volinia S, Cantley LC, Smerdon SJ, Gamblin SJ, Yaffe MB: Structural analysis of 14-3-3 phosphopeptide complexes identifies a dual role for the nuclear export signal of 14-3-3 in ligand binding. Mol Cell 4: 153–166, 1999
11. Grozinger CM, Schreiber SL: Regulation of histone deacetylase 4 and 5 and transcriptional activity by 14-3-3-dependent cellular localization. Proc Natl Acad Sci USA 97: 7835–7840, 2000
12. Fu H, Subramanian RR, Masters SC: 14-3-3 proteins: Structure, function, and regulation. Annu Rev Pharmacol Toxicol 40: 617–647, 2000
13. Rothman A, Kulik TJ, Taubman MB, Berk BC, Smith CW, Nadal-Ginard B: Development and characterization of a cloned rat pulmonary arterial smooth muscle cell line that maintains differentiated properties through multiple subcultures. Circulation 86: 1977–1986, 1992
14. Haribabu B, Hook SS, Selbert MA, Goldstein EG, Tomhave ED, Edelman AM, Snyderman R, Means AR: Human calcium-calmodulin dependent protein kinase I: cDNA cloning, domain structure and activation by phosphorylation at threonine-177 by calcium-calmodulin dependent protein kinase I kinase. Embo J 14: 3679–3686, 1995
15. Cruzalegui FH, Means AR: Biochemical characterization of the multifunctional Ca^{2+}/calmodulin-dependent protein kinase type IV expressed in insect cells. J Biol Chem 268: 26171–26178, 1993
16. Srinivasan M, Edman CF, Schulman H: Alternative splicing introduces a nuclear localization signal that targets multifunctional CaM kinase to the nucleus. J Cell Biol 126: 839–852, 1994
17. White SL, Low RB: Identification of promoter elements involved in cell-specific regulation of rat smooth muscle myosin heavy chain gene transcription. J Biol Chem 271: 15008–15017, 1996
18. Wu H, Rothermel B, Kanatous S, Rosenberg P, Naya FJ, Shelton JM, Hutcheson KA, DiMaio JM, Olson EN, Bassel-Duby R, Williams RS: Activation of MEF2 by muscle activity is mediated through a calcineurin-dependent pathway. Embo J 20: 6414–6423, 2001
19. Zeng H, Valencia T, Bai Y, Grant SR: Calcim/calmodulin-dependent enzymes regulate cardiac hypertrophy-sensitive gene expression. Miami Nature Biotechnol Short Reports 10: 1999
20. Valencia TG, Roberts LD, Zeng H, Grant SR: Tetracycline-inducible CaM kinase II silences hypertrophy-sensitive gene expression in rat neonate cardiomyocytes. Biochem Biophys Res Commun 274: 803–810, 2000
21. McDonough PM, Stella SL, Glembotski CC: Involvement of cytoplasmic calcium and protein kinases in the regulation of atrial natriuretic factor secretion by contraction rate and endothelin. J Biol Chem 269: 9466–9472, 1994
22. Ramirez MT, Zhao XL, Schulman H, Brown JH: The nuclear deltaB isoform of Ca^{2+}/calmodulin-dependent protein kinase II regulates atrial natriuretic factor gene expression in ventricular myocytes. J Biol Chem 272: 31203–31208, 1997
23. Luo ZJ, Zhang XF, Rapp U, Avruch J: Identification of the 14.3.3 zeta domains important for self- association and Raf binding. J Biol Chem 270: 23681–23687, 1995
24. Chan TA, Hermeking H, Lengauer C, Kinzler KW, Vogelstein B: 14-3-3 Sigma is required to prevent mitotic catastrophe after DNA damage. Nature 401: 616–620, 1999
25. Liu D, Bienkowska J, Petosa C, Collier RJ, Fu H, Liddington R: Crystal structure of the zeta isoform of the 14-3-3 protein. Nature 376: 191–194, 1995
26. Obsil T, Ghirlando R, Klein DC, Ganguly S, Dyda F: Crystal structure of the 14-3-3zeta:serotonin N-acetyltransferase complex. A role for scaffolding in enzyme regulation. Cell 105: 257–267, 2001
27. Alessandrini A, Greulich H, Huang W, Erikson RL: Mek1 phosphorylation site mutants activate Raf-1 in NIH 3T3 cells. J Biol Chem 271: 31612–31618, 1996
28. Huang W, Kessler DS, Erikson RL: Biochemical and biological analysis of Mek1 phosphorylation site mutants. Mol Biol Cell 6: 237–245, 1995
29. Hurley JH, Dean AM, Sohl JL, Koshland DE Jr, Stroud RM: Regulation of an enzyme by phosphorylation at the active site. Science 249: 1012–1016, 1990
30. Mansour SJ, Candia JM, Matsuura JE, Manning MC, Ahn NG: Interdependent domains controlling the enzymatic activity of mitogen-activated protein kinase kinase 1. Biochemistry 35: 15529–15536, 1996
31. Orr JW, Newton AC: Requirement for negative charge on 'activation loop' of protein kinase C. J Biol Chem 269: 27715–27718, 1994

Molecular and Cellular Biochemistry **242:** 163–171, 2003.

Regulation of MAPK pathways in response to purinergic stimulation of adult rat cardiac myocytes

Thomais Markou,[1] Guy Vassort[2] and Antigone Lazou[1]

[1]*Laboratory of Animal Physiology, Department of Zoology, School of Biology, Aristotle University of Thessaloniki, Thessaloniki, Greece;* [2]*INSERM U-390, Physiopathologie Cardiovasculaire, CHU Arnaud de Villeneuve, Montpellier, France*

Abstract

We investigated the activation of mitogen-activated protein kinases (MAPKs) pathways by purinergic stimulation in cardiac myocytes from adult rat hearts. ATPγS increased the phosphorylation (activation) of the extracellular signal regulated kinase 1 and 2 (ERK1/2) and p38 MAPK. ERK1/2 and p38 MAPK activation was differential, ERK1/2 being rapid and transient while that of p38 MAPK slow and sustained. Using selective inhibitors, activation of ERK1/2 was shown to involve protein kinase C and MEK1/2 while that of p38 MAPK was regulated by both protein kinase C and protein kinase A. Furthermore, we show that purinergic stimulation induces the phosphorylation of the MAPK downstream target, mitogen- and stress-activated protein kinase 1 (MSK1), in cardiac myocytes. The time course of MSK1 phosphorylation closely follows that of ERK activation. Inhibitors of the ERK and p38 MAPK pathways were tested on the phosphorylation of MSK1 at two different time points. The results suggest that ERKs initiate the response but both ERKs and p38 MAPK are required for the maintenance of the complete phosphorylation of MSK1. The temporal relationship of MSK1 phosphorylation and $cPLA_2$ translocation induced by purinergic stimulation, taken together with previous findings, is an indication that $cPLA_2$ may be a downstream target of MSK1. (Mol Cell Biochem **242**: 163–171, 2003)

Key words: purinergic agonists, mitogen-activated protein kinases, ERKs, p38 MAPK, MSK1, rat cardiomyocytes

Abbreviations: ERK – extracellular signal-regulated kinase; MAPK – mitogen-activated protein kinase; MSK1 – mitogen- and stress-activated protein kinase-1; PKA – protein kinase A; PKC – protein kinase C; PKI – cAMP dependent protein kinase inhibitor; PLA_2 – phospholipase A_2

Introduction

Extracellular purines have been recognized for many years to exert various and potent actions on the cardiovascular system [1]. ATP is released by both sympathetic and parasympathetic nerve terminals under physiological conditions, as well as by cardiomyocytes and various vascular cell types in cardiac ischemia. ATP modulates several ionic channels and exchangers; it acts on a number of intracellular effector systems and thereby elicits diverse responses in the heart. ATP increases contractility [2] and slows down the sinus pacemaker resulting in a negative chronotropic and dromotropic effect on the sinoatrial node. In ischemic hearts, ATP could also be a source of arrhythmia [1, 3].
Extracellular ATP exerts its effects by binding to two classes of P_2 receptors, the metabotropic receptor (P_{2Y}) and those receptors that are intrinsic inotropic channels (P_{2X}) [4, 5]. In most cell types including cardiomyocytes, the metabotropic

Address for offprints: A. Lazou, Laboratory of Animal Physiology, Department of Zoology, School of Biology, Aristotle University of Thessaloniki, Thessaloniki 54006, Greece (E-mail: lazou@bio.auth.gr)

purinoreceptors of the P_{2Y} subclass initiate their biological actions by G protein dependent activation of phospholipase C leading to IP_3 formation and subsequent elevation of intracellular calcium levels. P_{2Y} receptors have also been shown to be linked to the activation of PKA [6] PKC, MAPK and tyrosine kinases [1, 7].

It has been recently shown that, in isolated rat cardiac myocytes, extracellular ATP activates a TREK-like K^+ via the activation of $cPLA_2$ and the release of arachidonic acid [8]. The purinergic-induced $cPLA_2$ activation was suggested to require the simultaneous activation of both ERK1/2 and p38 MAPK, probably mediated by another downstream kinase, the mitogen- and stress-activated kinase-1 (MSK1). Considerable efforts have been made in recent years to study the mechanism of activation of $cPLA_2$ and the subsequent release of AA. $cPLA_2$ activity is regulated both by phosphorylation and by increase in intracellular calcium which drive its translocation from the cytosol toward phospholipid membranes. In a variety of cell types, $cPLA_2$ phosphorylation is achieved by p42/p44 mitogen activated protein kinases (ERK1/2) and/or p38 MAPK [9–11]. The MAPKs are a superfamily of proline directed Ser/Thr protein kinases, which are components of a critical signaling pathway linking membrane receptors to cytoplasmic and nuclear effectors. Three distinct MAPK subfamilies are best characterized in mammals: the extracellular signal regulated kinases (ERKs), the c-Jun N-terminal kinases (JNKs) and the p38 MAPK. MAPK pathways can be activated by a wide variety of different stimuli acting through diverse receptor families and including hormones, growth factors, vasoactive peptides, inflammatory cytokines and environmental stresses such as osmotic shock and ionizing radiation [12, 13]. The nature of these cascades allows considerable potential for signal integration and amplification as well as the possibility for cross talk between pathways because of the apparent overlapping substrate specificities of some of the components and of the upstream signaling molecules. There is considerable evidence of the immense potential significance of the regulation of MAPK pathways in the myocardium with respect to its reactions to pathological stresses (e.g. hypoxia, ischemia, reperfusion injury). However, little is known about the regulation of this signaling pathway by P_{2Y} receptors. In contrast to receptor tyrosine kinases, it is believed that the intermediate steps linking G protein coupled receptors to the activation of MAPKs vary among cell types and individual receptors [14].

In this study, we sought to further elucidate the molecular mechanisms whereby purinergic stimulation activates MAPK signaling cascades in adult rat cardiac myocytes. We show that ATP differentially activates ERKs and p38 MAPK, which in turn phosphorylate MSK1. Although ERKs seem to initiate MSK1 phosphorylation, both ERKs and p38 MAPK are required to maintain complete phosphorylation.

Materials and methods

Materials

ATPγS, RpcAMP, dimethyl sulfoxide, dithiothreitol, leupeptin, trans-epoxy succinyl-L-leucylamido-(4-guanidino)butane (E64) and phenyl methyl sulphonyl fluoride (PMSF) were obtained from Sigma Chemical Co. (St. Louis, MO, USA). SB203580, PD98059, GF109203X, U0126 and H89 were obtained from Calbiochem (La Jolla, CA, USA). Bradford protein assay reagent was from Bio-Rad (Hercules, CA, USA). Nitrocellulose (0.45 μm) was obtained from Schleicher & Schuell (Keene N.H. 03431, USA). Rabbit polyclonal antibodies specific for the phosphorylated MSK1 (Ser376) and the dually phosphorylated p38 MAPK were obtained from Cell Signaling (Beverly, MA, USA). A mouse monoclonal antibody to phosphorylated ERK1/2 was also from Cell Signaling (Beverly, MA, USA). Antibody to $cPLA_2$ and Western blotting chemiluminescence reagent kit were from Santa Cruz Biotechnology (Santa Cruz, CA, USA). Prestained molecular mass markers were from New England Biolabs (Beverly, MA, USA). Secondary antibodies were from DAKO (High Wycombe, Buckinghamshire, UK). X-OMAT AR film was purchased from Eastman Kodak Co. (New York, USA). General laboratory reagents were from Sigma Chemical Co. (St. Louis, MO, USA) or Merck (Darmstadt, Germany).

Cell isolation

Ventricular cardiac myocytes were isolated from 200–250 g adult male Wistar rats by retrograde aortic perfusion as previously described [15], with minor modification. Briefly, the hearts were perfused for 5 min with Krebs-Henseleit medium (hereafter referred to as incubation medium) containing in mM: 25 $NaHCO_3$, 4.7 KCl, 118.5 NaCl, 1.2 $MgSO_4 7H_2O$, 1.2 KH_2PO_4, 10 glucose and 5 μM added Ca^{2+} as $CaCl_2$. Perfusion continued for 40 min with incubation medium containing 0.5 mg/ml collagenase (CLS1, Biochrom KG) and 50 μM Ca^{2+}. The heart was gently dissociated through the bore of a large-tip pipette, followed by two decantations to separate dead cells. Cells were finally resuspended in incubation medium, in which added Ca^{2+} was gradually increased to 1 mM. Preparations were considered satisfactory only if the yield of rod-shaped cells was more than 70%.

Cell treatment and lysis

Freshly isolated ventricular myocytes were bathed at 37°C in incubation buffer containing ATPγS, inhibitors or vehicle for various times, lysed in ice-cold buffer (lysis buffer) containing 20 mM HEPES, pH 7.5, 20 mM β-glycerophosphate, 50 mM NaF, 2 mM EDTA, 0.2 mM Na_3VO_4, 5 mM dithiothreitol (DTT), 10 mM benzamidine, 200 μM leupeptin, 10 μM trans-epoxy succinyl-L-leucylamido-(4-guanidino)butane (E64), 300 μM phenyl methyl sulphonyl fluoride (PMSF), 1% (v/v) Triton X-100) and extracted on ice for 30 min. Cell lysates were centrifuged (10,000 g, 5 min, 4°C) and the supernatants were boiled with 0.33 vol of SDS/PAGE sample buffer (0.33 M Tris/HCl, pH 6.8, 10% (w/v) SDS, 13% (v/v) glycerol, 133 mM DDT, 0.2% (w/v) bromophenol blue).

For the preparation of subcellular fractions, cells were lysed in lysis buffer without Triton X-100 and extracted on ice for 5 min. Extracts were centrifuged (10,000 g, 15 min, 4°C) to separate the membranous from the cytosolic fractions. The pellet was washed with lysis buffer and resuspended in lysis buffer containing 1% Triton X-100. Protein concentrations were determined using the BioRad Bradford assay.

Immunoblot analysis

Proteins were separated by SDS-PAGE on 10% (MAPKs) or 8% (w/v) (MSK1, $cPLA_2$) acrylamide, 0.275% (w/v) bis-acrylamide slab gels and transferred electrophoretically onto nitrocellulose membranes (0.45 μm). Membranes were blocked with TBS-T (20 mM Tris-HCl, pH 7.5, 137 mM NaCl, 0.1% (v/v) Tween 20) containing 5% (w/v) nonfat milk power for 30 min at room temperature, and then incubated with the appropriate antibody (1:1000 dilution in TBS-T containing 5% w/v BSA) at 4°C overnight. Proteins were detected with horseradish peroxidase (HRP) conjugated secondary antibody (1:5000 dilution in TBS-T containing 1% (w/v nonfat milk powder, 1 h, room temperature) and were visualized by enhanced chemiluminescence. Scanning densitometry was used for semiquantitative analysis of the data.

Statistics

Data are presented as mean ± S.E.M. of n independent experiments. Statistical analyses (ANOVA with Dunnet post-test or two-tailed Student's *t*-test where appropriate) were performed using Instat (Graph Pad Software, San Diego, CA, USA) with significance taken as being established at $p < 0.05$.

Results

Extracellular ATP-stimulated ERK1/2 and p38 MAPK activation in cardiac myocytes

To investigate the activation of ERK1/2 and p38 MAPK by purinergic receptors in cardiac myocytes, we determined the phosphorylation of these kinases by immunoblotting with antibodies specific for the dually phosphorylated forms of each of the subfamilies after exposure to ATPγS over a 0.5–20 min period. ATP stimulated the rapid and transient activation of ERK1/2 (Figs 1A and 1B). Maximal activation (~ 6-fold compared with the controls) of ERK1/2 was observed within 2 min of ATP addition and returned to basal levels within 20 min. A different time-course was observed for the phosphorylation and activation of p38 MAPK (Figs 1C and 1D). p38 MAPK phosphorylation by ATP was of similar magnitude but delayed compared to that of ERK1/2, showing a maximal increase (~ 6.5-fold compared with the controls) at 10 min and remaining considerably elevated up to 20 min.

Mechanisms of ATP-induced ERK1/2 and p38 MAPK activation

To investigate the mechanism of purinergic activation of ERK1/2 we first examined the effect of a specific inhibitor of MEK1/2, PD98059 [16]. The ERK activation is believed to proceed through Raf, which phosphorylates and activates MEK1/2. PD98059 inhibits the activation of MEK1 and to a lesser extent MEK2, both of which are involved in ERK1/2 activation. As shown in Figs 2A and 2B, PD98059 (10 μM) completely abolished ERK1/2 activation by ATP suggesting that MEK1/2 are involved in ATP-induced ERK1/2 activation. Similar results were obtained when another structurally distinct inhibitor of MEK1/2, U0126 [17], was used (data not shown). SB203580, a p38 MAPK inhibitor, had no effect on ERK1/2 phosphorylation by ATP.

To assess the involvement of PKC in ATP-induced ERK1/2 activation, cardiac myocytes were pretreated with the specific PKC inhibitor GF109203X [18]. GF109203X at a concentration of 1 μM completely blocked ATP-induced ERK1/2 phosphorylation. When RpcAMP, a known PKA inhibitor, was used, ERK1/2 phosphorylation was not affected. In all cases, controls with the inhibitors alone were included (Figs. 2A and 2B).

We also examined the effects of these drugs on the activation of p38 MAPK to determine whether the same upstream mechanisms are involved in the activation of ERKs and p38 MAPK in cardiac myocytes. GF109203X markedly reduced p38 MAPK phosphorylation induced by ATP (Figs 2C and 2D). In contrast to its lack of effect on ERK1/2 phosphorylation, the PKA inhibitor RpcAMP partially inhibited p38 MAPK phosphorylation reducing it to 50% of its control ATP-stimulated value at 10 min. This was further confirmed using two other PKA inhibitors, H89 and PKI, which both showed similar partial inhibition (data not shown). Furthermore, pre-exposure of cardiac myocytes to SB203580 completely abolished the p38 MAPK increase in phosphorylation.

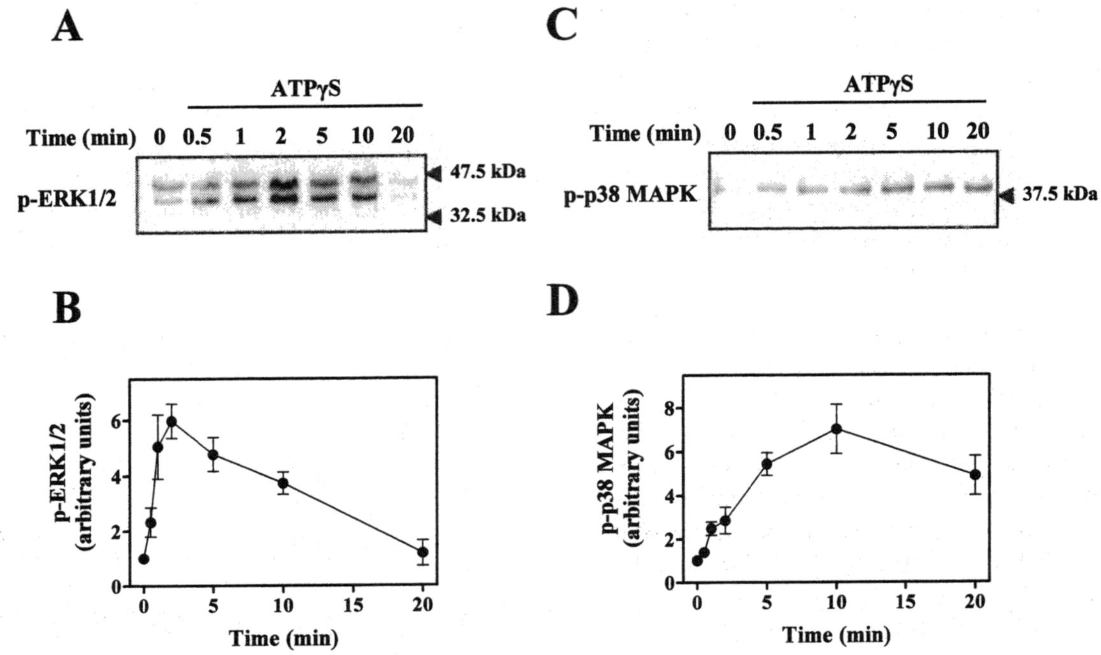

Fig. 1. Phosphorylation of MAPKs by ATPγS. Cardiac myocytes were exposed to 30 μM ATPγS for the times indicated. Extracts were subjected to SDS/PAGE and immunoblotted with an antibody selective for phospho-ERK1/2 (A) or phospho-p38 MAPK (C). Immunoblotted ERK1/2 and p38 MAPK were quantified by laser scanning densitometry (B and D). Results are means ± S.E.M. of 3 independent experiments.

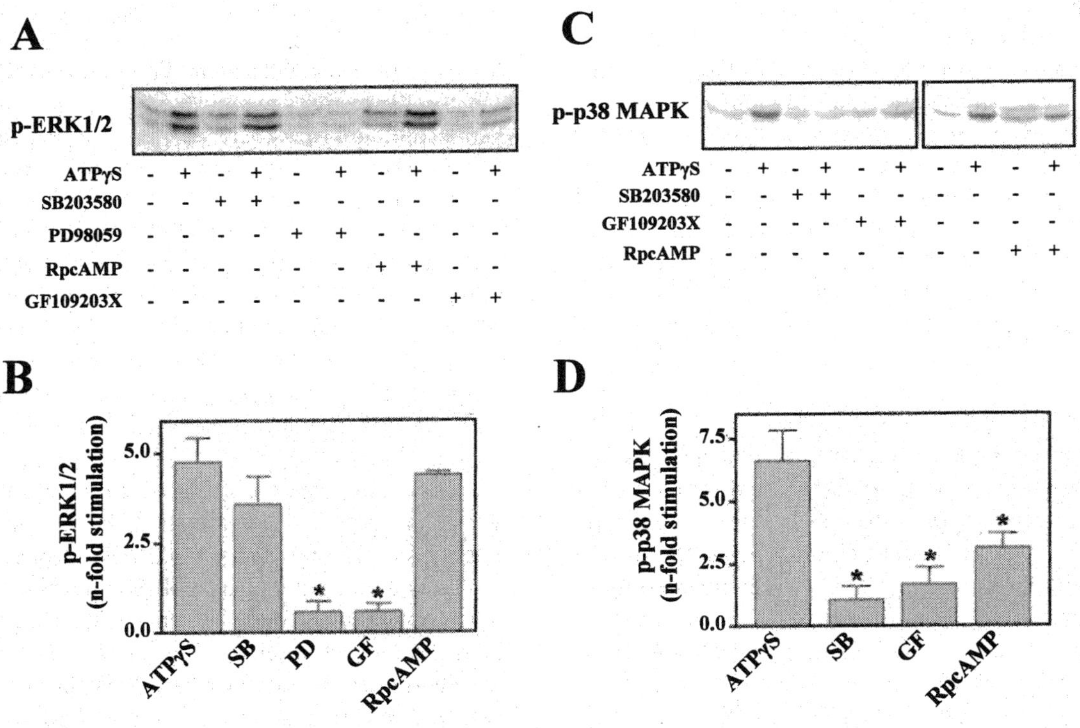

Fig. 2. Effect of various inhibitors on MAPK phosphorylation by ATPγS. Cardiac myocytes were either left untreated or preincubated with SB203580 (10 μM), PD98059 (10 μM), GF109203X (1 μM) or Rp-cAMP (100 μM). Then they were incubated in the absence of agonists or in the presence of 30 μM ATPγS, for 5 min (A, B) or 10 min (C, D). Cell extracts were immunoblotted for phosphorylated ERK1/2 (A) or p38 MAPK (C). These experiments were repeated 3 times with comparable results. ERK1/2 and p38 MAPK phosphorylation was quantified by laser scanning densitometry and plotted (B, D). The results are presented as mean ± S.E.M. of 3 independent experiments. *Significant inhibition of ERK1/2 or p38 MAPK phosphorylation ($p < 0.05$, ANOVA with Dunnet post-test) when compared with identically treated cells in the absence of inhibitors.

Although it was originally reported that SB203580 inhibition is through the binding to the ATP pocket, thus inhibiting the activity of the kinase [19], this finding was not unexpected. There are a number of recent reports where it is shown that SB203580 also inhibits the phosphorylation of p38 MAPK [20, 21].

Extracellular ATP stimulates MSK1 phosphorylation

The MAPKs regulate a variety of different biological processes through their actions in the cell cytoplasm and nucleus. Recently MSK1 has been identified as a downstream target of ERK and p38 MAPK pathways [22]. We therefore examined the time course of phosphorylation of MSK1 in response to ATP stimulation. Exposure of cardiac myocytes to ATP, increased the phosphorylation level of MSK1 as assessed by immunoblotting with an antibody specific for the kinase when phosphorylated at Ser376 (Fig. 3). The activation was rapid and maximal levels (~ 7-fold compared with the controls) were attained at 5 min after exposure to agonist. MSK1 phosphorylation was no more detected 20 min after stimulation.

Temporal differences in MSK1 response to inhibitors

Because both ERK1/2 and p38 MAPK are capable of phosphorylating and activating MSK1 [22] we used the MEK (PD98059 and U0126) and p38 MAPK (SB203580) inhibitors whose specificity is demonstrated in Fig. 2, to evaluate the relative roles of these MAP kinases in MSK1 activation in cardiac myocytes. We also used the PKC inhibitor GF109203X and the PKA inhibitor RpcAMP. Since we established that the time course of the ATP-induced activation was different for ERK1/2 and p38 MAPK (Fig. 1), the effects of the inhibitors were assessed 5 and 10 min after the purinergic stimulation. Interestingly, the effects of SB203580 and RpcAMP on the phosphorylation level of MSK1 varied with the time of stimulation (Fig. 4). After 5 min, phosphorylation was almost completely inhibited by PD98059 or U0126, but it was only partially prevented by SB203580 (Figs 4A and 4B). Similarly, prior treatment of cardiac myocytes with GF109203X resulted in a marked reduction of MSK1 phosphorylation. When both PD98059 and SB203580 were included in the incubation medium, the phosphorylation of MSK1 was completely inhibited. However, at this time point RpcAMP (or H89) had no effect on the activation of MSK1. After 10 min of stimulation with ATP, SB203580 inhibited to a greater extent the phosphorylation of the kinase; however to a value not significantly different from the extent of inhibition by PD98059 (Figs 4C and 4D). Similarly, preincubation of cardiomyocytes with RpcAMP before ATP stimulation, resulted in 82% inhibition of MSK1 phosphorylation.

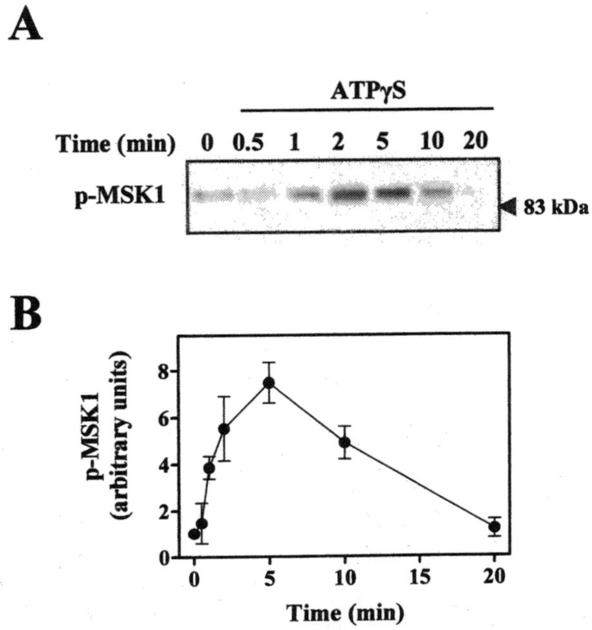

Fig. 3. Phosphorylation of MSK1 by ATPγS. Cardiac myocytes were exposed to 30 μM ATPγS for the times indicated. Cell extracts were subjected to SDS/PAGE and immunoblotted with an antibody which detects phosphorylated MSK1 (Ser376) (A). Phosphorylation of MSK1 by ATPγS was quantified by laser scanning densitometry (B). Results are means ± S.E.M. of 4 independent experiments.

Time course of the ATP-stimulated $cPLA_2$ activation

MSK1 was recently implicated as the responsible kinase for the phosphorylation and activation of $cPLA_2$ after purinergic stimulation of cardiac myocytes [8]. In view of the temporal differences in the maximal activation of ERK1/2, p38 MAPK and MSK1 (Figs 1 and 3), we sought to establish the time-course of the ATP-induced $cPLA_2$ activation and compare it with that of the kinases. $cPLA_2$ is fully activated by both phosphorylation and increases in intracellular calcium that drive its translocation from the cytosol toward the membranes. The reduced electrophoretic mobility of the enzyme due to its phosphorylation on Ser505, has also been used to assess activation of $cPLA_2$. We could not detect any change in electrophoretic mobility of $cPLA_2$ in response to purinergic stimulation of cardiomyocytes (data not shown). However, it has been recently shown that $cPLA_2$ can also be phosphorylated at other sites, in particular Ser727, without impact on its gel mobility [23]. We thus determined the redistribution of $cPLA_2$ from the cytosol to membranes in response to ATP in order to assess $cPLA_2$ activation. As shown in Fig. 5, purinergic stimulation triggered a rapid translocation of $cPLA_2$ with the maximal effect being observed at 10 min. These results fit with the time course of MSK1 activation. This type of temporal relationship suggests a cause and effect relationship between the activation of MSK1 and that

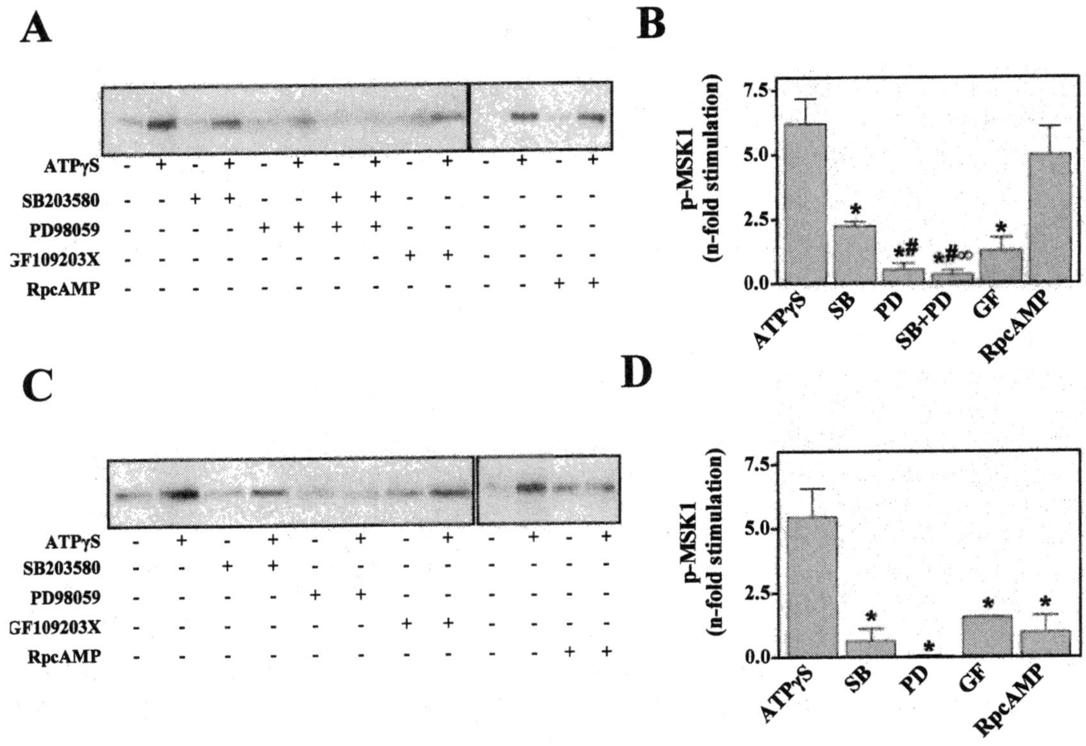

Fig. 4. Effect of various inhibitors on MSK1 phosphorylation by ATPγS. Cardiac myocytes were either left untreated or preincubated with SB203580 (10 μM), PD98059 (10 μM), GF109203X (1 μM), or Rp-cAMP (100 μM). They were incubated in the absence of agonists or in the presence of 30 μM ATPγS for 5 min (A, B), or 10 min (C, D). Proteins were separated by SDS/PAGE and phospho-MSK1 was detected via Western blot analysis (A, C). Blots were quantified by laser scanning densitometry (B, D). Results are means ± S.E.M. of 3 independent experiments. *Significant inhibition of MSK1 phosphorylation ($p < 0.05$, ANOVA with Dunnet post-test) when compared with identically treated cells in the absence of inhibitors. #Significant difference in MSK1 phosphorylation when compared with cells treated with SB203580. ∞Significant difference in MSK1 phosphorylation when compared with cells treated with RpcAMP.

of PLA_2. Further support for this conclusion had been obtained by the inhibition of ATP induced PLA_2 and AA release by Ro312080, which inhibits MSK1 activation [8].

Discussion

Extracellular ATP has a diverse range of biological effects in the cardiovascular system both in physiological and pathological conditions. ATP is released from the nerve terminals as a cotransmitter with norepinephrine and acetylcholine and affects both the choronotropy and inotropy of the heart. In addition, high extracellular ATP concentrations, such as those found locally during ischemia modulate the activity of exchangers and various transmembrane ionic currents leading to acidosis, Ca^{2+} overload and cell depolarization. These mechanisms could account for the early electrical disturbances observed under this pathologic condition. Recent studies in other cell systems also suggest that extracellularly applied ATP may trigger more slowly acting signal transduction cascades to mediate changes in cellular proliferation, growth and differentiation and apoptosis [24, 25]. Furthermore, in cardiac myocytes, ATP increases the expression of immediate-early genes such as c-fos and jun-B although it does not induce cell hypertrophy [26]. In fact, it has been shown that ATP inhibits adrenergic agonist induced hypertrophy in neonatal cardiac myocytes [27]. Activation of MAPK pathways could be important in the regulation of transcriptional events. MAPKs are activated in response to a variety of different stimuli and participate in various intracellular signaling pathways that control a wide spectrum of cellular processes including cell growth, differentiation and stress responses. It has been shown that ATP activates ERKs in neonatal cardiac myocytes although this activation has been dissociated from the hypertropic response [28]. In addition ERKs and p38 MAPK are activated in adult rat cardiac myocytes [8, 28] leading to the activation of PLA_2 and arachidonic acid release.

In the present study we demonstrate that, although both ERKs and p38 MAPK were phosphorylated to about the same extent in adult rat cardiac myocytes in response to purinergic stimulation, the pattern of activation of these two kinases is different. Phosphorylation of ERK1/2 is rapid and transient whereas p38 MAPK phosphorylation is induced more slowly and is sustained for at least 20 min. Specific inhibitors of ERK and p38 MAPK pathways have been a powerful tool in the

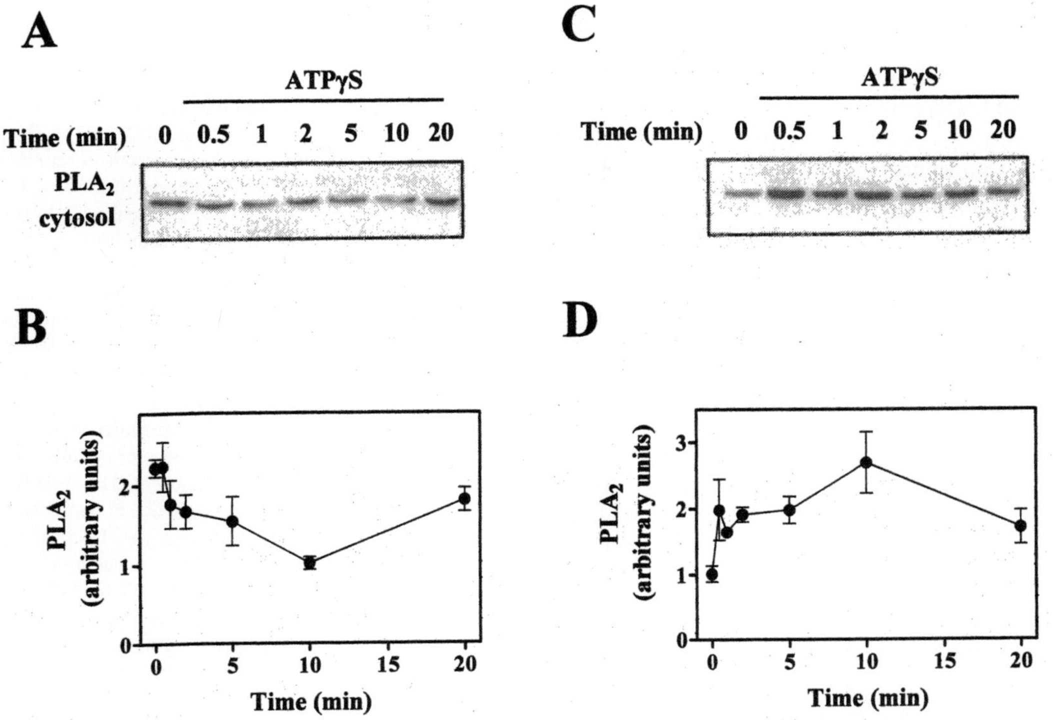

Fig. 5. $cPLA_2$ translocation to membranes after exposure of adult rat cardiac myocytes to ATPγS. Cardiac myocytes were exposed to 30 μM ATPγS for the time indicated. Cytosolic and membrane fractions were subjected to SDS/PAGE and immunoblotted with anti-$cPLA_2$ antibody (A, C). $cPLA_2$ from cytosolic (B) and membrane fractions (D) were quantified by laser scanning densitometry. Each point represents the mean ± S.E.M. of 5 independent experiments.

delineation of MAPK signaling in response to numerous stimuli. PD98059 and U0126 are selective inhibitors of MEK1/2 [16, 17] and have been used extensively to implicate MEK1/2 in regulating ERK1/2. Both drugs completely inhibited phosphorylation and activation of ERK1/2 (Figs 2A and 2B) implicating that the kinase is phosphorylated predominantly via the Raf/MEK pathway. Furthermore, the use of the specific PKC inhibitor, GF109203X, revealed that activation of both ERKs and p38 MAPK by ATP is PKC-dependent (Fig. 2). However, it is not clear whether the effect of PKC is independent of Raf activation or not. Similar results have been reported with α_1-adrenergic stimulation of neonatal cardiac myocytes where activation of both ERK1/2 and p38 MAPK by phenylephrine was PKC-dependent [29, 30].

Purinergic stimulation of cardiomyocytes increases cAMP content through a Gs-mediated activation of adenyl cyclase isoform V [31]. Here we confirm previous results [8] that the PKA inhibitor has an inhibitory effect on p38 MAPK phosphorylation while it does not affect ERK phosphorylation (Fig. 2). These results are further supported by the use of two other PKA inhibitors H89 and PKI. In the case of H89, the inhibitor at the concentration used (10 μM), had a slight stimulatory effect on ERKs. This is consistent with the notion that cAMP has a stimulatory and inhibitory effect on B-Raf and c-Raf-1, respectively [32] and previous findings that only c-Raf is present in rat neonatal cardiomyocytes [33].

To date, little is known on the regulation of the downstream molecular targets of ERK1/2 and p38 MAPK in the adult heart. Recently a new MAPK target protein, MSK1, has been identified [22]. In HEK293 cells, MSK1 is activated in an ERK-dependent manner by stimulation with phorbol esters or EGF whereas activation induced by UV irradiation, oxidative stress and other cell damaging stimuli is mediated through p38 MAPK. Here, we show that purinergic stimulation, which activates both ERKs and p38 MAPK in cardiac myocytes, also induces the phosphorylation of MSK1 (Fig. 3). The different temporal patterns of activation of ERKs and p38 MAPK by ATP would seem to indicate a distinct role for each of these MAPKs in MSK1 activation. Activation of MSK1 is rapid and closely follows the phosphorylation of ERK1/2. The parallel kinetics of ERKs and MSK1 phosphorylation could imply that phosphorylation by this MAPK is compulsory for MSK1 activation. This is further supported by the observation that the MEK inhibitors, PD98059 and U0126, almost completely inhibited MSK1 phosphorylation (Fig. 4). However, the inhibition of MSK1 phosphorylation observed in the presence of SB203580 (Fig. 4) implies that activation of the kinase may not be solely maintained by ERKs but by p38 MAPK as well. It is of interest that after 10 min stimulation with ATP, when p38 MAPK is fully activated, MSK1 phosphorylation was greatly reduced in the presence of SB203580 or RpcAMP. These results clearly indicate that p38 MAPK plays a role in MSK1 phosphorylation. However, at 20 min after purinergic stimulation, when p38 MAPK activation is still quite high but ERK activity has returned to basal levels (Fig. 1), MSK1 phosphorylation cannot be detected (Fig. 3) indicating that p38 MAPK alone cannot maintain phosphorylation of MSK1. Taken together the above results suggest that ERK1/2 act to trigger the phosphorylation of MSK1 and that both ERK1/2 and p38 MAPK appear to be required for full activation of MSK1 in cardiac myocytes in response to purinergic agonists.

The physiological role of activated MSK1 in cardiac myocytes is unknown. In other cell types, activated MSK1 has been shown to phosphorylate nucleosomal components such as histone H3/HMG-14 [34] as well as transcription factors such as cAMP-response element binding protein (CREB) and ATF1 [22, 35], indicating that it may play a role in the regulation of gene expression. In a previous study [8], it has been shown that Ro318220, a specific MSK1 inhibitor [22] markedly reduced $cPLA_2$ translocation and I_{TREK} activation in adult rat cardiac myocytes, as did inhibition of ERKs and p38 MAPK. These results suggest that MSK1 integrates signals from ERKs and p38 MAPK and then activates $cPLA_2$. In this study, we showed that the temporal increase in $cPLA_2$ translocation to the membranes (Fig. 4) closely followed MSK1 phosphorylation. These results further support the notion that $cPLA_2$ is a downstream target of MSK1.

In conclusion, we showed that purinergic stimulation of rat cardiomyocytes activates ERKs and p38 MAPK with a distinct temporal pattern. ERKs activation proceeds through the Raf/MEK pathway and its is PKC-dependent whereas both PKC and PKA regulate p38 MAPK signaling. We also demonstrated that under ATP stimulation of cardiomyocytes, MSK1 is phosphorylated with kinetics parallel to that of ERKs phosphorylation. The results from the use of various inhibitors of the ERK and p38 MAPK pathways on the phosphorylation of MSK1 suggest that ERKs initiate the response but both types of MAP kinases, ERKs and p38 MAPK, are required for the maintenance of the MSK1 phosphorylation. The temporal relationship of MSK1 phosphorylation and $cPLA_2$ translocation to the membranes, taken together with previous findings [8], is an indication that $cPLA_2$ may be a downstream target of MSK1.

Acknowledgements

The work was supported by a grant from the Greek General Secretariat of Research and Technology under the programme for Bilateral Scientific and Technological Cooperation with France (PLATON).

References

1. Vassort G: Adenosine 5′-triphosphate: A P2-purinergic agonist in the myocardium. Physiol Rev 81: 767–806, 2001
2. Legssyer A, Poggioli J, Renard D, Vassort G: ATP and the other adenine compounds increase mechanical activity inositol triphosphate production in rat heart. J Physiol 401: 185–199, 1988
3. Kuzmin AI, Lakomkin VL, Kapenko VI, Vassort G: Interstitial ATP level and degradation in control and postmyocardial infracted rats. Am J Physiol 275, C766–C771, 1998
4. Fredholm BB, Abbracchio MP, Burnstock G, Dubyak GR, Harden TK, Jacobson KA, Schwabe U, Williams M: Towards a revised nomenclature for P1 and P2 receptors. Trends Pharmacol Sci 18: 79–82, 1997
5. Ralevic V, Burnstock G: Receptors for purines and pyrimidines. Pharmacol Rev 50: 415–492, 1998
6. Communi D, Govaerts C, Parmentier M, Boeynaems JM: Cloning of a human purinergic P2Y receptor coupled to phospholipase C and adenyl cyclase. J Biol Chem 272: 31969–31973, 1997
7. Puceat M, Roche S, Vassort G: Src family tyrosine kinase regulates intracellular pH in cardiomyocytes. J Cell Biol 141: 1637–1646, 1998
8. Aimond F, Fauzier J-M, Bony C, Vassort G: Simultaneous activation of p38 MAPK and p44/42 MAPK by ATP stimulates the K^+ current I_{TREK} in cardiomyocytes. J Biol Chem 275: 39110–39116, 2000
9. Leslie CC: Properties and regulation of cytosolic phospholipase A_2. J Biol Chem 272: 16709–16712, 1997
10. Gijon MA, Leslie CC: Regulation of arachidonic acid release and cytosolic phospholipase A2 activation. J Leukoc Biol
11. Gijon MA, Spencer DM, Kaiser AL, Leslie CC: Role of phosphorylation sites and the C2 domain in regulation of cytosolic phospholipase A1. J Cell Biol 145: 1219–1232, 1999
12. Sugden PH, Clerk A: Cellular mechanisms of cardiac hypertrophy. J Mol Med 76: 725–746, 1998
13. Widmann C, Gibson S, Jarpe MB, Johnson GL: Mitogen-activated protein kinase: Conservation of a three kinase module from yeast to human. Physiol Rev 79: 143–180, 1999
14. Della Roca GJ, van Viesen T, Daaka Y, Luttrell DK, Luttrell LM, Lefkowitz RJ: Ras-dependent mitogen-activated protein kinase activation by G protein-coupled receptors. Convergence of Gi- and Gq-mediated pathways on calcium/calmodulin, Pyk2 and Src kinase. J Biol Chem 272: 19125–19132, 1997
15. Lazou A, Fuller JS, Bogoyevitch AM, Orfali AK, Sugden HP: Characterization of stimulation of phosphoinositide hydrolysis by α_1-adrenergic agonists in adult rat hearts. Am J Physiol 267: H970–H978, 1994
16. Alessi DR, Cuenda A, Cohen P, Dudley DT, Saltiel AR: PD098059 is a specific inhibitor of the activation of mitogen-activated protein kinase kinase *in vitro* and *in vivo*. J Biol Chem 270: 27489–27494, 1995
17. Favata MF, Horiuchi KY, Manos EJ, Daulerio AJ, Stradley DA, Feeser WS, Van Dyk DE, Pitts WJ, Earl RA, Hobbs F, Copeland RA, Magolda RL, Scherle PA, Trzaskos JM: Identification of a novel inhibitor of mitogen-activated protein kinase kinase. J Biol Chem 273: 18623–18632, 1998
18. Toullec D, Pianetti P, Coste H, Bellevergue P, Grand-Perret T, Ajakane M, Baudet V, Boissin P, Boursier E, Loriolle F, Duhamel L, Charon D, Kirilovsky J: The bisindolylmaleimide GF109203X is a potent and selective inhibitor of protein kinase C. J Biol Chem 266: 15771–15781, 1991
19. Lee JC, Laydon JT, McDonnell PC, Gallagher TF, Kumar S, Green D, McNulty D, Blumenthal MJ, Heys JR, Landvatter SW, Strickler JE, McLaughlin MM, Siemens IR, Fisher SM, Livi GP, White JR, Adams

JL, Young PR: A protein kinase involved in the regulation of inflammatory cytokine biosynthesis. Nature 372: 739–746, 1994
20. Frantz B, Klatt T, Pang M, Parsons J, Rolando A, Williams H, Tocci MJ, O'Keefe SJ, O'Neill MA: The activation state of p38 mitogen-activated protein kinase determines the efficiency of ATP competition for pyridinylimidazole inhibitor binding. Biochemistry 37: 13846–13853, 1998
21. Yamboliev IA, Hedges JC, Mutnick JL, Adam LP, Gerthoffer WT: Evidence for modulation of smooth muscle force by the p38 MAP kinase/Hsp27 pathway. Am J Physiol 278: H1899–H1907, 2000
22. Deak M, Clifton AD, Lucocq JM, Alessi DR: Mitogen- and stress-activated protein kinase-1 (MSK1) is directly activated by MAPK and SAPK2/p38 and may mediate activation of CREB. EMBO J 17: 4426–4441, 1998
23. Hefner Y, Borsch-Haubold G, Murakami M, Wilde JI, Pasquet S, Schieltz, Ghomashchi F, Yates JR III, Armstrong CG, Paterson A, Cohen P, Fukunaga R, Hunter T, Kudi I, Watson SP, Gelb MH: Serine 7272 phosphorylation and activation of cytosolic phospholipase A2 by MNK1-related protein kinases. J Biol Chem 275: 37542–37551, 2000
24. Antonysamy MA, Moticka EJ, Ramkurar V: Adenosine acts as an endogenous modulator of IL-2 dependent proliferation of cytotoxic T lymphocytes. J Immunol 15: 2813–2831, 1995
25. Abbracchio MP, Burnstock G: Purinergic signaling: Pathophysiological roles. Jpn J Pharmacol 78: 113–145, 1998
26. Zheng JS, Boluyt MO, O'Neil L, Crow MT, Lakatta EG: Extracellular ATP induces immediate-early gene expression but not cellular hypertrophy in neonatal cardiac myocytes. Circ Res 74: 1034–1041, 1994
27. Zheng JS, Boluyt MO, Long X, O'Neil L, Lakatta EG, Crow MT: Extracellular ATP inhibits adrenergic agonist-induced hypertrophy of neonatal cardiac myocytes. Circ Res 78: 525–535, 1996
28. Post GR, Goldstein D, Thuerauf DJ, Glembotski CC, Brown JH: Dissociation of p44 and p42 mitogen activated protein kinase activation from receptor-induced hypertrophy in neonatal rat ventricular myocytes. J Biol Chem 271: 8452–8457, 1996
29. Bogoyevitch MA, Glennon PE, Andersson MB, Clerk A, Lazou A, Marshall JC, Parker PJ, Sugden PH: Endothelin-1 and fibroblast growth factors stimulate the mitogen-activated protein kinase signaling cascade in cardiac myocytes. The potential role of the cascade in the integration of two signaling pathways leading to myocyte hypertrophy. J Biol Chem 269: 1110–1119, 1994
30. Clerk A, Michael A, Sugden PH: Stimulation of the p38 mitogen-activated protein kinase pathway in neonatal rat ventricular myocytes by the G protein-coupled receptor agonists, endothelin-1 and phenylephrine: A role in cardiac myocyte hypertrophy? J Cell Biol 142: 523–535, 1998
31. Puceat M, Bony C, Jaconi M, Vassort G: Specific activation of adenyl cyclase V by purinergic agonists. FEBS Lett 431: 189–194, 1998
32. Gao Z, Chen T, Weber MJ, Linden J: A2B adenosine and $P2Y_2$ receptors stimulate mitogen-activated protein kinase in human embryonic kidney-293 cells. Cross-talk between camp and protein kinase C pathways. J Biol Chem 274: 5972–5980, 1999
33. Bogoyevitch MA, Marshall CJ, Sugden PH: Hypertrophic agonists stimulate the activities of the protein kinases c-Raf and A-Raf in cultured ventricular myocytes. J Biol Chem 270: 26303–26310, 1995
34. Thomson S, Clayton AL, Hazzalin CA, Rose S, Barrat MJ, Mahadevan LC: The nucleosomal response associated with immediate-early gene induction is mediated via alternative MAP kinase cascades: MSK1 as a potential histone H3/HMG-14 kinase. EMBO J 18: 4779–4793, 1999
35. Arthur JSC, Cohen P: MSK1 is required for CREB phosphorylation in response to mitogens in mouse embryonic stem cells. FEBS Lett 482: 44–48, 2000

Molecular and Cellular Biochemistry **242:** 173–180, 2003.

Thyroid hormone and cardioprotection: Study of p38 MAPK and JNKs during ischaemia and at reperfusion in isolated rat heart

Constantinos Pantos,[1] Vassiliki Malliopoulou,[1] Ioannis Paizis,[1] Panagiotis Moraitis,[1] Iordanis Mourouzis,[1] Stylianos Tzeis,[1] Evangelia Karamanoli,[1] Demosthenes D. Cokkinos,[1] Hariclia Carageorgiou,[1] Dennis Varonos[1] and Dennis V. Cokkinos[2]

[1]*Department of Pharmacology, University of Athens, Goudi, Athens;* [2]*1st Cardiology Department, Onassis Cardiac Surgery Center, Kallithea, Athens, Greece*

Abstract

It has been recently shown that long-term thyroxine administration increases the tolerance of the heart to ischaemia. The present study investigated whether thyroxine induced cardioprotection involves alterations in the pattern of p38 mitogen activated protein kinase (p38MAPK) and c-Jun NH_2-terminal kinases (JNKs) activation during ischaemia-reperfusion. L-thyroxine (T4) was administered in Wistar rats (25 μg/100 g/day, subcutaneously) for 2 weeks (THYR), while normal animals served as controls (NORM). NORM and THYR isolated rat hearts were perfused in Langendorff mode and subjected to 10 or 20 min of zero-flow global ischaemia only and also to 20 min of ischaemia followed by 10, 20 or 45 min of reperfusion. Postischaemic recovery of left ventricular developed pressure at 45 min of reperfusion was expressed as % of the initial value. Activation of p38 MAPK and JNKs was assessed at the different times of the experimental setting by standard Western blotting techniques using a dual phospho p38MAPK and phospho JNKs (p46/p54) antibodies. Activation of p38 MAPK was significantly attenuated during ischaemia and reperfusion in thyroxine treated hearts compared to normal hearts. JNKs were found to be activated only during the reperfusion period. The levels of phospho JNKs were found to be lower in thyroxine treated hearts as compared to untreated hearts, though not at a statistically significant level. Postischaemic functional recovery was higher in THYR as compared to NORM, $p < 0.05$. In summary, in hearts pretreated with thyroxine, p38 MAPK was attenuated during ischaemia and at reperfusion and this was associated with improved postischaemic recovery of function. (Mol Cell Biochem **242**: 173–180, 2003)

Key words: thyroid hormone, ischaemia, p38 MAPK, JNK, rat heart, ischaemic preconditioning

Introduction

Over the past years various therapeutic approaches have been suggested for the protection of the heart against ischaemic injury. Interventions targeting the preischaemic period are thought to be an effective therapeutic strategy that potentially can be applicable in clinical practice. Ischaemic preconditioning and pharmacological mimicking of this phenomenon are shown to be one of the most powerful means of protection currently available [1].

Thyroid hormone has been recently found to play an important role in the adaptive response of the heart to various ischaemic insults. In fact, long-term thyroxine pretreatment renders the heart more resistant to sustained ischaemic injury [2] and this effect was found to be potentiated by cardioprotective means such as ischaemic preconditioning [3]. Several

Address for offprints: C. Pantos, Department of Pharmacology, University of Athens, 75 Mikras Asias Avenue, 11527 Goudi, Athens, Greece (E-mail: cpantos@cc.uoa.gr)

studies show that the heart is one of the most thyroid hormone-responsive tissues in the body and therefore it is expected that thyroxine treatment could potentially exert various physiological effects on the myocardium [4]. It is already known that thyroxine administration can regulate the transcription of a variety of myocyte specific genes and it has been suggested that thyroxine could be a suitable inotropic agent that can be used in various clinical conditions [4]. Furthermore, recent studies have demonstrated that thyroxine can also regulate cardiac metabolism as well as intracellular molecules that are found to be involved in cardioprotection, effects that potentially could alter the adaptive response of the heart to ischaemia [2, 5, 6]. In fact, long-term thyroxine administration was found to be associated with increased induction of hsp70 mRNA and attenuation of p38 Mitogen Activated Protein Kinase (p38MAPK) activity in response to ischaemia and reperfusion [2]. Such intracellular changes could potentially account for the cardioprotection conferred by thyroxine pretreatment. However, the exact mechanisms of thyroxine induced cardioprotection is a topic that remains to be further investigated.

Recent research concerning the underlying mechanisms of the adaptive response of the heart against various ischaemic insults has emphasized the important role of the mitogen activated protein kinases in that response. In fact, it has been shown that part of the cellular response to ischaemic stress involves activation of several members of the mitogen activated protein kinase family such as p38 MAPK and c-Jun NH_2-terminal kinases (JNKs) [7]. In the isolated perfused heart, p38 MAPK can be activated by global ischaemia and this activation is maintained during reperfusion [8, 9]. In the same model, JNKs are not found to be activated by ischaemia, whereas they can be activated by reperfusion following ischaemia [8–10]. However, it remains unclear whether activation of these kinases is detrimental for cell survival or it is a part of a protective mechanism against ischaemia or ischaemia and reperfusion. Current evidence indicates that the duration and/or intensity of the stimulus possibly determines the pattern of their activation and their effect, beneficial or deleterious, on cell survival against various stresses [11, 12]. In fact, transient activation of p38 MAPK was found not to have a deleterious effect on the cell, whereas sustained activation of p38 MAPK induces apoptosis that can be attenuated by inhibition of p38 MAPK activation [12]. Along with these lines, a few studies have shown that in cardioprotective manipulations such as ischaemic and pharmacological preconditioning, the activation of p38 MAP kinase during the following prolonged ischaemia is found to be attenuated [13, 14]. Furthermore, recent studies showed that interventions that could interfere with intracellular events not only during the preischaemic or ischaemic period but also during the reperfusion phase can determine the extent of ischaemic injury [14, 15]. However, the pattern of activation of p38 MAPK and JNKs particularly and its possible correlation to protection has been poorly investigated in the context of cardioprotective interventions targeting the preischaemic period.

In the present study, we extended our previous work on thyroxine induced cardioprotection by examining the activation pattern of p38 MAPK and JNKs during ischaemia and reperfusion in thyroxine pretreated and normal hearts as well as in ischaemically preconditioned hearts and its possible correlation to postischaemic functional recovery.

Materials and methods

Animals

Eighty Wistar male rats (280–320 g) were used for this study. The rats were handled in accordance with the Guide for the Care and Use of Laboratory Animals published by the US National Institutes of Health (NIH Publication No 85-23, revised 1985).

Thyroxine administration

L-Thyroxine (T4) (Sigma Chemicals, St. Louis, MO, USA) (25 μg/100 g body wt) was given subcutaneously once daily for 14 days. This treatment results in a long-term moderate hyperthyroidism. Normal rats were treated with normal saline given once daily for 14 days [2, 3, 6, 16, 17].

Isolated heart preparation

A non-ejecting isolated rat heart preparation was perfused at constant coronary flow according to the Langendorff technique, as previously described [2, 3, 6, 17]. In this model, coronary flow per gram of cardiac tissue was similar in all the experimental groups. Rats were anaesthetized with intraperitoneal injection of ketamine hydrochloric acid and heparin 1000 IU/kg was given intravenously before thoracotomy. The hearts were perfused with oxygenated (95% O_2/5% CO_2) Krebs-Henseleit buffer at a constant temperature of 37°C and were paced at 320 bpm with a Harvard pacemaker. The pacemaker was turned off during the period of ischaemia. Postischaemic function was assessed by the recovery of the left ventricular developed pressure that was measured at the end of the reperfusion period and was expressed as % of the initial value (LVDP%). Severity of ischaemic contracture (increase in minimal value of ventricular pressure during ischaemia) was assessed by the time to peak contracture, T_{max} in min and by the magnitude of peak contracture, C_{max} in mmHg [2, 3, 6, 17].

Total protein preparation, SDS-PAGE and immunoblotting

Protein analysis was performed as previously described [2]. Briefly, 0.2 g frozen tissue was homogenized in ice-cold Tris-sucrose buffer (0.35 M sucrose, 10 mM Tris-HCl pH = 7.5, 1 mM ethylenediaminetetraacetic acid (EDTA), 0.5 mM dithiothreitol (DTT), 0.1 mM phenylmethylsulfonylfluoride (PMSF)) and the resulting homogenate was centrifuged at 15000 g for 20 min at 4°C. Protein concentrations were determined by the bicinchoninic acid (BCA) method using bovine serum albumin (BSA) as a standard.

After boiling for 5 min in Laemmli sample buffer, protein aliquots (40 μg) were loaded onto 12% (w/v) acrylamide gels and subjected to SDS-PAGE. After Western blotting, filters were probed with specific antibodies against dual phospho-p38 MAPK, total p38 MAPK (1:1000, New England Biolabs, Hitchin, Herts, UK), phospho-JNK, total JNK (1:1000, New England Biolabs, Hitchin, Herts, UK) and phospho-MKK3/6 (1:1000, New England Biolabs, Hitchin, Herts, UK) and immunoreactivity was detected by enhanced chemiluminescence. Immunoreactivity was expressed as ratio of the optical densities of dual phospho-p38 and phospho-JNK to total p38 MAP kinase and JNK optical densities, respectively. Phospho-MKK3/6 immunoreactivity was expressed as ratio of the optical density of phospho-MKK3/6 to total JNK optical density.

Experimental protocol

Normal and thyroxine treated hearts were stabilized for 20 min without being subjected to any ischaemic stress and were used for baseline measurements: NORM(base), n = 4 and THYR-(base), n = 5.

Normal and thyroxine treated hearts after initial stabilization period were subjected to (i) 10 min of zero-flow global ischaemia; NORM(10I), n = 4, THYR(10I), n = 5, (ii) 20 min of zero-flow global ischaemia; NORM(20I), n = 4, THYR-(20I), n = 4, and (iii) 20 min of zero-flow global ischaemia followed by 10 min of reperfusion; NORM(20I/10R), n = 5, THYR(20I/10R), n = 5, (iv) 20 min of zero-flow global ischaemia followed by 20 min of reperfusion; NORM(20I/20R), n = 5, THYR(20I/20R), n = 5, (v) 20 min of zero-flow global ischaemia followed by 45 min of reperfusion; NORM-(20I/45R), n = 9, THYR(20I/45R), n = 6.

Normal hearts after initial stabilization period were subjected to four cycles of brief episodes of ischaemia and reperfusion consisting of 3 min of zero-flow global ischaemia, 5 min of reperfusion, and three cycles of 5 min of ischaemia and 5 min of reperfusion followed by (i) 10 min of zero-flow global ischaemia; NORM-Pc(10I), n = 5, (ii) 20 min of zero-flow global ischaemia; NORM-Pc(20I), n = 4, (iii) 20 min of zero-flow global ischaemia followed by 10 min of reperfusion; NORM-Pc(20I/10R), n = 5, (iv) 20 min of zero-flow global ischaemia followed by 20 min of reperfusion; NORM-Pc(20I/20R), n = 5, (v) 20 min of zero-flow global ischaemia followed by 45 min of reperfusion; NORM-Pc(20I/45R), n = 6.

Statistical analysis

Values are presented as mean (S.E.M). Unpaired *t*-test and Mann-Whitney test were used for differences between groups. A two-tailed test with a p value less than 0.05 was considered significant.

Results

Parameters of cardiac function

Left ventricular developed pressure at the end of the stabilization period (LVDP) and at the end of reperfusion (LVDP45), left ventricular end diastolic pressure at 45 min of reperfusion (LVEDP45), LVDP%, T_{max} and C_{max} for all groups are shown in Table 1.

Activation of p38 MAPK during sustained ischaemia

Activation of p38 MAPK during sustained ischaemia was assessed by the dual phosphorylation of p38 MAPK and the levels of phospho-p38 at different time points of the ischaemic phase are shown in Fig. 1A. In normal hearts, p38 MAPK was found to be activated at 10 min of ischaemia and this activation was further increased at the end of ischaemia (Fig. 1A). In thyroxine treated hearts, p38 MAPK was also found to be activated at 10 min of ischaemia, whereas returned to baseline levels at the end of ischaemia. A similar pattern of activation of p38 MAPK was found in preconditioned hearts (Fig. 1A).

At 10 min of ischaemia, phospho-p38 MAPK was 1.5 fold more in NORM(10I) as compared to THYR(10I) hearts, $p < 0.05$, and 1.2 fold more as compared to NORM-Pc(10I), $p = 0.1$. At 20 min of ischaemia, phospho-p38 MAPK was found to be 1.9 fold more in NORM(20I) as compared to THYR-(20I) hearts, $p < 0.05$, and 2.0 fold more as compared to NORM-Pc(20I), $p < 0.05$ (Fig. 1A).

In addition, in thyroxine treated hearts, the upstream activators of p38 MAPK, MKK3/6, were also found to be less activated during sustained ischaemia as compared to normal hearts. In fact, phospho-MKK3/6 were 1.5 and 2 fold more in normal than in thyroxine-treated hearts at 10 and 20 min of ischaemia respectively, $p < 0.05$.

Table 1. Left ventricular developed pressure at the end of the stabilization period (LVDP in mmHG) and at 45 min of reperfusion (LVDP45 in mmHG), left ventricular end-diastolic pressure at 45 min of reperfusion (LVEDP45 in mmHg), recovery of left ventricular developed pressure (LVDP%), time to peak contracture (T_{max} in min) and magnitude of peak contracture (C_{max} in mmHg) in NORM(20I/45R), THYR(20I/45R) and NORM-Pc(20I/45R) groups. The values are mean (S.E.M.).

Group	LVDP baseline	LVDP 45 min reperfusion	LVDP %	LVEDP 45 min reperfusion	T_{max}	C_{max}
NORM(20I/45R) (n = 9)	123.8 (4.6)	54.1 (6.5)	42.9 (4.1)	66.6 (6.3)	–	65.3 (2.5)
THYR(20I/45R) (n = 6)	117.0 (4.8)	78.6* (4.7)	67.6* (4.2)	45.8* (4.3)	10.5 (0.9)	98.5 (7.8)*
NORM (Pc+20I/45R) (n = 6)	129 (8.1)	97.1* (10.9)	75.2* (8.3)	35.7* (7.8)	14.5 (1.9)	74.4 (11.2)

*p < 0.05 vs. NORM(20I/45R).

Activation of p38 MAPK during reperfusion

In normal hearts, p38 MAPK was found to be activated at 10 min of reperfusion and this activation was further increased at 20 min of reperfusion, while declined at the end of reperfusion (Fig. 1B). In thyroxine treated hearts, p38 MAPK was also found to be activated at 10 min of reperfusion, further increased at 20 min of reperfusion but returned to baseline levels at the end of reperfusion (Fig. 1B). In preconditioned hearts, the activation of p38 MAPK followed a similar pattern as in thyroxine treated hearts (Fig. 1B).

The levels of phospho-p38 MAPK were not different between normal, thyroxine-treated and preconditioned hearts at 10 and 20 min of reperfusion (Fig. 1B). At the end of reperfusion, phospho-p38 MAPK was found to be 1.7 fold more in normal hearts as compared to thyroxine-treated hearts, $p < 0.05$ and 1.3 fold more in normal hearts as compared to preconditioned hearts, $p = 0.09$ (Fig. 1B).

Activation of JNKs during reperfusion

Activation of JNKs during reperfusion was assessed by the dual phosphorylation of JNKs and the levels of p54 and p46 JNK at different time points of reperfusion phase are shown in Fig. 2. JNKs were not found to be activated during ischaemia (data not shown). JNKs were found to be activated at 10 min of reperfusion and this activation was further increased at 20 min, and declined at the end of reperfusion, though it remained above the baseline levels.

At 10 min of reperfusion, phospho-p54 JNK was found to be 1.4 fold more in NORM(20I/10R) than in THYR(20I/10R) hearts, $p < 0.05$ and 1.4 fold more in NORM(20I/10R) than in NORM-Pc(20I/10R) hearts, $p < 0.05$. Phospho-p46 JNK was found to be 1.6 fold more in NORM(20I/10R) than in THYR(20I/10R) hearts, $p = 0.05$ and 1.9 fold more in NORM(20I/10R) than in NORM-Pc(20I/10R) hearts, $p < 0.05$. At 20 min of reperfusion JNKs were found to be equally activated in all groups. At 45 min of reperfusion, phospho-p54 JNK was 1.8 fold more and phospho-p46 JNK 1.9 fold more in NORM(20I/45R) than in NORM-Pc(20I/45R) hearts,

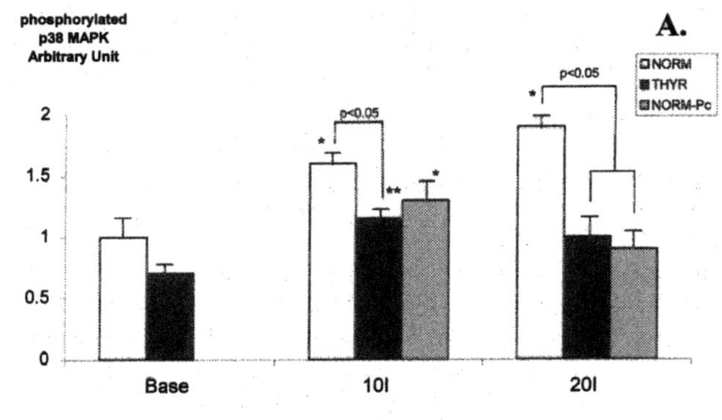

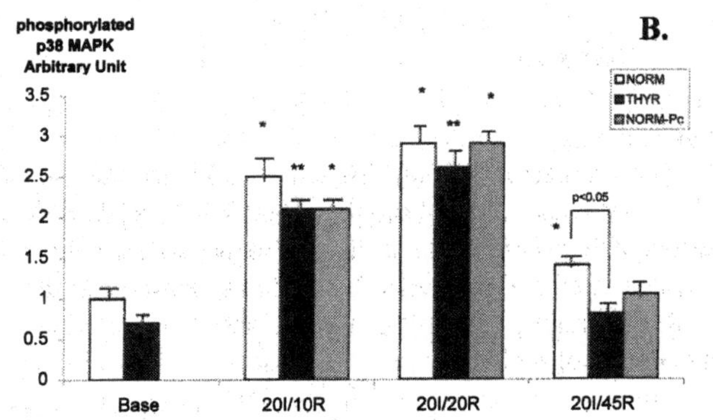

Fig. 1. Phosphorylated p38 MAP kinase protein expression in normal (NORM), thyroxine-treated (THYR) and ischaemically preconditioned hearts (NORM-Pc) at different time points during ischaemia (A) and during reperfusion (B). Base – baseline; I – ischaemia; R – reperfusion (columns are means of optical ratios, bar = S.E.M.).

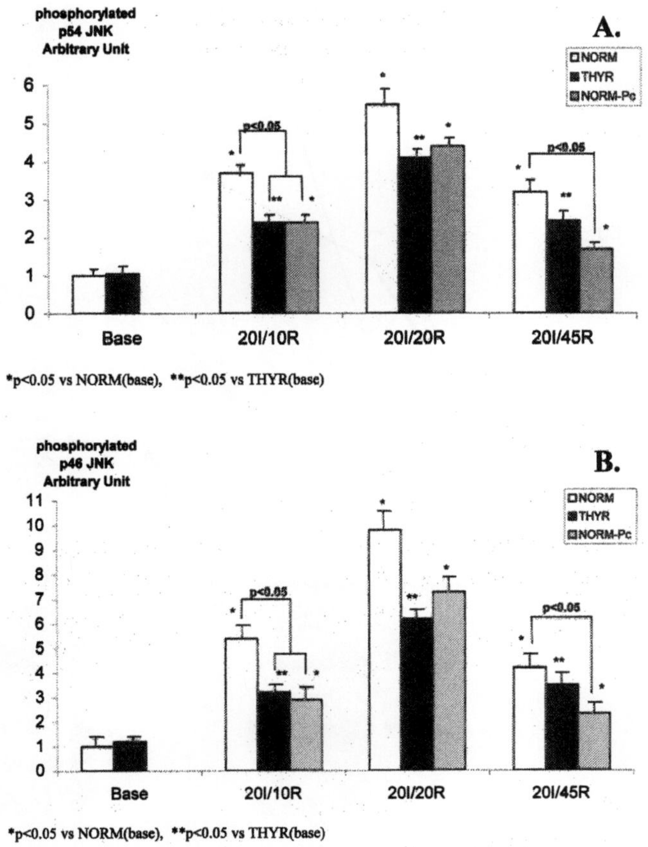

Fig. 2. Phosphorylated p54 JNK (A) and p46 JNK (B) protein expression in normal (NORM), thyroxine-treated (THYR) and ischaemically preconditioned hearts (NORM-Pc) at different time points during reperfusion. Base – baselines; I – ischaemia; R – reperfusion (columns are means of optical ratios, bar = S.E.M.).

$p < 0.05$ while there was not a statistical difference between NORM(20I/45R) and THYR(20I/45R) hearts.

Discussion

In the present study we evaluated the pattern of activation of two subfamilies of the mitogen activated kinase family, the p38 MAPK and JNKs throughout the entire experimental setting of zero-flow global ischaemia and reperfusion in thyroxine treated and non treated isolated rat hearts in order to shed some light on the role of these kinases in thyroxine induced cardioprotection.

Activation of p38 MAPK during sustained ischaemia

Activation of p38 MAPK (as evidenced by the dual phosphorylation of this kinase) was found to occur at the early phase of ischaemia (10 min) in both normal and thyroxine treated hearts. The phosphorylated p38 MAPK though was significantly less in thyroxine treated hearts than in normal hearts. Furthermore, at the end of the 20 min of ischaemia, p38 MAPK activation was found to decline and return to baseline levels in thyroxine treated hearts, whereas activation of p38 MAPK continued to increase in normal hearts. Thus, the levels of the phosphorylated p38 MAPK were significantly less in thyroxine treated than in untreated hearts. Furthermore, the upstream activators of p38 MAPK, MKK3/6, were also shown to have a similar pattern of phosphorylation to that of p38 MAPK. This pattern of activation was associated with improved postischaemic recovery of function in thyroxine treated than in non treated hearts. Similar results were also found for the preconditioned hearts. In fact, p38 activation was attenuated at the end of ischaemia following a four cycle preconditioning protocol, while it was less but not at a statistical significance at the 10 min of ischaemia as compared to normal hearts.

Several studies have clearly demonstrated that attenuation of p38 activation during sustained ischaemic stress is correlated with increased tolerance of the myocardium against an ischaemic insult. In cell based models, it has been shown that during lethal ischaemic stress sustained activation of p38 MAPK occurs and this results in cell death, while inhibition of p38 MAPK activation during the ischaemic insult by a p38 MAPK inhibitor such as SB203580 was found to increase cell survival [12, 13]. Furthermore, interventions targeting the preischaemic period such as ischaemic preconditioning [14, 18], SB202190 administration [18], isoproterenol administration [19], or overexpression of active PKCδ in cell based models [20] were shown to attenuate ischaemia induced p38 MAPK activation and this was associated with less ischaemic injury. On the basis of these data, it could be suggested that the pattern of p38 MAPK activation that was observed to occur in thyroxine treated hearts during ischaemia might be an essential element of thyroxine induced cardioprotection.

Ischaemic contracture and activation of p38 MAPK during ischaemia

It is known that, during global ischaemia, a rapid decrease of myocardial contractility occurs, whereas diastolic pressure is increased, a phenomenon known as ischaemic contracture. In previous studies as well as in this study we have found that thyroxine treated hearts display an abnormal pattern of contracture [3, 17]. In fact, ischaemic contracture occurs earlier in those hearts than in normal hearts. This early occurrence of contracture has been related to earlier energy depletion that occurs during ischaemia due to the lower levels of preischaemic myocardial glycogen that are found in thyroxine treated as compared to normal hearts [21]. Interestingly, early occurrence of contracture has been observed in various treatments that target the preischaemic period such as ischaemic preconditioning [22], dobutamine [23], norepinephrine [24], and even the administration of SB202190, a p38 MAPK inhibitor [18]. All of these interventions have also been found to increase the tolerance of the heart to ischaemic stress [18, 22–24].

It is likely that this early energy deprivation that occurs during ischaemia and has a profound effect on the mode of the mechanical function of the heart, could potentially affect energy dependent intracellular processes such as the activation of stress-induced mitogen activated kinases. In fact, Gabai *et al.* [25] have recently shown that, in a cell based model, during ATP depletion, activities of stress kinases are decreased to almost an undetectable level (because stress kinase could not be phosphorylated by upstream kinases without ATP), whereas restoration of ATP increased significantly their activity. On the basis of those data one could suggest that the attenuated and non sustained p38 MAPK activation that is observed in thyroxine treated hearts during ischaemia is probably the result of the early energy deprivation that occurs in those hearts as compared to normal hearts. This is also supported by the fact that a similar pattern of p38 MAPK activation was found to occur in preconditioned hearts in which ischaemic contracture was also shown to be accelerated (Fig. 3).

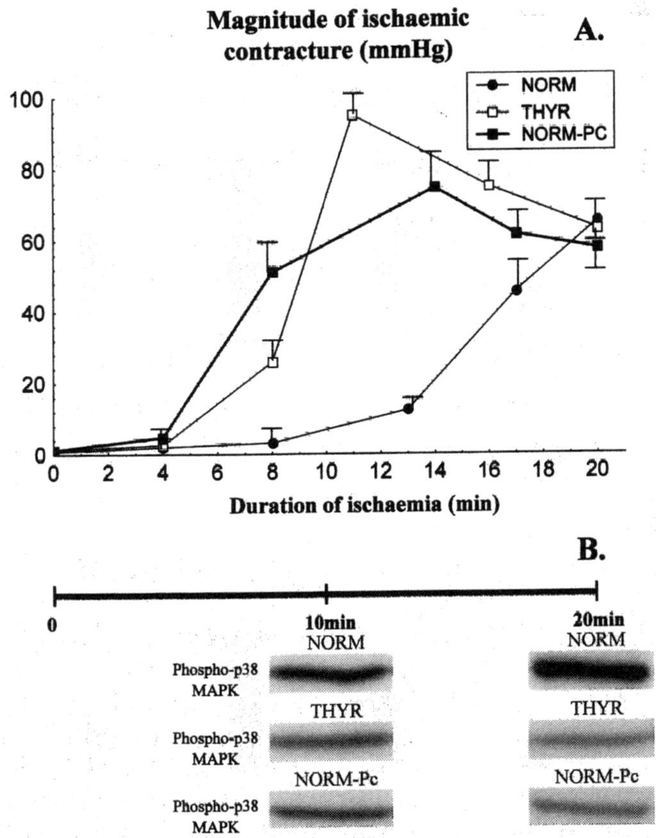

Fig. 3. (A) Ischaemic contracture profiles of normal (NORM), thyroxine treated (THYR) and ischaemically preconditioned hearts (NORM-Pc) subjected to 20 min of ischaemia (bar = S.E.M.). (B) Western blots showing the expression of phospho-p38 MAPK at 10 and 20 min of ischaemia in NORM, THYR and NORM-Pc.

The effects of reperfusion on p38MAPK and JNKs

The reperfusion phase has been poorly investigated in relation to preischaemic cardioprotective interventions. Moreover, the importance of the various events that occur during reperfusion regarding cardioprotection remains largely unknown. In the present study, in order to identify the possible role of reperfusion phase in thyroxine induced cardioprotection, we have investigated the effect of thyroxine pretreatment on the pattern of activation of p38 MAPK and JNKs during reperfusion. This study was also extended to hearts that were subjected to another means of protection such as ischaemic preconditioning.

Our study showed that during the early (10 min) and middle (20 min) phase of reperfusion increased activation of p38 MAPK occurred to almost a similar degree in thyroxine and in preconditioned hearts as compared to normal hearts. Furthermore, at the end of reperfusion p38 MAPK activation declined towards the baseline levels more rapidly though in thyroxine treated hearts than in preconditioned and normal hearts.

JNKs were also found to be activated at middle reperfusion to a similar degree in thyroxine and preconditioned hearts as compared to normal hearts. However, JNKs activation was found to be delayed in thyroxine treated and preconditioned hearts at 10 min of reperfusion as compared to normal hearts. This finding could probably reflect a delayed energy restoration that might occur at the onset of reperfusion in hearts that displayed a greater energy deprivation (profound exacerbation of ischaemic contracture in preconditioned and thyroxine treated hearts) during the ischaemic phase. At the end of reperfusion, JNKs activation declined but more rapidly in preconditioned hearts than in thyroxine treated or normal hearts. Attenuated activation of JNK1 has been also reported to occur in preconditioned hearts at the end of reperfusion following sustained ischaemia [26].

It appears that, at reperfusion, preischaemic interventions such as thyroxine pretreatment as well as preconditioning can modify the duration of activation of certain types of mitogen activated kinases probably by altering their dephosphorylation process. This process is shown to be dependent on the type of the cardioprotective intervention: increased decay of p38 MAPK inactivation was more profound in thyroxine-treated hearts whereas JNKs were more rapidly inactivated in preconditioned hearts. The contribution though of these particular changes to cardioprotection remains largely unknown. However, it is worth noting that pharmacological interventions that interfere with the dephosphorylation of MAP kinases, such as vanadate (a phosphatase inhibitor), were found to increase the tolerance of the cell to ischaemic stress [27].

The present study shows that a specific pattern of activation of p38 MAPK occurs during ischaemia and reperfusion in thyroxine treated hearts that possibly could account for the conferred protection in those hearts. However, future studies with pharmacological interventions targeting specifically the mitogen activated kinases dependent pathways at different stages (ischaemia and/or reperfusion) will probably be needed for better understanding of the underlying mechanisms of thyroxine induced cardioprotection. This would be of clinical and therapeutical importance since there is ongoing clinical evidence that thyroid hormone alterations can play an important role in the response of the heart to ischaemia. In fact, during acute myocardial infarction thyroid hormone levels were found to be reduced and this was associated with more severe infarcts and increased mortality [28, 29]. Interestingly, similar results were confirmed by experimental studies of acute myocardial infarction in which it was shown that thyroid hormone levels are reduced acutely and long-term administration of increased dose of thyroxine at postischaemic period could reverse the extent of ischaemic injury [30]. Furthermore, thyroid hormones are found to be altered in other clinical conditions with increased incidence of coronary events such as diabetes and heart failure [4].

In summary, our results indicate that in thyroxine treated hearts p38 MAPK activation was attenuated both during ischaemia and reperfusion and this was associated with increased postischaemic recovery of function.

Acknowledgements

This research has been supported by G. Karelias and Bodosakis Foundation. Evangelia Karamanoli is a recipient of a State Scholarship's Foundation fellowship. Ioannis Paizis is a recipient of a Public Benefit Foundation Alexandros S. Onassis fellowship.

References

1. Nakano A, Cohen MV, Downey JM: Ischemic preconditioning: From basic mechanisms to clinical applications. Pharmacol Ther 86: 263–275, 2000
2. Pantos C, Malliopoulou V, Mourouzis I, Karamanoli E, Tzeis SM, Carageorgiou H, Varonos D, Cokkinos DV: Long-term thyroxine administration increases HSP70 mRNA expression and attenuates p38 MAP kinase activity in response to ischaemia. J Endocrinol 170: 207–215, 2001
3. Pantos C, Cokkinos DD, Tzeis S, Malliopoulou V, Mourouzis I, Carageorgiou H, Limas C, Varonos D, Cokkinos D: Hyperthyroidism is associated with preserved preconditioning capacity but intensified and accelerated ischaemic contracture in rat heart. Basic Res Cardiol 94: 254–260, 1999
4. Klein I, Ojamaa K: Thyroid hormone – targeting the heart. Endocrinology 142: 11–12, 2001
5. Fryer LGD, Holness MJ, Decock JB, Sugden MC: Cardiac protein kinase C expression in two models of cardiac hypertrophy associated with an activated cardiac renin-angiotensin system: Effect of experimental hyperthyroidism and genetic hypertension (the mRen-2 rat). J Endocrinol 158: 27–33, 1998
6. Pantos C, Malliopoulou V, Mourouzis I, Karamanoli E, Paizis I, Steimberg N, Varonos D, Cokkinos DV: Long-term thyroxine administration protects the heart in a similar pattern as ischaemic preconditioning. Thyroid 12: 325–329, 2002
7. Han J, Lee J-D, Bibbs L, Ulevitch RJ: A MAP kinase targeted by endotoxin and hyperosmolarity in mammalian cells. Science 265: 808–811, 1994
8. Bogoyevitch M, Gillespie-Brown J, Ketterman A, Fuller S, Ben-Levy R, Ashworth A, Marshall CJ, Sugden PH: Stimulation of the stress-activated mitogen-activated protein kinase subfamilies in perfused heart: p38/RK mitogen-activated protein kinases and c-jun N-Terminal kinases are activated by ischemia/reperfusion. Circ Res 79: 162–173, 1996
9. Yin T, Sandhu G, Wolfgang C, Burrier A, Randy LW, Rigel DF, Tsonwin H, Whelan J: Tissue-specific pattern of stress kinase activation in ischemic/reperfused heart and kidney. J Biol Chem 272: 19943–19950, 1997
10. Knight RJ, Buxton DB: Stimulation of c-Jun kinase and mitogen-activated protein kinase by ischaemia and reperfusion in the perfused rat heart. Biochem Biophys Res Commun 218: 83–88, 1996
11. Chen Y-R, Wang X, Templeton D, Davis RJ, Tan T-H: The role of c-Jun N-terminal kinase (JNK) in apoptosis induced by ultraviolet C and γ radiation. J Biol Chem 271: 31929–31936, 1996
12. Mackay K, Mochly-Rosen D: An inhibitor of p38 mitogen-activated protein kinase protects neonatal cardiac myocytes from ischemia. J Biol Chem 274: 6272–6279, 1999
13. Nagarkatti D, Sha'afi R: Role of p38 MAP kinase in myocardial stress. J Mol Cell Cardiol 30: 1651–1664, 1998
14. Marais E, Genade S, Huisamen B, Strijdom JG, Moolman JA, Lochner A: Activation of p38 MAPK induced by a multicycle ischaemic preconditioning protocol is associated with attenuated p38 MAPK activity during sustained ischaemia and reperfusion. J Mol Cell Cardiol 33: 769–778, 2001
15. Ma XL, Kumar S, Gao F *et al.*: Inhibition of p38 mitogen-activated protein kinase decreases cardiomyocyte apoptosis and improves cardiac function after myocardial ischemia and reperfusion. Circulation 99: 1685–1691, 1999
16. Pantos CI, Tzilalis V, Giannakakis S, Cokkinos DD, Tzeis SM, Malliopoulou V, Mourouzis I, Asimacopoulos P, Carageorgiou H, Varonos DD, Cokkinos DV: Phenylephrine induced aortic vasoconstriction is attenuated in hyperthyroid rats. Int Angiol 20: 181–186, 2001
17. Pantos C, Mourouzis I, Tzeis S, Malliopoulou V, Cokkinos DD, Asimakopoulos P, Carageorgiou H, Varonos DD, Cokkinos DV: Propranolol diminishes cardiac hypertrophy but does not abolish acceleration of the ischemic contracture in hyperthyroid hearts. J Cardiovasc Pharmacol 36: 384–389, 2000
18. Schneider S, Chen W, Hou J, Steenbergen C, Murphy E: Inhibition of p38 MAPK α/β reduces ischemic injury and does not block protective effects of preconditioning. Am J Physiol Heart Circ Physiol 280: H499–H508, 2001
19. Marais E, Genade S, Strijdom JG, Moolman JA, Lochner A: p38 MAPK activation triggers pharmacologically-induced beta-adrenergic preconditioning, but not ischaemic preconditioning. J Mol Cell Cardiol 33: 2157–2177, 2001

20. Saurin AT, Martin JL, Heads RJ, Foley C, Mockridge W, Wright MJ, Wang Y, Marber MS: The role of differential activation of p38-mitogen-activated protein kinase in preconditioned ventricular myocytes FASEB J 14: 2237–2246, 2000
21. Van Der Vusse GJ, Coumans WA, Ulrich M, Van Bilsen M: Thyroxine induced alteration in cardiac energy metabolism. J Moll Cell Cardiol 30: A110, 1998
22. Kolocassides KG, Galinanes M, Hearse DJ: Dichotomy of ischaemic preconditioning. Improved postischaemic contractile function despite intensification of ischaemic contracture. Circulation 93: 1725–1733, 1996
23. Asimakis GK, Conti VR: Preconditioning with dobutamine in the isolated rat heart. Life Sci 57: 177–187, 1995
24. Hearse DJ, Sutherland FJ: Catecholamines and preconditioning: Studies on contraction and function in isolated rat hearts. Am J Physiol 277: H136–H143, 1999
25. Gabai VL, Meriin AB, Yaglom JA, Wei JY, Mosser DD, Sherman MY: Suppression of stress kinase JNK is involved in HSP72-mediated protection of myogenic cells from transient energy deprivation: HSP72 alleviates the stress-induced inhibition of JNK dephosphorylation. J Biol Chem 275: 38088–38094, 2000
26. Sato M, Cordis G, Maulik N, Das DK: SAPKs regulation of ischemic preconditioning. Am J Physiol Heart Circ Physiol 279: H901–H907, 2000
27. Mackay K, Mochly-Rosen D: Involvement of a p38 mitogen-activated protein kinase phosphatase in protecting neonatal rat cardiac myocytes from ischemia. J Moll Cell Cardiol 32: 1585–1588, 2000
28. Kimura T, KandaT, Kotajima N, Kuwabara A, FukumuraY, Kobayashi I: Involvement of circulating interleukin-6 and its receptor in the development of euthyroid sick syndrome in patients with acute myocardial infarction. Eur J Endocrinol 143: 179–184, 2000
29. Friberg L, Drvota V, Bjelak AH, Eggertsen G, Ahnvc S: Association between increased levels of reverse triiodothyronine and mortality after acute myocardial infarction. Am J Med 111: 699–703, 2001
30. Ojamaa K, Kenessey A, Shenoy R, Klein I: Thyroid hormone metabolism and cardiac gene expression after acute myocardial infarction in the rat. Am J Physiol 279: E1319–E1324, 2000

Molecular and Cellular Biochemistry **242:** 181–187, 2003.

Involvement of mitogen-activated protein kinases and reactive oxygen species in the inotropic action of ouabain on cardiac myocytes. A potential role for mitochondrial K_{ATP} channels

Jiang Tian,[1] Jiang Liu,[1] Keith D. Garlid,[2] Joseph I. Shapiro[1] and Zijian Xie[1]

[1]*Departments of Pharmacology and Medicine, Medical College of Ohio, Toledo, OH;* [2]*Department of Biochemistry and Molecular Biology, OGI School of Science and Engineering, Oregon Health & Sciences University, Beaverton, OR, USA*

Abstract

Binding of ouabain to Na^+/K^+-ATPase activated multiple signal transduction pathways including stimulation of Src, Ras, p42/44 MAPKs and production of reactive oxygen species (ROS) in rat cardiac myocytes. Inhibition of either Src or Ras ablated ouabain-induced increase in both $[Ca^{2+}]_i$ and contractility. While PD98059 abolished the effects of ouabain on $[Ca^{2+}]_i$, it only caused a partial inhibition of ouabain-induced increases in contractility. On the other hand, pre-incubation of myocytes with N-acetyl cysteine (NAC) reduced the effects of ouabain on contractility, but not $[Ca^{2+}]_i$. Furthermore, 5-hydroxydecanoate (5-HD) blocked ouabain-induced ROS production and partially inhibited ouabain-induced increases in contractility in cardiac myocytes. Pre-incubation of myocytes with both 5-HD and PD98059 completely blocked ouabain's effect on contractility. Finally, we found that opening of mitochondrial K_{ATP} channel by diazoxide increased intracellular ROS and significantly raised contractility in cardiac myocytes. These new findings indicate that ouabain regulates cardiac contractility via both $[Ca^{2+}]_i$ and ROS. While activation of MAPKs leads to increases in $[Ca^{2+}]_i$, opening of mitochondrial K_{ATP} channel relays the ouabain signal to increased ROS production in cardiac myocytes. (Mol Cell Biochem **242**: 181–187, 2003)

Key words: Na^+/K^+-ATPase, ouabain, contractility, $[Ca^{2+}]_i$, Ras/MAPK, mitochondrial K_{ATP} channel, reactive oxygen species

Introduction

Na^+/K^+-ATPase is an energy-transducing ion pump in most mammalian cells [1, 2]. It carries out the active transport of Na^+ and K^+ across the plasma membrane using the energy generated from hydrolysis of ATP. In the heart, this enzyme also serves as a functional receptor for digitalis compounds such as digoxin and ouabain [3–6]. Binding of ouabain to cardiac Na^+/K^+-ATPase inhibits the ion pumping function of the enzyme and increases myocyte contractility in the heart. This effect on cardiac contractility serves as the basis for the therapeutic use of digitalis drugs in the management of congestive heart failure [3–6].

In recent years we have demonstrated that binding of ouabain to the Na^+/K^+-ATPase can also convert the enzyme into a signal transducer [7–14]. It appears that ouabain promotes the interaction of the Na^+/K^+-ATPase with Src, resulting in activation of the kinase. The activated Src in turn binds to and transactivates the epidermal growth factor receptor (EGFR), leading to recruitment of adaptor protein Shc and subsequent stimulation of Ras [12, 13]. Downstream from Ras ouabain stimulates p42/44 mitogen-activated protein kinases (MAPKs) and increases mitochondrial production of reactive oxygen species (ROS) [10, 11]. Interestingly, activation of some of these signal transduction pathways by ouabain is independent of ouabain-induced changes in intracellular ion concentrations

Address for offprints: Z. Xie, Department of Pharmacology, Medical College of Ohio, 3035 Arlington Avenue, Toledo, OH 43614-5804, USA (E-mail: Zxie@mco.edu)

as well as in contractility of cardiac myocytes [13]. Significantly, we have recently shown that the classic effects of ouabain on intracellular calcium ($[Ca^{2+}]_i$) also depend on the signal transducing function of the Na^+/K^+-ATPase [14]. Inhibition of either protein tyrosine kinases or Ras or p42/44 MAPKs, but not ROS production diminishes ouabain-induced increases in $[Ca^{2+}]_i$. These findings led us to extend the above investigation and test the role of the signal transducing function of the Na^+/K^+-ATPase in ouabain-induced regulation of contractility in cardiac myocytes. We report here that the effects of ouabain on cardiac contraction not only depend on activation of MAPKs and the subsequent increase in $[Ca^{2+}]_i$, but also require opening of mitochondrial K_{ATP} channels (mitoK_{ATP}), which causes an increase in intracellular ROS.

Materials and methods

Materials

Collagenase Type II was from Worthington (Freehold, NJ, USA). Indo-1-AM and CM-DCFH diacetate were obtained from Molecular Probes (Eugene, OR, USA). Diazoxide and 5-HD were from Sigma (Saint Louis, MO, USA). PP2 was purchased from Calbiochem (San Diego, CA, USA).

Cell preparation and culture

The same protocols were used to prepare Ca^{2+}-tolerant adult rat ventricular myocytes as described in our previous work [14, 15]. In brief, Sprague–Dawley rats weighing between 250–300 g were anesthetized with sodium pentobarbital (60 mg/kg i.p.). The hearts were rapidly removed, attached to an aortic cannula, and retrograde perfused for 15 min with Joklik medium to wash out the blood, followed by 5 min perfusion with a nominally Ca^{2+}-free Joklik medium supplemented with 20 mM creatine and 60 mM taurine. The heart was then perfused with collagenase type II until the heart became soft and flaccid. Myocytes were dissociated from the left ventricle, harvested, and plated onto laminin-coated coverslips as previously described [14]. Medium was changed 2 h post plating. Over 95% of myocytes were quiescent, and they were used for the experiments after an overnight culture.

Fluorescence microscopic measurements of $[Ca^{2+}]_i$, contractility and ROS

Myocytes cultured on coverslips were perfused and paced at 0.5 Hz. $[Ca^{2+}]_i$ was measured by indo-1 as previously described [14]. Myocytes were loaded with 10 μM indo-1-AM for 30 min. Indo-1 fluorescence was recorded using a microscope-based fluorescence system (Photon Technology International, Monmouth Junction, NJ, USA). The probe was excited at 365 nm, and fluorescence emitted at 405 and 485 nm was recorded at 60 Hz in real time. $[Ca^{2+}]_i$ was calculated based on the fluorescence ratio and the Ca^{2+}calibration curve [14]. Myocyte contractility was measured as cell shortening using an edge detector as previously described [16]. Under each experimental condition signals were obtained from about 12 single cells from 3–5 different preparations. Intracellular ROS concentration was measured in cells loaded with 10 μM CM-DCFH diacetate as previously described [11]. Under each experimental condition about 15 single myocytes were imaged with an Attofluor imaging system, and CM-DCF fluorescence was measured at an excitation wavelength of 480 nm and an emission wavelength of 520 nm.

Preparation of replication-defective adenoviruses and adenovirus infection of cardiac myocytes

Replication-defective adenoviruses expressing a dominant negative Asn^{17} Ras were generated, amplified, purified, and used for the infection of myocytes as described before [10]. An identical virus containing the β-galactosidase gene (β-Gal), instead of the Asn^{17} Ras, was used as the control [10].

Analysis of data

Data are given as the mean ± S.E. Statistical analysis was performed using the Student's *t*-test, and significance was accepted at $p < 0.05$. Each presented immunoblot is representative of the similar results from at least three separate experiments.

Results

Ouabain regulation of cardiac contractility requires activation of Src, Ras, and MAPKs

We showed previously that binding of ouabain to the Na^+/K^+-ATPase activated Src, resulting in transactivation of the EGFR and subsequent stimulation of the Ras/MAPK cascade in cardiac myocytes [12, 14]. Inhibition of either Src or Ras or MAPKs blocked ouabain-induced increases in $[Ca^{2+}]_i$ [14]. Because increases in $[Ca^{2+}]_i$ are essential for ouabain-induced rise in contractility [17, 18], we postulated that the effects of ouabain on myocyte contractility must also be mediated via the above pathways. As depicted in Fig. 1, ouabain, at non-

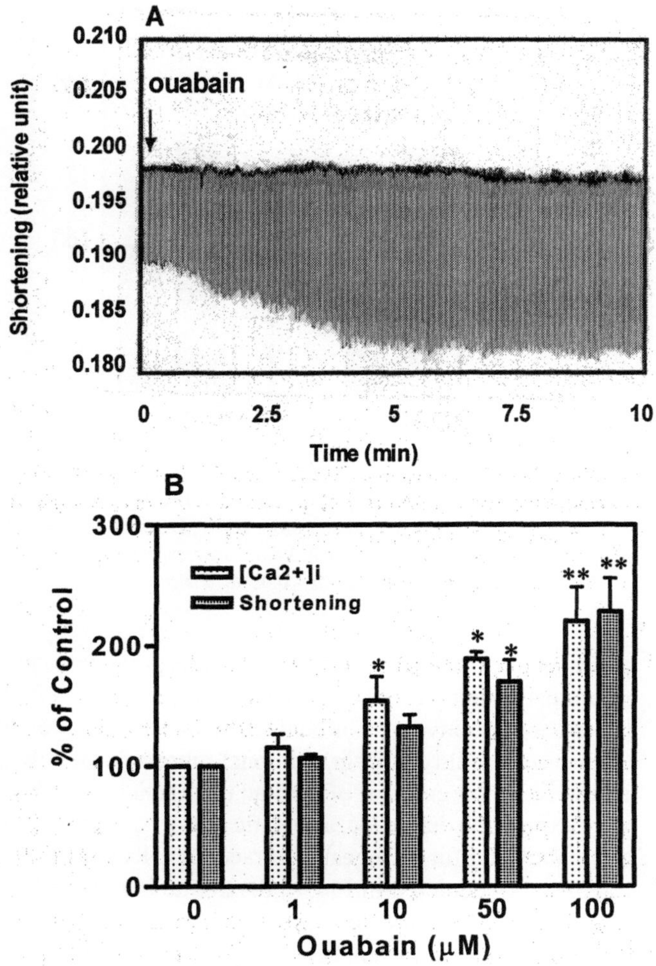

Fig. 1. Effects of ouabain on contractility and $[Ca^{2+}]_i$ in adult rat cardiac myocytes. To measure myocyte contractility and $[Ca^{2+}]_i$, cells were perfused and paced at 0.5 Hz as described under Materials and methods. Cell contractility was measured as cell shortening using an edge detector, and $[Ca^{2+}]_i$ was determined based on the indo-1 fluorescence ratio of 405 and 485 nm. Panel A shows a representative trace of contractility in a single cell. Ouabain (100 μM) was added to the medium at the time as indicated by the arrow. Panel B shows that ouabain increases both contractility and $[Ca^{2+}]_i$ in a dose dependent manner. Values are presented as mean ± S.E. of 15 single cells from 4 different experiments. $*p < 0.05$ and $**p < 0.01$ vs. control.

toxic concentrations, increased contractility in cultured adult rat cardiac myocytes in a time- and dose-dependent manner. The effects of ouabain on contractility correlated well with the rise in $[Ca^{2+}]_i$. As expected, inhibition of Src by PP2 completely blocked the effects of ouabain on contractility (Fig. 2). In addition, in cells expressing dominant negative mutant of Asn[17] Ras, ouabain also failed to stimulate myocyte contraction (Fig. 2). These data are consistent with the findings that inhibition of either Src or Ras blocks ouabain-induced increases in $[Ca^{2+}]_i$ [14]. Surprisingly, while pre-incubation of myocytes with PD 98059 to inhibit MAPK completely

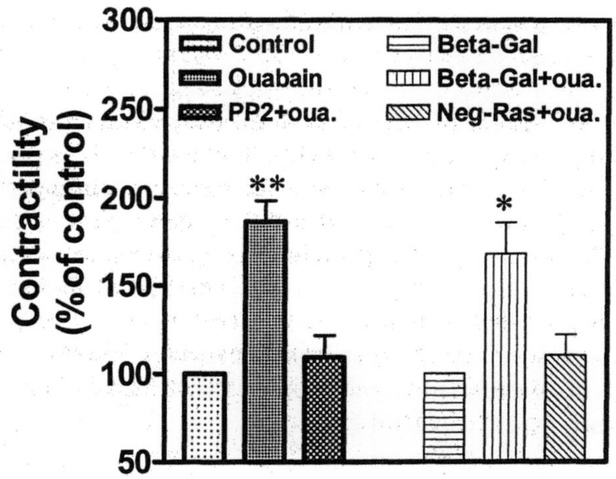

Fig. 2. Effects of PP2 and dominant negative Ras on ouabain-induced increase in contractility. Myocytes were pre-incubated with 1 μM PP2 for 20 min or transduced with adenoviruses expressing a dominant negative Ras for 12 h. (Cells transduced with the same amount of β-gal viruses were used as viral control.) Both treated and control myocytes were then exposed to 100 μM ouabain for 10 min and contractility was measured as in Fig. 1. Values are presented as mean ± S.E. of 10 single cells. $*p < 0.05$ and $**p < 0.01$ vs. control.

abolished ouabain-induced increases in $[Ca^{2+}]_i$, it only caused a partial inhibition of ouabain-induced rise in contractility (Fig. 3). These findings indicate that factors other than increases in $[Ca^{2+}]_i$ also contribute to ouabain regulation of cardiac contractility.

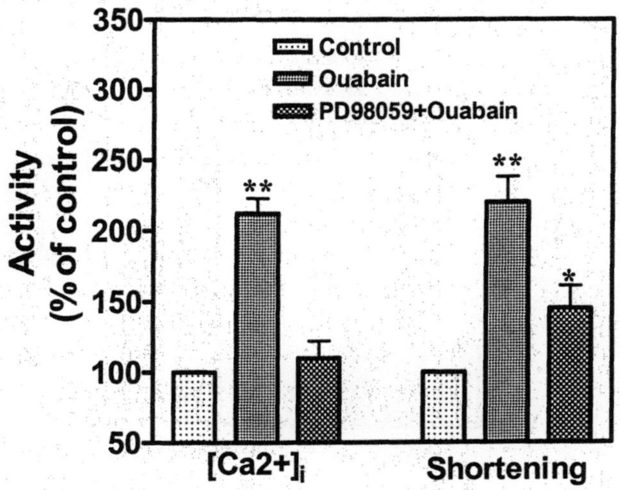

Fig. 3. Effects of PD 98059 on ouabain-induced increases in contractility and $[Ca^{2+}]_i$. Myocytes were pre-incubated with 30 μM PD98059 for 30 min, and then exposed to 100 μM ouabain for 10 min. Contractility and $[Ca^{2+}]_i$ were measured as in Fig. 1. Values are presented as mean ± S.E. of 15 single cells from 4 different experiments. $*p < 0.05$ and $**p < 0.01$ vs. control.

Involvement of ROS in ouabain-induced increases in contractility

Since activation of Ras by ouabain also increased mitochondrial production of ROS [13, 14], the above findings led us to examine whether ROS are involved in ouabain-induced regulation of cardiac contractility. As depicted in Fig. 4, ouabain increased ROS production in cardiac myocytes. Pre-incubation of myocytes with 10 mM NAC abolished ouabain-induced rise in intracellular ROS (Fig. 4) as previously noted in neonatal cardiac myocytes. Interestingly, NAC also caused a significant inhibition of ouabain-induced increases in contractility.

Involvement of mitoK$_{ATP}$ in the effects of ouabain on contractility

MitoK$_{ATP}$ opening plays a pivotal role in cardioprotection by K$_{ATP}$ channel openers and ischemic preconditioning [19–22]. In particular, mitoK$_{ATP}$ opening is required to trigger cardioprotective signaling pathways [23], and we have proposed that this function is mediated by inducing mitochondrial ROS production [24]. Accordingly, we assessed the role of inhibition of mitoK$_{ATP}$ in ouabain-induced ROS production and rise in contractility. As depicted in Fig. 5, pre-incubation of myocytes with 200 μM 5-HD inhibited the effects of ouabain on intracellular ROS. Interestingly, 5-HD also suppressed ouabain-induced increases in contractility under the same experimental conditions (Fig. 5). These findings support a proposal that opening of mitoK$_{ATP}$ and subsequent rise in ROS production are involved in ouabain-induced regulation of cardiac contractility. To gain additional support that mitoK$_{ATP}$ is involved in regulation of cardiac contraction, myocytes were exposed to diazoxide, a specific mitoK$_{ATP}$ agonist [22], and monitored for changes in intracellular ROS and contractility. As depicted in Fig. 6, diazoxide increased both intracellular ROS and contractility in cardiac myocytes. Interestingly, pre-incubation of myocytes with either 10 mM NAC or 200 μM 5-HD caused a complete inhibition of diazoxide-induced increases in both ROS and myocyte con-

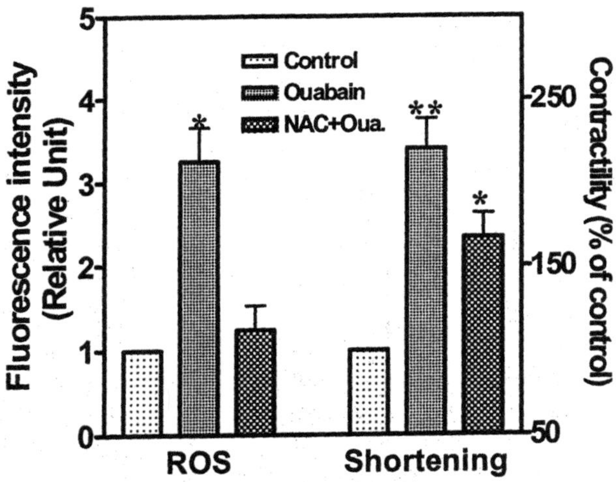

Fig. 4. Effects of NAC on ouabain-induced increases in intracellular ROS and contractility in adult rat cardiac myocytes. Myocytes were loaded with 5- (and 6-)chloromethyl-2′,7′-dichlorofulorescin diacetate, and treated with 100 μM ouabain for 10 min in the presence or absence of 10 mM NAC. Both ROS and myocytes contractility were measured as described under Materials and methods. Values are mean ± S.E. of 12 cells from 4 independent experiments. *p < 0.05, **p < 0.01 vs. control.

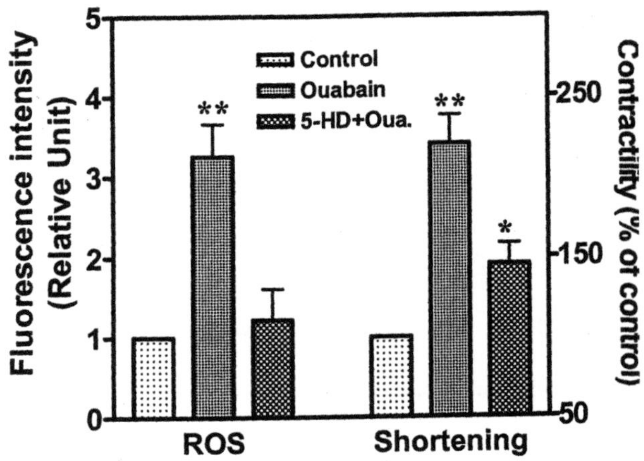

Fig. 5. Effects of 5-HD on oubain-induced increases in intracellular ROS and myocyte contractility. Myocytes were preincubated with 200 μM 5-HD for 30 min, and then exposed to 100 μM ouabain. Intracellular ROS and contractility was measured as in Fig. 4. Values are presented as mean ± S.E. of 10–15 single cells. *p < 0.05, **p < 0.01 vs. control

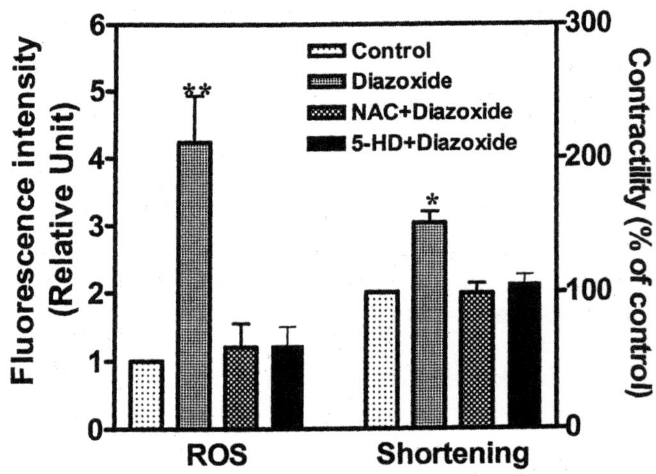

Fig. 6. Effects of diazoxide on intracellular ROS and myocytes contractility. Myocytes were treated with 10 μM diazoxide in the presence or absence of either 10 mM NAC or 200 μM 5-HD. Intracellular ROS and contractility were then measured as in Fig. 4. Values are presented as mean ± S.E. *p < 0.05, **p < 0.01 vs. control.

traction (Fig. 6). Since activation of MAPKs and opening of mitoK_{ATP} are required for ouabain-induced increases in $[Ca^{2+}]_i$ and ROS, respectively, we reasoned that inhibition of these two pathways should completely abolish the effects of ouabain on myocyte contraction. Indeed, as shown in Fig. 7, pre-incubation of myocytes with both PD98059 and 5-HD blocked the ouabain-induced increases in contractility.

Discussion

It has been known for long time that cardiac glycosides including ouabain increase $[Ca^{2+}]_i$ and contractility in cardiac myocytes by binding to the Na^+/K^+-ATPase [3–6]. In rat cardiac myocytes we showed that 10–100 μM ouabain caused about 20–50% inhibition of Na^+/K^+-ATPase in a dose-dependent manner [8, 15]. It is important to note that under our experimental conditions, ouabain at concentrations up to 100 μM did not cause arrhythmic contraction and Ca^{2+} overload in 15 min. It also had no effect on diastolic cell length. On the other hand, these non-toxic concentrations of ouabain caused a rapid stimulation of multiple signal transduction pathways including activation of Ras, p42/44 MAPKs and mitochondrial production of ROS (Fig. 4 and [14]). Concomitantly, it also raised $[Ca^{2+}]_i$ and contractility in a time- and dose-dependent manner in these cells. When the relation between the signal transducing function of the enzyme and the pharmacological effects of ouabain on $[Ca^{2+}]_i$ and contractility

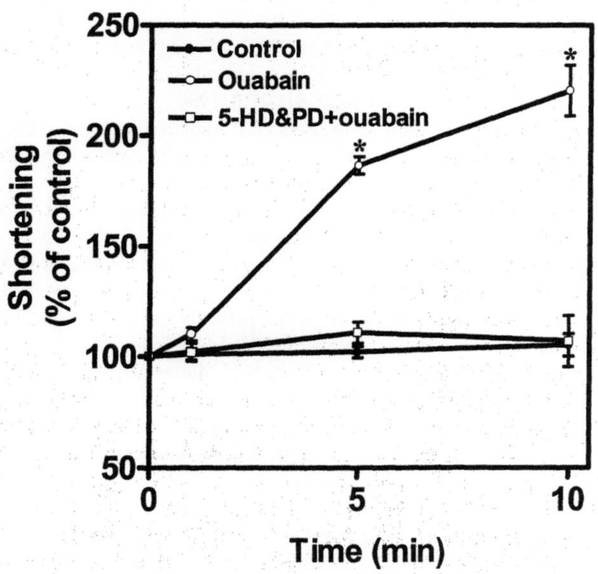

Fig. 7. Combination of PD98059 and 5-HD abolished ouabain-induced increases in contractility. Myocytes were simultaneously pre-incubated with 200 μM 5-HD and 30 μM PD98059 for 30 min, then exposed to 100 μM ouabain for 5 or 10 min and assayed for contractility as in Fig. 1. Values are mean ± S.E. of 10 measurements from 3 different preparations. $*p < 0.05$, $**p < 0.01$ vs. control.

were determined (Figs 1–5), we showed that inhibition of either Src by PP2 or Ras by expression of Asn[17] Ras ablated ouabain-induced increases in both $[Ca^{2+}]_i$ and contractility. Clearly, the factors that relay extracellular ouabain to increases in $[Ca^{2+}]_i$ and contractility must be the effectors of Ras.

It is well established that $[Ca^{2+}]_i$ is the central regulator of cardiac contractility [17, 18]. Since ouabain raised $[Ca^{2+}]_i$ via a MAPK-dependent pathway, we postulated that inhibition of MAPKs by PD98059 should abolish the effects of ouabain on both $[Ca^{2+}]_i$ and contractility in cardiac myocytes. Surprisingly, while pre-incubation of myocytes with PD 98059 completely blocked ouabain-induced rise in $[Ca^{2+}]_i$, it only caused a partial inhibition of the effects of ouabain on contractility. These findings led us to propose that factors other than MAPKs and $[Ca^{2+}]_i$ are also involved in ouabain-induced increases in contractility. Because inhibition of Ras completely blocked the effects of ouabain on contractility, we reasoned that the additional regulatory element(s) must be the other effector(s) of Ras. We showed previously that downstream from Ras ouabain also increased mitochondrial production of ROS in cardiac myocytes [13, 14]. Early studies of others also indicated that antioxidant α-tocopherol reduced the positive inotropic action of digitalis on atria muscle [25]. Therefore, we postulated that ROS might work in concert with $[Ca^{2+}]_i$ in regulation of contractility in response to ouabain. This notion was supported by the experiments shown in Fig. 4, in which pre-incubation of myocytes with NAC caused a significant inhibition of the effects of ouabain on contractility. Since NAC showed no effect on ouabain-induced changes in Na/K-ATPase activity (data not shown), $[Ca^{2+}]_i$ [14], and c-fos expression [11], its effect on ouabain-induced increases in contractility strongly suggest an involvement of ROS in ouabain regulation of cardiac contraction.

We showed previously that ouabain stimulated mitochondrial production of ROS via a Ras-dependent pathway [13]. We noted that the ouabain signaling pathway [7–14] and the cardioprotective signaling pathway [26] contain many elements in common, including mitochondrial ROS production. Moreover, cardioprotection by diazoxide [19] has been proposed to begin with mitoK_{ATP}-dependent stimulation of mitochondrial ROS production [23, 24]. Therefore, we reasoned that opening of mitoK_{ATP} might relay the signal from Ras to ROS production. This hypothesis is supported by the findings shown in Fig. 5 in which inhibition of mitoK_{ATP} by 5-HD blocked ouabain-induced increases in intracellular ROS. Furthermore, 5-HD also caused a significant inhibition of ouabain-induced rise in contractility, and pre-incubation of myocytes with both PD98059 and 5-HD completely abolished the effects of ouabain on contractility in cardiac myocytes. Interestingly, opening mitoK_{ATP} by incubation of myocytes with diazoxide was sufficient to raise intracellular ROS concentration and increase contractility in myocytes. These effects were blocked by either 5-HD or NAC (Fig. 6). Clearly, a modest increase in intracellular ROS via opening of mitoK_{ATP} by either ouabain or diazoxide can cause a significant rise in contractility in cardiac myocytes.

Although the mechanism by which ROS regulate cardiac contractility remains to be resolved, it is appropriate to consider the following two alternatives. First, the effects of ROS may be mediated through sensitizing myofilament to $[Ca^{2+}]_i$ by its direct action on contractile proteins or regulation of protein phosphorylation via activation of protein kinases such as PKC [27, 28]. Interestingly, activation of PKC was also found to be responsible for endothelin-1-induced increases in myofilament Ca^{2+} sensitivity [29]. In addition, early studies had suggested a role of PKC in ouabain-induced inotropy in the heart [30, 31], and we have recently demonstrated that ouabain activates multiple isozymes of PKC in cardiac myocytes under our experimental conditions [32]. Because there is evidence that increases in ROS are sufficient to activate PKC [33], it is quite possible that ROS may exert its effect on myocyte contraction by increasing myofilament Ca^{2+} sensitivity through activation of PKC. Alternatively, since mitoK_{ATP} plays an important role in regulation of mitochondrial energy metabolism [21], it is also possible that opening of mitoK_{ATP} and the subsequent rise in ROS production may prime mitochondria into a higher working state so that ATP can be delivered more efficiently to the myofibrils for contraction. Interestingly, early studies did suggest that digitalis affect mitochondrial energy metabolism [34, 35].

In summary, we demonstrated here that ouabain regulates cardiac contractility via activation of at least two pathways. Activation of p42/44 MAPKs and inhibition of the ion pumping function of the Na^+/K^+-ATPase by ouabain increased $[Ca^{2+}]_i$ whereas opening of mitoK_{ATP} stimulated the production of ROS. Both $[Ca^{2+}]_i$ and ROS, in turn, worked in concert, resulting in increases in contractility in cardiac myocytes.

Acknowledgements

This work was supported by National Institutes of Health Grants HL-36573 and HL-63238 awarded by the National Heart, Lung and Blood Institute (to Z.X.), and by GM-55324 awarded by the National Institute of General Medical Sciences (to K.D.G.), United States Public Health Service, Department of Health and Human Services.

References

1. Skou JC: The Na,K-pump. Meth Enzymol 156: 1–29, 1988
2. Lingrel JB, Kuntzweiler T: Na^+,K^+-ATPase. J Biol Chem 269: 19659–19662, 1994
3. Akera T, Ng YC: Digitalis sensitivity of Na^+,K^+-ATPase, myocytes and the heart. Life Sci 48: 97–106, 1991
4. Schwartz A, Grupp G, Wallick E, Grupp IL, Ball WJ Jr: Role of the Na^+K^+-ATPase in the cardiotonic action of cardiac glycosides. Prog Clin Biol Res 268B: 321–338, 1988
5. Eisner DA, Smith TW: The Heart and Cardiovascular System. In: H.A. Fozzard, E. Haber, R.B. Jennigs, A.M. Katz, H.E. Morgan (eds). Raven Press, New York, 1992, pp 863–902
6. Barry WH, Hasin Y, Smith TW: Sodium pump inhibition, enhanced calcium influx via sodium–calcium exchange, and positive inotropic response in cultured heart cells. Circ Res 56: 231–241, 1985
7. Xie Z: Ouabain interaction with cardiac Na/K-ATPase reveals that the enzyme can act as a pump and a signal transducer. Cell Mol Biol 47: 383–390, 2001
8. Peng M, Huang L, Xie Z, Huang W-H, Askari A: Partial inhibition of Na^+/K^+-ATPase by ouabain induces the Ca^{2+}-dependent expressions of early-response genes in cardiac myocytes. J Biol Chem 271: 10372–10378, 1996
9. Huang L, Li H, Xie Z: Ouabain-induced hypertrophy in cultured cardiac myocytes is accompanied by changes in expressions of several late response genes. J Mol Cell Cardiol 29: 429–437, 1997
10. Kometiani P, Li J, Gnudi L, Kahn BB, Askari A, Xie Z: Multiple signal transduction pathways link Na^+/K^+-ATPase to growth-related genes in cardiac myocytes: The roles of ras and mitogen-activated protein kinases. J Biol Chem 273: 15249–15256, 1998
11. Xie Z, Kometiani P, Liu J, Li J, Shapiro JI, Askari A: Intracellular reactive oxygen species mediate the linkage of Na^+/K^+-ATPase to hypertrophy and its marker genes in cardiac myocytes. J Biol Chem 274: 19323–19328, 1999
12. Haas M, Askari A, Xie Z: Involvement of Src and epidermal growth factor receptor in the signal transducing function of Na^+/K^+-ATPase. J Biol Chem 275: 27832–27837, 2000
13. Liu J, Tian J, Haas M, Shapiro JI, Askari A, Xie Z: Ouabain interaction with cardiac Na^+/K^+-ATPase initiates signal cascades independent of changes in intracellular Na^+ and Ca^{2+} concentrations. J Biol Chem 275: 27838–27844, 2000
14. Tian J, Gong X, Xie Z: Signal-transducing function of Na^+/K^+-ATPase is essential for ouabain's effect on $[Ca^{2+}]_i$ in rat cardiac myocytes. Am J Physiol 281: H1899–H1907, 2001
15. Xie Z, Wang Y, Askari A, Huang W-H, Klaunig JE, Askari A: Studies on the specificity of the oxygen free radical effects on cardiac sodium pump. J Cell Mol Cardiol 22: 911–920, 1990
16. Bassani RA, Bassani JWM, Bers DM: Mitochondrial and sarcolemmal Ca^{2+} transport reduce $[Ca^{2+}]_i$ during caffeine contractures in rabbit cardiac myocytes. J Physiol (Lond) 453: 591–608, 1992
17. Bers DM: Calcium fluxes involved in control of cardiac myocyte contraction. Circ Res 87: 275–281, 2000
18. Satoh H, Ginsburg KS, Qing K, Terada H, Hayashi H, Bers DM: KB-R7943 block of Ca(2+) influx via Na(+)/Ca(2+) exchange does not alter twitches or glycoside inotropy but prevents Ca(2+) overload in rat ventricular myocytes. Circulation 101: 1441–1446, 2000
19. Garlid KD, Paucek P, Yarov-Yarovoy V, Murray HN, Darbenzio RB, D'Alonzo AJ, Lodge NJ, Smith MA, Grover GJ: Cardioprotective effect of diazoxide and its interaction with mitochondrial ATP-sensitive K^+ channels: Possible mechanism of cardioprotection. Circ Res 81: 1072–1082, 1997
20. Grover GJ, Garlid KD: ATP-sensitive potassium channels: A review of their pharmacology. J Mol Cell Cardiol 32: 677–695, 2000
21. Kowaltowski AJ, Seetharaman S, Paucek P, Garlid KD: Bioenergetic consequences of opening the ATP-sensitive K(+) channel of heart mitochondria. Am J Physiol Heart Circ Physiol 280: H649–657, 2001
22. Garlid KD, Paucek P, Yarov-Yarovoy V, Sun X, Schindler PA: The mitochondrial K_{ATP} channel as a receptor for potassium channel openers. J Biol Chem 271: 8796–8799, 1996
23. Pain T, Yang XM, Critz SD, Yue Y, Nakano A, Liu GS, Heusch G, Cohen MV, Downey JM: Opening of mitochondrial K(ATP) channels

triggers the preconditioned state by generating free radicals. Circ Res 87: 460–466, 2000
24. Garlid KD: Opening mitochondrial K(ATP) in the heart – what happens, and what does not happen. Basic Res Cardiol 95: 275–279, 2000
25. Tawfik H, Schlieper P: The influence of a-tocopherol-nicotinate (Renascin) on cardiac glycoside effects. In: E. Erdmann, K. Greff, J.C. Skou (eds). Cardiac Glycosides 1785–1985. 1986, pp 109–117
26. Baines CP, Cohen MV, Downey JM: Signal transduction in ischemic preconditioning: The role of kinases and mitochondrial K(ATP) channels. J Cardiovasc Electrophysiol 10: 741–754, 1999
27. Steinberg SF, Goldberg M, Rybin VO: Protein kinase C isoform diversity in the heart. J Mol Cell Cardiol 27: 141–153, 1995
28. Baudet S, Weisser J, Janssen AP, Beulich K, Bieligk U, Pieske B, Noireaud J, Janssen PML, Hasenfuss G, Prestle J: Increased basal contractility of cardiomyocytes overexpressing protein kinase C and blunted positive inotropic response to endothelin-1. Cardiovasc Res 50: 486–494, 2001
29. Wang H; Endoh M: Chelerythrine and genistein inhibit the endothelin-1-induced increase in myofilament Ca^{2+} sensitivity in rabbit ventricular myocytes. Eur J Pharmacol 424: 91–96, 2001
30. Gotoh H, Kamiyama A, Shibayama R, Sawada M, Kashimoto T: Involvement of phosphoinositide turnover in ouabain inotropism. Biochem Biophys Res Commun 194: 72–78, 1993
31. Otani B, Otani H, Morita M, Das DK: Mol Cell Biochem 90: 111–120, 1989
32. Mohammadi K, Kometiani P, Xie Z, Askari A: Role of protein kinase C in the signal pathways that link Na^+/K^+-ATPase to ERK1/2. J Biol Chem 276: 42050–42056, 2001
33. Konishi H, Tanaka M, Takemura Y, Matsuzaki H, Ono Y, Kikkawa U, Nishizuka Y: Activation of protein kinase C by tyrosine phosphorylation in response to H_2O_2. Proc Natl Acad Sci USA 94: 11233–11237, 1997
34. Lee KS; Klaus W: The subcellular basis for the mechanism of inotropic action of cardiac glycosides. Pharmacol Rev 23: 193–229, 1971
35. Wollenberger A: The energy metabolism of the failing heart and the metabolic action of the cardiac glycosides. Pharmacol Rev 1: 311–344, 1949

Molecular and Cellular Biochemistry **242:** 189–191, 2003.

Index to Volume 242

Developments in Molecular and Cellular Biochemistry

Series Editor: Naranjan S. Dhalla, Ph.D., M.D. (Hon.), FACC

1. V.A. Najjar (ed.): *Biological Effects of Glutamic Acid and Its Derivatives*. 1981 ISBN 90-6193-841-4
2. V.A. Najjar (ed.): *Immunologically Active Peptides*. 1981 ISBN 90-6193-842-2
3. V.A. Najjar (ed.): *Enzyme Induction and Modulation*. 1983 ISBN 0-89838-583-0
4. V.A. Najjar and L. Lorand (eds.): *Transglutaminase*. 1984 ISBN 0-89838-593-8
5. G.J. van der Vusse (ed.): *Lipid Metabolism in Normoxic and Ischemic Heart*. 1989 ISBN 0-7923-0479-9
6. J.F.C. Glatz and G.J. van der Vusse (eds.): *Cellular Fatty Acid-Binding Proteins*. 1990 ISBN 0-7923-0896-4
7. H.E. Morgan (ed.): *Molecular Mechanisms of Cellular Growth*. 1991 ISBN 0-7923-1183-3
8. G.J. van der Vusse and H. Stam (eds.): *Lipid Metabolism in the Health and Diseased Heart*. 1992 ISBN 0-7923-1850-1
9. Y. Yazaki and S. Mochizuki (eds.): *Cellular Function and Metabolism*. 1993 ISBN 0-7923-2158-8
10. J.F.C. Glatz and G.J. van der Vusse (eds.): *Cellular Fatty-Acid-Binding Proteins, II*. 1993 ISBN 0-7923-2395-5
11. R.L. Khandelwal and J.H. Wang (eds.): *Reversible Protein Phosphorylation in Cell Regulation*. 1993 ISBN 0-7923-2637-7
12. J. Moss and P. Zahradka (eds.): *ADP-Ribosylation: Metabolic Effects and Regulatory Functions*. 1994 ISBN 0-7923-2951-1
13. V.A. Saks and R. Ventura-Clapier (eds.): *Cellular Bioenergetics: Role of Coupled Creatine Kinases*. 1994 ISBN 0-7923-2952-X
14. J. Slezák and A. Ziegelhöffer (eds.): *Cellular Interactions in Cardiac Pathophysiology*. 1995 ISBN 0-7923-3573-2
15. J.A. Barnes, H.G. Coore, A.H. Mohammed and R.K. Sharma (eds.): *Signal Transduction Mechanisms*. 1995 ISBN 0-7923-3663-1
16. A.K. Srivastava and J.-L. Chiasson (eds.): *Vanadium Compounds: Biochemical and Therapeutic Applications*. 1995 ISBN 0-7923-3763-8
17. J.M.J. Lamers and P.D. Verdouw (eds.): *Biochemistry of Signal Transduction in Myocardium*. 1996 ISBN 0-7923-4067-1
18. E.-G. Krause and R. Vetter (eds.): *Biochemical Mechanisms in Heart Function*. 1996 ISBN 0-7923-4118-X
19. R. Vetter and E.-G. Krause (eds.): *Biochemical Regulation of Myocardium*. 1996 ISBN 0-7923-4259-3
20. G.N. Pierce and W.C. Claycomb (eds.): *Novel Methods in Molecular and Cellular Biochemistry of Muscle*. 1997 ISBN 0-7923-4387-5
21. F.N. Gellerich and S. Zierz (eds.): *Detection of Mitochondrial Diseases*. 1997 ISBN 0-7923-9925-0
22. P.K. Singal, V. Panagia and G.N. Pierce (eds.): *The Cellular Basis of Cardiovascular Function in Health and Disease*. 1997 ISBN 0-7923-9974-9
23. S. Abdel-aleem and J.E. Lowe (eds.): *Cardiac Metabolism in Health and Disease*. 1998 ISBN 0-7923-8104-1
24. A.K. Srivastava and B. Posner (eds.): *Insulin Action*. 1998 ISBN 0-7923-8113-0
25. V.A. Saks, R. Ventura-Clapier, X. Leverve, A. Rossi and M. Rigoulet (eds.): *Bioenergetics of the Cell: Quantitative Aspects*. 1998 ISBN 0-7923-8118-1
26. G.N. Pierce, H. Rupp, T. Izumi and A. Grynberg (eds.): *Molecular and Cellular Effects of Nutrition on Disease Processes*. 1998 ISBN 0-7923-8171-8
27. K. Ahmed, E. Chambaz and O.G. Issinger (eds.): *Molecular and Cellular View of Protein Kinase CK2*. 1998 ISBN 0-7923-8208-0

Developments in Molecular and Cellular Biochemistry

28. M.V. Cohen, J.M. Downey, R.J. Gelpi and J. Slezák (eds.): *Myocardial Ischemia and Reperfusion*. 1998 ISBN 0-7923-8173-4
29. D.A. Bernlohr and L. Banaszak (eds.): *Lipid Binding Proteins within Molecular and Cellular Biochemistry*. 1998 ISBN 0-7923-8223-4
30. R. Albarez-Gonzalez (ed.): *ADP-Ribosylation Reactions: From Bacterial Pathogenesis to Cancer*. 1998 ISBN 0-7923-8235-8
31. S. Imai and M. Endo (eds.): *Muscle Physiology and Biochemistry*. 1998 ISBN 0-7923-8264-1
32. D.K. Das (ed.): *Stress Adaptation, Prophylaxis and Treatment*. 1999 ISBN 0-7923-8406-7
33. H. Rupp and B. Maisch (eds.): *Control of Gene Expression by Catecholamines and the Renin-Angiotensin System* ISBN 0-7923-7981-0
34. X. Shi, V. Castranova, V. Vallyathan and W.G. Perry (eds.): *Molecular Mechanisms of Metal Toxicity and Carcinogenesis*. 2001 ISBN 0-7923-7498-3
35. K. Ahmed, J.E. Allende and O.-G. Issinger (eds.): *Protein Kinase CK2 – From Structure to Regulation*. 2001 ISBN 0-7923-7666-8
36. R.K. Sharma (ed.): *Guanylate Cyclase*. 2002 ISBN 0-7923-7682-8
37. V. Vallyathan, V. Castranova and S. Shi (eds.): *Oxygen/Nitrogen Radicals: Cell Injury and Disease*. 2002 ISBN 1-4020-7085-3
38. J.F.C. Glatz (ed.): *Cellular Lipid Binding Proteins*. 2002 ISBN 1-4020-7220-1
39. E. Kardami, L. Hryshko and N. Mesaeli (eds.): *Cardiac Cell Biology*. 2003 ISBN 1-4020-7296-1

KLUWER ACADEMIC PUBLISHERS – DORDRECHT / BOSTON / LONDON

MIX
Papier aus verantwortungsvollen Quellen
Paper from responsible sources
FSC® C105338

FSC
www.fsc.org

If you have any concerns about our products,
you can contact us on
ProductSafety@springernature.com

In case Publisher is established outside the EU,
the EU authorized representative is:
Springer Nature Customer Service Center GmbH
Europaplatz 3, 69115 Heidelberg, Germany

Printed by Libri Plureos GmbH
in Hamburg, Germany